Tibor Müller Harmund Müller

Modelling in Natural Sciences

With Contributions by

G. Adrian, U. Bielert, W. Breitung, T. Bromage, C. Hertler, S. Heusch, J. Kazmierczak, M. Kelm, S. Kempe, B. Kienzler, A. Kotchourko, M. Lempert, J. Lützenkirchen, D. Majewski, M. Ostrowski, F. Schrenk, C. Theis

Springer-Verlag Berlin Heidelberg GmbH

Tibor Müller
Harmund Müller

Modelling in Natural Sciences

Design, Validation and Case Studies

with 133 Figures and 16 Tables

Springer

DR. TIBOR MÜLLER
Boulevard Clovis 49a
1000 Bruxelles
Belgium

DR. HARMUND MÜLLER
Roßbergring 90a
64354 Reinheim-Zeilhard
Germany

ISBN 978-3-642-05516-4 ISBN 978-3-662-05304-1 (eBook)
DOI 10.1007/978-3-662-05304-1

Cataloging-in-Publication Data applied for

Bibliographic information published by Die Deutsche Bibliothek
Die Deutsche Bibliothek lists this publication in the Deutsche Nationalbibliografie; detailed bibliographic data is available in the Internet at <http://dnb.ddb.de>.

http://www.springer.de

Originally published by Springer-Verlag Berlin Heidelberg New York in 2003
Softcover reprint of the hardcover 1st edition 2003

Camera ready by authors
Cover design: E. Kirchner, Heidelberg
Printed on acid-free paper 32/3141/as 5 4 3 2 1 0

Preface

Zwar weiß ich viel,
doch möcht' ich alles wissen[1]
Goethe, Faust I, Vers 601

Man has always recognized his limitations as a challenge to strive for new horizons. Today, technical progress enables him to realize this ambition; one of the means are models granting new insights into phenomena or problems which cannot be observed or (otherwise) explained. Depending on the standpoint of the expert, the model is either mainly retrospective – like Darwin's theory of natural selection as a model to explain the evolution of species – or it concentrates prospectively on the future by trying to predict events, e.g. catastrophes such as floods or droughts. Naturally, all these models are not perfect as they are man-made, but they do help to solve problems. Politicians should draw consequences from these observations; but as they cannot be expected to evaluate such models, they need highly qualified advisers.

This exposition already indicates that the term *model* encompasses many different facets with far-reaching consequences. We quote several examples in order to demonstrate the rather indefinite interpretation of the term and the various purposes models are to serve; in fact we come to the conclusion that there are literally 'models everywhere'.

Diverse as models are, they all share some common ideas such as the structural aspects of the modelling process. We approach the study from the viewpoint of system theory: a model is a system reflecting a section of reality which is a system itself and the process of modelling is defined by mappings between these systems. We base our considerations upon systems analysis and systems design as well as on the theory of *morphisms*, mappings conserving certain structures immanent in that part of 'reality' mirrored by a model.

---

[1] *I've learnt a deal, made books my drink and meat,*
But cannot rest till knowledge is complete
translated by Philip Wayne 1949

By using these aspects the processes of generating and evaluating models are described straightforwardly. The structures of morphisms permit a very detailed characterization of models and help to describe their set-up. The evaluation of a model leads to the determination of its *range of validity* which depends on the evidence available to the evaluator. The availability of evidence characterizes this scrutiny as a stochastic process depending on interdisciplinary accepted facts 'supporting' a model or 'casting doubt on it'; at worst the decision whether a model is dependable or not may be entirely based on personal probabilities. By applying the probability approach it is possible to quantify the quality of a model by degrees.

The *validity* of a model is the generic quality criterion a model has to fulfil. A closer examination will lead to a subdivision into subordinate aspects which must be tested in adequate procedures; their significance depends on the respective field of research. The quantification of the dependability of a model by degrees of validity should not lead to the conclusion that this is an absolute system of measures. It rather allows the comparison of models as being more or less 'valid' and its components are determined by consensus among the members of the relevant scientific community – which sometimes may not be much more than 'guessing'. Validity may not imply much more than the consensus that a model 'is not invalid', i.e. that it serves its purposes, is in concordance with our observations and contributes to the solution of the underlying problems.

Our considerations lead to the conclusion that modelling is not only a highly specialized technical procedure, demanding the outstanding competence of experts in the relevant field of research, but also an art which requires creativity and imagination from the modeller. The real mechanic – skilled as he may be – must combine both elements or he will fight a losing battle.

We are greatly indebted to Mrs Christine Müller who read and corrected innumerable proofs; this book could never have been completed without her help. We sincerely appreciate the efforts of the authors of all the detailed case studies appended to this book, which are the concrete counter-balance to the first part and without which the book would not be complete. Last but not least, we would like to thank Springer-Verlag, particularly Dr. Christian Witschel and his team, for their support which made this publication possible.

CH–Erlinsbach, B–Brussel/Bruxelles, D–Reinheim-Zeilhard

Januar 2003 T.M. & H.M.

Contents

1 Models

1.1 Examples

The term *model* is used in many different meanings of the word. So, to start off, we want to outline the range covered by this expression. In the following examples we shall try to give various interpretations of what is meant when we talk about models. There will not yet be any structuring of this enumeration; we shall just put down the examples in the order they occur to us, and we do not claim completeness of the list.

Ex. 1.1 A model may be a three-dimensional representation – very often on a reduced scale – of an object, like a model-car, a model-aeroplane or a figurine.

Each of these models is constructed in order to serve its own special purpose. A model-car, for instance, used as a toy by small children will only roughly resemble a real car; its material, set-up and shape will be chosen so that no harm is done to the child playing with it. The representation is simplified and reflects just the most significant features of the idea of a 'real' car. These essentials seem to be the four wheels of an automobile. Further details are sometimes added to these models in order to make them more manageable for children. Therefore handles may be added to a model-car; of course, there is no part like this on a 'real' car, and often these model-cars rather look like a four-wheeled iron than a car known to an adult. Four wheels – and, maybe, a handle to operate the model – seem to be sufficient to represent an automobile to a small child.

A model-car designed for collectors will reflect as many details of the body of the original car as possible. Some parts of the interior fittings, like the seats, the steering wheel and the gearshift lever, may be reproduced in a collector's model-car, but there will be no further details of the drive, the breaks and so on.

If a car is modelled in order to study its aerodynamics in a wind-tunnel, some significant features of its shape and the structure of its surface will be represented by the model, while other technical details are non-essential. This model is to serve a different *purpose*. There will definitely not be a handle on top of it – as may be the case with a toy-car –, and there will not be any stickers and additional accessories the collector's model may be supplied with.

If a model is to demonstrate the functions of a car in a driving school, details of its outer appearance will be neglected, but the clutches, the breaks and other technical systems relevant to the way an automobile works will be represented in greater detail. Some extra attachments may even be incorporated into the model not present in its original, e.g. there may be an axis facilitating the rotation of the model and the inspection of it from different angles of vision.

Furthermore, there are model-cars representing non-existing objects. These are fancy-cars like 'Batman's batmobile'; they are modelled according to their creator's imagination. They somehow look like the cars we are accustomed to, but it may well be that their four wheels – like those of a toy – are the only and essential feature they have in common with a real car.

Finally, a model-car may be a design-study: the model of a car not yet existing, but possibly being produced in the future. In that case the model is a *prototype* (Greek *protos* and *tupos*: 'the first' and 'stamp') demonstrating its creator's ideas of what the real car to be developed should look like. This model may be a three-dimensional representation of the proportions of the car to be manufactured in the future, but it may as well only be a (two-dimensional) draft reflecting its designer's aesthetic perceptions.

With model-aeroplanes the situation may be different. A paper model, for instance, will have little in common with an existing aircraft. Its prime purpose is to simulate the performance of real planes.

Given a figurine, a doll or anything like that, it may look similar to its original, but on a different scale, or it may neither in shape nor in function resemble anything 'real' at all. Garden gnomes can be traced back to the appearance of miners, but some 'Godzilla'-like monsters do not look like anything real or reflect the performances of any existing object or person or of any object designed to be produced.

The kind of model described in Ex. 1.1 may also represent an entire *ensemble* of objects. Architectural models, for instance, reflect blocks of buildings and even whole cities already existing or only planned and still to be erected. ♦ [1]

Ex. 1.2 A model may be a figure to be reproduced in the same or in another material, on the same or on a different scale.

This type of model may represent a real object, but, maybe, on a different scale, or the figure is purely fictitious, and it is not similar to anything 'real'. In this respect there is no difference between these models and those described in Ex. 1.1, but they serve different *purposes*. The models of Ex. 1.2 are *prototypes* to be reproduced in large numbers. Another word for prototype is *archetype* (Greek *arkhaios* and *tupos*: 'ancient' or 'originally' and 'stamp'), but in general this term is used in another context (cf. Ex. 1.18).

The reproduction of a prototype can be done manually or automatically, for instance, by copying the figure as a piece turned on the lathe or by casting. If the figure is reproduced by casting, a further (intermediate negative) model is necessary: a *mould* will be taken from the figure (cf. Ex. 1.4). Within this mould each protruding part of the surface of the figure will be represented by a corresponding depression, and *vice versa*. ♦

Ex. 1.3 An exemplary person or thing or an idea to be imitated or copied is called a model. We also talk of photo-models and of models posing for artists; and yet, an exemplary model may also be regarded as a 'warning'; then it is referred to in order to deter people from copying it or some of its characteristics.

In this example the model is *prototypical* (in a positive or negative sense) and chosen according to the ideas of a single individual or of a group of persons. A model athlete will be a model of a specific discipline; thus, a model sprinter will look different from a model boxer. A model of morals of the Catholic will be different from the corresponding ideal of a Buddhist, and both may – or will – be different from the model political leader. The latter example of a political leader shows that the preferences also depend on the special circumstances of the period of time as well as on the historical and social backgrounds of those accepting the model in question as an example.

---

[1] The ♦-symbol designates the end of an example or of its discussion.

The artist's model is chosen in accordance with special individual aesthetic ideas. Rubens' models did not resemble those posing for Picasso. Besides, even if the same model sits for different artists, the resulting works of art doubtless look different.

The reproductions of these models are restricted to certain select features. We may try to copy the style of a tennis champion, but we may not want to adopt his mannerisms and social behaviour; we may even detest his morals. Often the way the copy mirrors the modelled prototype is not apparent to everyone; it may, for instance, be difficult to trace any similarity between a portrait painted by a modern artist and the model posing for it.

Yet, these models may just as well be constructed to serve the purpose of deterring from copying them or certain of their features. In that case they are counter-examples used, e.g. in education in order to warn people not to copy them, because that will be detrimental and lead to a dreadful ending. ♦

Ex. 1.4 Relief plates used by printers or by the designers of cloths, coining dies and baking tins are called models.

These models may be derived from a positive-form model as in Ex. 1.2 – the way a mould is formed – or they may be generated directly as a negative-form model. In the former case they are intermediate models used for copying a given object. They thus serve the purpose of reproducing a particular figure as described in Ex. 1.2. The lower die used for stamping coins is the *matrix*, the counter die is called *patrix*. The prime patrix is the proto- or archetype of a coin in every sense of the words *first stamp* of Ex. 1.1 and Ex. 1.2. ♦

Ex. 1.5 Model garments are expensive, designed by prominent couturiers and fabricated in limited numbers only.

In that case copying or the close imitation of the model is unwelcome, and there may be nothing 'real' represented by the model. The model garment is created according to the ideas and perceptions of the couturier. But it may well be that the creator of a model garment wants to influence fashion and set certain trends, then the garment serves the purpose of a prototype; or he may wish to produce something unique, entirely extravagant and not transferable to the development of fashion at all. ♦

Ex. 1.6 Unified series of technical products are models, like *FORD*'s *Tin Lizzy T-*Model or the *AIRBUS* model *A 380*.

Here the model represents certain technological ideas and standards. It is a set of objects which are interchangeable with respect to their technical conception, and the term *model* is also used to represent this technical conception itself.

The *fungibility* of products may serve as another example of unification and standardization. When trading in commodities, the individual specimens of these goods must be standardized in a way that one of them can replace any other one or can be replaced by any other specimen. In order to achieve fungibility, certain *types* of the commodities are defined, like types of grain or types of crude. ♦

Philosophical ideas and theories in the field of humanities are models, or they may be represented by models. Ideas of morals as well as of aesthetics belong to particular culture groups and periods of time. In this respect these examples have something in common with Ex. 1.3 where an ideal person is regarded as a model. The philosopher will evolve theories representing his ideas in detail. Theories are rather *dialectical* models and not physical, like, e.g. the model-cars in Ex. 1.1 or the paragons in Ex. 1.3.

Frequently, philosophical theories are very abstract, and it can be doubted that there are any *real* structures – material or not – meeting their demands. If they exist, these structures will be considered as *representing* the theory in question: they are *models* of the theory.

As these aspects cover a wide range of models, we will discuss two cases in the following examples Ex. 1.7 and Ex. 1.8 demonstrating the aspect of finding representations of philosophical theories. There will be further examples expounding the sometimes limited means of validating these models.

Ex. 1.7 The ideas of the *Enlightenment* – emphasizing reason, rationalism and individualism rather than religion and tradition – may serve as the first example: J.-J. Rousseau (1712-1778) was convinced that everything and everybody was good and ideal by nature, but was degenerated under the influence of society, and he also believed that morals declined because of the progress of civilization. He summarized these ideas in his well-known and then provocative appeal 'Rétournons à la Nature'. Rousseau created a pedagogical model based on the idea of an unrestricted development of the human personality and put it down in his novel 'Émile ou De l'éducation' (Rousseau 1964).

As nobody knows in advance whether a theory can be applied to everyday life, it has to be tested out. Rousseau's theory triggered off various experiments, *modelling* his ideas on education. As *one* example of a pedagogical model translating Rousseau's ideas into reality we mention Neill's *Summerhill*-programme – founded in 1924 (e.g. Neill 1968, 1996) – based on the principles of anti-authoritarian education[1] and influenced by Freud's and Reich's theories. We think the *Summerhill*-project failed as it was never accepted as a basis of teaching curricula used by state school systems. But the institution still exists, and they proudly announce – among other facts – that '... no pressure is applied to the child to start going to lessons' (http://www.first-ask.de/summerhill2000/About-Summerhill.html). Furthermore, there are some modern pedagogical conceptions trying to resurrect Neill's ideas under different labels, like anti-pedagogy (e.g. Braunmühl 1975, 1990) or the educational concept based on Steiner's anthroposophy.

If we regard Rousseau's pedagogical theory and its representation in his novel 'Émile ou De l'éducation' as a *model* of his ideas under the influence of the historical and social situation of that time, and if we call Neill's *Summerhill* a *model* of Rousseau's perceptions, the latter is a model of a model. So we realize that the term *modelling* may be *hierarchical* leading to models of models of models and so on. Besides, we learn from the above-mentioned example that Neill carried out *experiments* in order to *test* Rousseau's model.

When studying educational models, we discover the phenomenon of 'respiration': the models oscillate between periods governed by rather authoritarian principles and periods of a more unrestrained atmosphere at schools. There is no definite periodicity to these cycles, but they are noticeable (cf. Subsect. 2.2.6 and Subsect. 2.3.2.1). ♦

Ex. 1.8 As a second example we mention the Emperor Friedrich[2] II (Friedrich von Hohenstaufen, 1194-1250) a member of the dynasty of Staufen. As we think it demonstrates some of the characteristics of modelling in humanities we will go into that in more detail. From the modern point of view we would classify the Emperor's experiment as belonging to the field of developmental psychology.

---

[1] There have been earlier attempts to translate those ideas into pedagogical practice, dating from 1921 when Neill was a co-founder of an International School (Neue Schule) at Hellerau near Dresden in Germany.

[2] We shall justify the usage of the German spelling 'Friedrich' in the discussion of this example.

Salimbene de Adam (da Parma, 1221-1288), a Minor friar of the Order of the Franciscans, was a chronicler contemporary with Friedrich II, and his reports seem to be credible. He records a strange psychological experiment of the Emperor. According to Salimbene Friedrich II believed in some sort of *proto-language*, a language every human being would use for communication, even if he never learned to talk.

As the considerations of Friedrich II are somehow typical of some aspects of modelling we will describe that experiment with great attention to detail. We shall give the example by quoting the English translation of Salimbene's Chronicle (Baird et al. 1986); but as we think this version contains some shortcomings regarding those aspects of modelling we want to discuss, we shall quote Salimbene's original text, edited by Holder-Egger (1905-1913), in italics, first.

According to Salimbene de Adam Friedrich II followed certain superstitions [1], one of them being that he believed in the existence of a *proto-language*. The original report is (Holder-Egger 1905-1913): *Secunda eius superstitio fuit, quia voluit experiri, cuismodi linguam et loquelam haberent pueri, cum adolevissent, si cum nemine loquerentur. Et ideo precepit baiulis et nutricibus, ut lac infantibus darent, ut mammas sugerent, et belnearent et mundificarent eos, sed nullo modo blandirentur eis nec loquerentur. Volebat enim cognoscere, utrum Hebream linguam haberent, qui prima fuerat, an Grecam vel Latinam vel Arabicam aut certe linguam parentum suorum, ex quibus nati fuissent. Sed laborabat in cassum, quia pueri sive infantes moribantur omnes. Non enim vivere possent sine aplausu et gestu et letitia faciei et blanditiis baiularum et nutricum suarum. Unde fascenine, -narum, dicuntur carmina, que mulier dicit movendo cunas, ut sopiat puerum, sine quibus puer male posset dormire et quietem habere.* The English version (Baird et al. 1986) reads 'His second idiosyncracy was that he wanted to discover what language a child would use when he grew up if he had never heard anyone speak. Therefore, he placed some infants in the care of wet-nurses, commanding them to bathe and suckle the children, but by no means ever to speak to or fondle them. For he wanted to discover whether they would speak Hebrew, the first language, or Greek, Latin, Arabic, or the language of their parents. But he labored in vain, because all of the infants died. For they cannot live without the praise, fon-

---

[1] Superstition #1 reported by Salimbene was that the Emperor was somewhat idiosyncratic about the spelling of his name; he thus 'had the thumb of a certain notary cut off because he had written his name in a way different from the way the Emperor desired. The Emperor wanted an *i* in the first syllable of his name, in this manner: *Fridericus*, but the notary had spelled it with an *e*, writing *Fredericus*' (Baird et al. 1986).

dling, playfulness, and happy expressions of their nurses. Thus the songs which a woman sings while rocking the cradle are called *fascenine*, *-narum*, without which a child can scarcely quieten down and go to sleep.'

According to Salimbene de Adam Friedrich II believed this proto-language might, for instance, be Hebrew, because this was the common language (not only a *lingua franca*, but somehow the natural means of verbal communication in those days) used by everybody before the Babel of languages, or it might be Greek, Latin, Arabic or the mother tongue of the child's parents, a knowledge of which it may have inherited. In order to verify his theory, he performed an experiment and isolated some new-born children. They were nourished and taken care of, but their social contacts were limited to the members of this group of children and to the physical contacts with their wet-nurses. No adult ever talked to them or played with them. The children were expected to communicate in their proto-language and to grow up without being influenced by the surrounding cultural assets. The experiment did not lead to the expected result as the children wasted away. So the theory of Friedrich II was a model of the way people develop linguistic means of communication, and his *kindergarten* was a model, too: it was a means to *test* out the Emperor's theory by *experimenting* in order to find out whether it was true.

In Salimbene's opinion the experiment, quite naturally, had to fail. He obviously disapproved of it, and he put forward his own theory that children necessarily needed verbal communication and loving care, that 'thus the songs which a woman sings while rocking the cradle are called fascenine, -narum, without which a child can scarcely quieten down and go to sleep'.

We have to add that Friedrich II was one of the most erudite men of his time; he was – what we today would call – a universal genius. So he did not do his tests on a whim, but we think he was confronted with an open scientific question and wanted to find out by experiments. From our modern point of view we may share Salimbene's disgust. Baird's translation seems to be too complimentary when he talks about 'idiosyncracies' where Salimbene equals *superstitiones*, 'superstitions', with '...idle curiosity, lack of faith, perversity, tyranny, and accursedness ...' (Baird et. al 1986), *Porro alias superstitiones et curiositates et maledictiones et incredulitates et perversitates et abusiones habuit similiter Fridericus, ...* (Holder-Egger 1905-1913). But, as a matter of fact, Friedrich II only forbade verbal communication between the children and the adults caring for them. Certainly, there was some physical contact and some non-verbal communication – like smiling – between the children and their wet-nurses, and that contact seems to be of crucial importance for the development of babies (Spitz 1954, 1957). It may well be that the failure of the experiment did not discourage Friedrich II and refute his belief in a proto-language. Seidler (1964), for instance, thinks it may be possible that the children's deaths were caused by infections, easily spreading under hygienic conditions of the 13th century and the way the children were lodged, all of them living close together.

Of course, nobody would repeat this experiment today, and we would not expect children to 'speak Hebrew, or Greek, Latin, Arabic or the language of their parents' under the conditions described by Salimbene. On the other hand, we may expect a community of human beings to develop some means of acoustic communication. If we carried out the corresponding *Gedankenexperiment*[1], we would have to isolate a group of new-born children, keep them under optimum living conditions, but prevent any acoustic contact with them; we might even shield them from any background noises. The experiment would be restricted to the observation of one generation only; any children born by the members of this group would be taken away from them in order to prevent any acoustic feedback from the babies to the guinea-pigs. The experimenters might expect the members of this group to develop in the course of their lives some sort of verbal communication – with an estimated average lifetime of some 75 years –, nothing with a sophisticated grammar, but some sort of language. To make sure that this was a *proto-language*, the experiment should be carried out with various groups of testees developing verbal communication patterns not significantly different from each other.

Obviously, the question whether there is a *proto-language* is still open, and it might never be answered once and for all, because to answer it means to carry out the experiment described above, and who would dare to do so. There may be analogous experiments involving animals – apes, pigs, dolphins – kept under the conditions of the *Gedankenexperiment* and interpretations transferable to man, thus inducing some sort of solution of the problem of Friedrich II. Nevertheless, we see that theories have to be *tested out* in order to determine their credibility, that testing is tantamount to experimenting, and sometimes the experiments are nothing but *Gedankenexperimente*: in those cases we have to content ourselves with living with a theory relying on the available evidence. ♦

---

[1] This term is one of the few loan-words of German origin without an equivalent in the English language. A *Gedankenexperiment* is only fictitious, imagined and never carried out for various reasons (cf. Sect. 2.2).

In our view, both experiments (Neill's as well as that of Friedrich II) failed, but this may not reflect the opinions of Friedrich II or Neill. We conclude that, in general, there will be *tests evaluating* a model, and the model may turn out a 'good' model or a 'poor' model according to the outcomes of the experiment carried out. The interpretation of these outcomes may be controversial and far from being objective. There are, obviously, quite a few theoreticians of pedagogy trying to save the *Summerhill*-ideas. On the other hand, if we agree that both experiments failed, we have to ask what was wrong. Is the fact that both experiments failed a piece of circumstantial evidence that the theory of Friedrich II as well as Rousseau's model (or Freud's or Reich's theories) are 'poor'? Or does this fact rather show that Friedrich II and Neill carried out the 'wrong' experiments in order to verify the respective theory, or that they performed appropriate experiments, but were just unlucky? It may well be that those theories actually are 'good', but the experiments carried out in order to test them out were 'poor' models of the underlying theories or were done under unfavourable circumstances.

We think, there is another example of a philosophical model reflecting many of the ideas fundamental to our concepts:

Ex. 1.9 Plato's well-known 'Allegory of the Cave'

We do not think it necessary to relate it word by word, but will just report the central ideas, following Dickinson (1931). In his Republic, Book VII, Plato describes the following scenario:

'Picture men in an underground cave-dwelling, with a long entrance reaching up towards the light along the whole width of the cave; in this they lie from their childhood, their legs and necks in chains, so that they stay where they are and look only in front of them, as the chain prevents them turning their heads round. Some way off, and higher up, a fire is burning behind them, and between the fire and the prisoners is a road on higher ground. Imagine a wall built along this road, like the screens which showmen have in front of the audience, over which they show the puppets. Then picture also men carrying along this wall all kinds of articles which overtop it, statues of men and other creatures in stone and wood and other materials; naturally some of the carriers are speaking, others are silent. These men (in the cave) can see nothing but the shadows of objects carried along the wall and will suppose these to be the only reality. Further ... (there is) ... an echo reverberating back from the wall opposite the prisoners, so that any word spoken by the carriers will seem to them (the prisoners) to proceed from the shadows at which they are looking. ...'

Plato then speaks of one prisoner being released from his bonds, looking at 'the heavenly bodies' and seeing the light of the sun for the first time. Later on he returns to the cave and when reporting about the 'reality' he has seen outside, none of his fellow prisoners understands him.

The allegory demonstrates various levels of gaining recognition. At the same time we discover several aspects of modelling. First of all, there are 'all kinds of articles, ... statues of men and other creatures in stone ...'. These are, obviously, articles we would call models (the scale models of Ex. 1.1). Furthermore, these models are not the entities visible to the prisoners in the cave. They are projected again, and only their shadows are the phenomena visible to those dwelling in the cave. Thus these shadows are models of models, and again we are dealing with hierarchies in models. In that case it seems to be impossible to learn about reality, the real objects behind the models. The released prisoner is only capable of reaching a higher level of recognition by distinguishing one level of modelling – the statues – from the following one – the shadows of the statues.

The allegory (Greek *allos*: 'other' and *agora*: 'place of assembly') itself is a model transforming Plato's abstract perception of the gaining of recognition over consecutive stages into a similar scenario. In order to make it more comprehensible, the philosopher breaks his proposition down to its basic thoughts and reassembles them in an entirely different context – the allegory – but with the same message. The allegory serves the purpose of an analogy (Greek *analogia*: 'proportion'): the correspondence between the philosopher's perception and the described scenario. ♦

Ex. 1.10 Mathematical ideas and theories are models which may again be represented by models.

There are simple cultural techniques like, e.g. counting. When counting, we may *represent* each single object to be counted by an equivalent; this may, for instance, be one of our fingers. A barkeeper may keep a record of his customers' consumption representing each drink by a mark – like a slash – on their beer mats. The mathematicians have translated this way of counting into theories of sets, of denumerability, of natural numbers and so on. These methods can be traced back to the beginning of mankind: probably already primeval men, when counting their mammoths 'put out to pasture', represented each mammoth by a pebble or something similar. Mathematical theories, e.g. the theory of denumerability, thus model the way the human brain associates the elements of one set of entities with those of another one. These sets are mutual models of their counterparts. In our example the set of pebbles used for counting represents – within the theory of denumerability – the set of mammoths, and *vice versa.*

In the course of time mathematicians may have refined a given theory by incorporating many axioms. The question arising now is whether this is still 'reasonable', whether the theory is not overburdened with inconsistent axioms, and whether there is still anything 'real' representing it. If no such interpretation of the theory is possible, it deals with the empty set, and propositions about this set are nonsensical. If we state, for instance, 'if pigs could fly – automatically assuming that there are no flying pigs – we would give you a large amount of money', our generous offer to pay this sum is not much of an obligation, because we are talking about the empty set of flying pigs. This is due to the fact that we consider the two propositions 'this animal is a pig' and 'this animal can fly' to be contradictory; we are 'sure' it is impossible for a pig to fly. So the mathematician will have to look for *models*, entities representing – and thus *supporting* – his theory, ensuring that this theory is *valid* for his model and not contradictory.

If we continue with our analysis of the saying 'if pigs could fly ...', we will discover further possibilities of interpretation, though they may sound rather absurd. Suddenly we do not feel so confident after all, because nobody really knows if there will not be flying pigs one day, and then we will be obliged to pay, even though we tried to prevent this by demanding 'flying pigs' as the essential condition for the readiness to pay. The saying does not require that pigs fly by their own efforts like birds do. We know that pigs are quite intelligent and learn easily. Who knows, one day, there might well be a pig with a pilot's licence flying a plane. Maybe, after all, it is not really out of the question to watch pigs fly. The German equivalent saying is more definite; it is, 'if Easter Sunday and Whit Sunday coincide in one day ...'; this is definitely impossible by the definitions of Easter and Whitsun.

Mathematical theories – though consistent and confirmed by the aforementioned kind of model – are often very abstract and hard to understand. In that case the mathematician may try to *project* the purely theoretical structure onto a more concrete one possessing the same relations given within the original, thus *representing* the structure in question, but to be managed more easily.

Mathematical models may be models of entirely mathematical problems. In that case they are means to solve these problems. For instance, numerical methods, like Newton's algorithm or *regula falsi* used for the solution of equations, power series expansions used to determine values of the *Integral Sine* or the *Normal Distribu-*

tion Function, model the underlying mathematical entities – equations and functions. Their purpose is to solve equations approximately or to approximate the values of functions not explicitly representable.

We realize some of these mathematical models are developed to reflect something real, something happening in *reality* like counting, for example. Moreover, a model may be a means to make a complex theory more comprehensible. It then is a *representation* of the underlying theory. Mathematical models are also constructed in order to solve complex mathematical problems approximately, and they are means to prove the consistency of mathematical theories. In this respect they can be compared with the experiment of Friedrich II in Ex. 1.8, with the help of which he wanted to show that his theory was in accordance with 'real' life.

All these examples of mathematical models have in common that they represent imaging processes mapping an object of research into the model. ♦

Ex. 1.11 Mathematical representations of structures and processes in social and natural sciences and in humanities are models.

Mathematical calculi are used to represent complicated non-mathematical 'real' structures and processes. These representations constitute a different example of mathematical modelling. In general, they will be extremely simplifying, but they may lead to solutions of problems occurring within the original processes.

There is a vast range of applications of this sort of models. They represent physical phenomena observed in physics, chemistry and biology as well as statistical data models used in medicine, for instance, to determine the correlation between smoking habits and diseases of the respiratory organs. Besides, they are used for textual emendations in literature – Kaufmann (1967) has given an interesting example concerning this subject – as well as for the identification of fingerprints when combating crime. They serve to optimize economic decisions, e.g. in inventory theory (cf. e.g. Taha 1976).

In all those cases the mathematical calculus may lead to a mathematical solution of the mathematical optimization problem, and, maybe, this solution can be transferred to the underlying non-mathematical situation represented by the model. And if we are lucky, it will also yield a solution of the non-mathematical problems arising within these fields of research (cf. Ex. 2.4).

There are quite a few problems connected with each of these fields of application of modelling. One of them is to determine the *quality* of the set-up of the model. We shall analyse this problem in the following chapters. Nevertheless, we already had reason to reflect the adequacy of models when we discussed the model of Friedrich II and the realization of Rousseau's educational model by Neill (cf. Ex. 1.7 and Ex. 1.8). We summarize that models may be means to *mediate* between *complex phenomena* observed in *reality* and our understanding of them. ♦

Ex. 1.12 Non-mathematical scientific models.

Frequently scientific models are non-mathematical. The direction of modelling as an imaging process – as we have just set out – may be reverted: non-mathematical models can thus also be used to represent mathematical phenomena.

We start off with the *abacus*, a frame with beads threaded on wires. This instrument is used for arithmetic calculations within the ring of integers. The beads and their positions within the frame represent various powers of a given base which may vary among different groups of users. The rules of calculation are also defined by the users of the abacus. There are various designs used in different cultures (Menninger 1958). Abacuses as well as slide-rules, for example, facilitate rather complicated arithmetic calculations by using mechanical analogues. More complicated devices, like adding machines and computers, use different physical states in order to represent mathematical entities.

Apart from this example illustrating the way mathematical operations are modelled by physical processes, there may be phenomena observed in nature and models developed in order to explain them. Well-known examples are the phenomenon of biological evolution which can be explained by Darwin's theory of natural selection and the observation of the phenomenon of compounds being formed of different elements in chemistry; it is explained by the theory of valences and the model of the atom.

A more material example is the visualization of complex molecules by models consisting of table-tennis balls in different colours, each colour representing an element, each ball standing for an atom. In these models the valences are visualized by press studs glued to the surfaces of the balls. The positions of the connectors represent the respective angles of valency. Models of the atom are constructed in the same way with balls of various sizes visualizing the atomic nucleus and the electrons. The distances between the 'electrons' and the 'atomic nucleus' within these models represent the respective energy levels.

In physics, when we try to determine the trajectories of bodies rotating round their common centre of gravity, we may portray the problem in simplified terms by concentrating the mass of each of the bodies in one point of space – their centre of mass. This is the model of material points.

All these models are means to facilitate the manipulation of mathematical problems the way the *abacus* does – akin to Newton's algorithm, *regula falsi* and the power series expansions of Ex. 1.10 – or the mass-point-model. They may either be *theories* trying to explain phenomena observed in reality, or they may be means to visualize complex theories difficult to understand. In these cases the model mediates between something difficult or incomprehensible and our limited capabilities of understanding or dealing with it. This difficult or incomprehensible matter may be a mathematical entity, a phenomenon observed in reality or a theory developed in order to describe such a phenomenon.

Sometimes the objects of research are even phenomena hardly comprehensible, like, for instance, the fact that the dinosaurs disappeared after they had existed for some 100,000,000 years during the Mesozoic Era. There are several theories trying to explain this phenomenon. They model abrupt changes in the living conditions of the reptiles, but none of them can account for the phenomenon of their sudden extinction; they all are speculative, and there have been no means yet to *verify* them. ♦

Ex. 1.13 Measurements, Characterizations

When we wish to characterize a body, we may do so by taking photographs or by *modelling* it in a different material, but we can also measure its physical properties. We may describe it, e.g. as a rectangular solid and measure its dimensions, and we may determine, e.g. its mass (or weight) or its electrical conductivity. So, eventually, we get a set of data (and non-numerical descriptions) representing the given body.

There are non-physical means of numerical characterizations. Cars, for instance, can be identified by car numbers. The employees of a firm are identifiable by personnel numbers or, for instance, by Social Security Card numbers. ♦

We shall now discuss a further group of mathematical and non-mathematical models. The way they represent 'reality' is by no means obvious at first glance, and the interpretation of the models, in general, depends on agreement within the groups using them.

Ex. 1.14 Symbols, Icons, Metaphors

We have already studied counting and the representation of natural numbers by marks on a beer mat (Ex. 1.10), and we have talked about calculations performed on an abacus (Ex. 1.12). In these examples the natural numbers are mapped into sets made up of symbols or ensembles of objects – like beads threaded on a wire or pebbles representing a set of mammoths as we have seen in Ex. 1.10.

One way of interpreting natural numbers is to regard them as cardinal numbers. In this case we may decide to represent each such number by a set of marks of the corresponding cardinality. Thus, the sets of vertical lines, respectively of dots, in Fig. 1.1 stand for the natural numbers one to six.

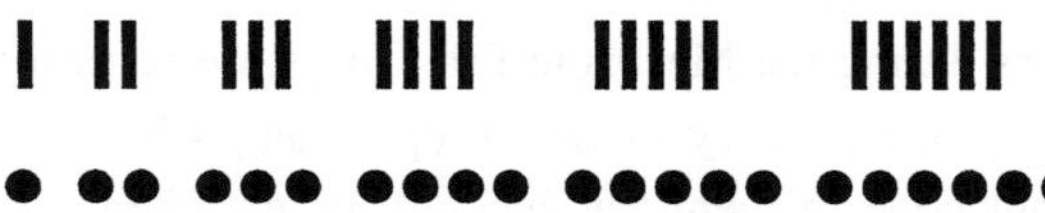

Fig. 1.1. Symbolic representation of the natural numbers 1, ..., 6

These sets of signs represent the cardinal numbers in a rather *obvious* way, and most people who are able to count will interpret Fig. 1.1 correctly as representations of the numbers one to six. Any other sign, e.g. ♣, ♠, ♥ or ♦, used as an element of the sets given in Fig. 1.1, will serve the same purpose.

For better legibility we shall now rearrange the symbols, e.g. those in the top row of Fig. 1.1. A common way of rearrangement is to cross out four consecutive parallel marks by the following fifth as shown in Fig. 1.2. This representation of natural numbers is frequently used in tally-charts.

Fig. 1.2. Tally-chart representation of the natural numbers 1, ..., 6

This is a still comprehensible way of counting by representing each natural number – interpreted as a cardinal number – by a set of this cardinality. And even most of those not familiar with tally-charts and this way of arranging the marks within each symbolic representation of the numbers five and six will probably interpret Fig. 1.2 correctly.

I II III IV V VI ... X ... L ... C ... D ... M ...

Fig. 1.3. Roman numerals

As soon as we represent the natural numbers by the Roman numerals, as in Fig. 1.3, the situation gets more complicated. We may guess that **V** stands for the number five and **L** for fifty and so on if we possess a complete list of these numerals ranging from one to, say, some 1,100. But there is no obvious way of interpreting this representation of natural numbers when we look at them out of order. These symbols cannot be interpreted without a knowledge of their historical development. One possible explanation is depicted in Fig. 1.4 where **V** symbolizes the thumb and the little finger of a hand representing the number five. **X** is the combination of two **V**s, **C** stands for *Centum* – the Latin word for 'hundred' – and **M** for *Mille* – the Latin 'thousand'. **L**, finally, is the lower half of **C**, the way we present it in Fig. 1.4, and **D** is the right half of **M** in this representation. **VI** for six is obviously **V** plus **I**. On the other hand, **IV** representing the number four as **V** minus **I** has to be agreed upon, and there is still the more comprehensible Roman representation of four by **IIII**, too.

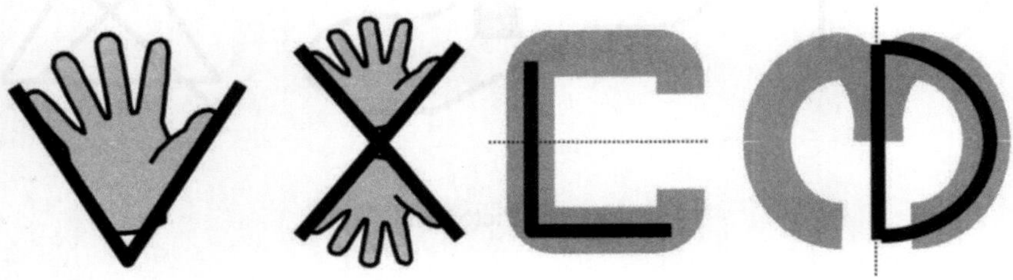

Fig. 1.4. Interpretation of the Roman numerals

The Arabic numerals can be traced back to similar 'obvious' abstractions. Today we just agree on taking the number symbols **1**, ..., **9** as representations of the numbers one to nine and **0** for nought. Furthermore, we agree on building up the numbers exceeding the numeral **9** by *decimal* grouping, but different cultures have developed their own rules of constructing numerals. The English 'twenty-one' is 'vingt et un' (twenty and one) in French, but it is 'einundzwanzig' (one and twenty) – the other way round - in German; 'ninety-one' is 'quatre-vingt-onze' (four (times) twenty (plus) eleven) in French, to give some examples.

The situation is even more complicated when we regard further abstractions, like the introduction of the symbol ∞ not representing a number, but rather any numerical process exceeding any given numerical limit. The Chinese abstract in a similar way by compressing singular forms to abstractions. As we have pointed out we might have used any mark in order to represent the elements of the sets visualizing the cardinal numbers in Fig. 1.1. If, thus, we had used signs representing a tree, a hair or a woman, any triple of these signs would have stood for the number 'three', but by abstraction they also represent the Chinese words 'forest', 'hair' and 'gossip' (Menninger 1958). These abstractions may still be comprehensible, but in the course of time the various symbols used in the singular form, thus representing one 'tree', one 'hair' or one 'woman', have gradually changed their graphic reproductions and have adopted rather abstract forms: they have become *icons* (Greek *eikon*: 'image'), and it is almost impossible to grasp their meanings and to relate them to the familiar *real* trees, hairs and women, the objects of the mappings: you must simply know them.

Fig. 1.5. Pictograms

Pictograms are sometimes even more difficult to interpret. They are iconic, too, and if we look at Fig. 1.5, we may think pictogram *A* on the left symbolizes something like the order 'turn left', but it could also be the *logo* of a company, or it may be incomprehensible to any person who does not interpret the union of a square and a triangle as an arrow pointing to the left. Pictogram *B* represents an escalator to most of those who have seen an escalator before. In order to comprehend this pictogram, we have to know what an escalator is and what it is used for. The information we can get from this pictorial representation can be converted into linguistic analogues, like 'there is a means of transport to get from one floor of this building to the next one'. The third pictogram C is even more complicated to understand. We all know it from the instructions of textile manufacturers, and we have learned that it means 'no bleaching'. Actually, we do not really see what a triangle has to do with bleaching, but we do admit that the ✕ – crossing out the triangle – looks like a warning better not to 'triangle' our textiles showing this symbol. The rest of the pictogram, the triangle, irritates us a lot, because this symbol makes us helpless and confused at the same time.

If we regard symbols – icons and pictograms – as models, the images of something 'real', like numbers, directions, escalators, washing instructions, we realize that their correct interpretation depends on *evidence*. This available evidence is founded on our experience and knowledge, on our cultural background and on a *consensus* among the members of a group of persons using these models for communication (a *community*).

The linguistic analogue of an icon is a metaphor (Greek *meta*: 'over' and *phorein*: 'to carry', i.e. 'to transform'). It is a reduction onto the basics, transforming a complex train of thought into a word or a saying reproducing the underlying ideas like 'to be a tower of strength'. ♦

Ex. 1.15 Euphemisms, Swearwords, Camouflage

Some people rather talk about *Beelzebub* than about the Devil, they say a person *passed away* instead of saying he died, and they regard his corpse as his *remains*. These are *euphemisms* (Greek *eu*-: 'good' and *phanai*: 'to speak'), expressions less unpleasant than the facts they describe. In a similar way the Red Indians used to talk about 'going to the happy hunting-grounds' when they wanted to avoid speaking of death. Here again the correct interpretation of the words depends on the consensus among those using them. Maybe, we still guess what is meant by *Beelzebub* if we are well-versed in the Bible, but it is more difficult to understand what the word *remains* means when it is used as a euphemism. And we think it is impossible to know what *really* happened to someone who went to the 'happy hunting-grounds' unless we are informed about the particular meaning of this idiom.

A euphemism will sometimes serve as a *camouflage* (French *camoufler*: 'to disguise'). In that case it may be used in order to enhance the status of something or someone possibly regarded as inferior. It may be the projection of an ideal and used to serve the purpose of a *nom de guerre*. A prostitute, for instance, may call herself a *model*, thus euphemistically avoiding the correct term used to describe her profession. Here the term *model* – as used in our Ex. 1.3 – itself is used for modelling, and it serves the purpose of improving the social position and thus the social reputation of the 'model'. On the other hand, in that case the *camouflage* is only intended half-heartedly, because the person calling herself a *model* does not really want to deny her proper profession completely. She would certainly be disappointed if there were no other clients than those asking her to pose for a painting. So, there must be a *consensus* between these professionals and their clients about the meaning of the word 'model'.

The same is often true for swearwords as we all know from experience. If we call somebody an ass, we do not want to insinuate he is the animal; we rather want to express that he is a stupid person. Talking about asses when talking about foolish persons is founded on *consensus*, too: there is nothing stupid, foolish or ridiculous about a real ass, on the contrary, *real* asses seem to be rather intelligent and far from being stupid. Likewise, the acceptance of a swearword depends on consensus. There are words just tolerated by members of one social class, while a mortal insult in another. The usage of the word 'bloody' as a swearword may serve as an example. While it is an absolute taboo for the members of the middle and upper class in Great Britain, it is commonly used by lower class people in order to emphasize their anger; it serves as a synonym for words like 'utterly' or 'extremely' when talking about, e.g. 'a bloody waste of time'. The same is true of different civilizations where such words may socially be just acceptable in one cultural area while a mortal insult in another one. ♦

Ex. 1.16 Reconstructions

Models may be gained by reconstruction. Archaeologists design models of ancient cities destroyed thousands of years ago and covered by layers of sediments, like Pompeii or Troy as examples. These reconstructions are sometimes based on only few remains of the original city. In the same way palaeontologists try to reconstruct a specimen of a certain species such as dinosaurs, from some excavated bones.

As the modelled entities do no longer exist these models convey the imagination of their creators reflecting individual experience. The skin colour of dinosaurs, for instance, is permanently assumed to be a greyish green, but up to now nobody knows whether they ever looked like that. The moment the scientist gets a piece of information concerning the colour of the skin he must correct his model accordingly. Thus, the reconstructions have to be constantly adapted to the growing insight into the relevant field of research.

Yet, there are also examples of an extremely dubious way of reproducing models – from our point of view. In this context we refer to the reports of Phlegon 'paradoxographus Traillanus'. He was a Greek freedman of the Emperor Hadrian (AD 117-138, * 76) and came from Tralleis (Latin *Tralles*), an ancient Carian city on the Meander River. Phlegon was a paradoxographer; he compiled paradoxes and published them in his 'Book of Marvels' (Hansen 1996). Among these 'marvels' there are several reports on giant bones found within the ancient world. These reports agree with the reports of other ancient historians; obviously they were fossils, skeletons of mammoths, elephants, whales and others (Mayor 2000). According to Phlegon (and other ancient historians), 'there was an earthquake' in the reign (AD 14-37) of Emperor Tiberius Claudius Nero (Caesar Augustus *BC 42) 'in which many ... cities of Asia Minor disappeared. ... In the cracks in the earth huge bodies appeared that the local inhabitants were hesitant to remove, although as a sample they sent to Rome (to Tiberius) a tooth of one of the bodies. It was not just a foot long but even greater than this measure. ... Tiberius ... avoided the sacrilege of the robbing of the dead. He summoned a certain 'geometer' (mathematician), Pulcher by name ..., and bade him to fashion a face in proportion to the size of the tooth. The geometer ... made a construction. Tiberius ... sent the tooth back to where it had come from.' (Hansen 1996). 'The facts are historically and scientifically sound, ... the size of the giant tooth (about 12 inches) that was sent to Tiberius corresponds to the length of the molar of a woolly or steppe mammoth. ... Pulcher's replica (was) presumably a grotesque humanoid bust of clay or wax ...' (Mayor 2000). This 'reconstruction' will certainly not stand up to modern scientific theories of the 'existence of giants'; nevertheless, it is an interesting example of modelling. Obviously, ancient people had evidence that there had been or even still existed creatures of gigantic size, much larger than human beings, and they concluded that these 'giants' looked or must have looked like human beings, but on a larger scale. The 'replica', ordered by Tiberius, contains two different kinds of models. First, there is the reproduction of the tooth, the original was sent back 'to its home soil for reburial' (Mayor 2000). This reproduction in a different material was supposed to be an exact copy of the fossil. But there was also the 'reconstruction' of the giant's face and of all the missing parts of the original body not present in the collection of the found fossils. This reconstruction shows how the scientist – a contemporary of Tiberius – represented the picture he had of the 'giant'; this picture was based on the then *available evidence*: it was just the human body which he simply reproduced on a larger scale. Furthermore, the members of the scientific community in the days of Tiberius obviously agreed on the fossil being part of a giant and accepted Pulcher's reproduction as adequate. There were no means to scrutinize and *test* these ideas, they just seemed perfectly *plausible*.

Other reproducing models are phantom portraits of criminals wanted by the police. They are (re-) produced using the descriptions of witnesses, and they are improved with the help of further evidence. The reproduction of the course of a criminal offence is another example. In that case a whole sequence of conditions and actions is modelled by an assumed plot taking findings and available evidence into account. We accept this reconstruction as true until we have to correct it according to the growing knowledge of further details.

Crash-tests performed with newly designed automobiles may serve as a further example. The dummies used in these tests are models of the human body, and the results of these tests serve as evidence for certain reactions of the man-machine-system to external stresses. The scenario of severe accidents will thus be simulated, and the evaluation of the damages to the cars and the dummies will lead to the recognition of risks car-drivers are exposed to in reality. ♦

The models in Ex. 1.16 are related to our next example. It may be regarded as a more complex generalization of the aforementioned reproductions.

Ex. 1.17 Specimens

A scientist may find a specimen of a species of animals or plants extinct or never seen before. This specimen may either be dead, but well-preserved (e.g. petrified) or still alive. It is a model of its species, and its appearance is regarded as representative of the whole. In case there is a small number of these specimens, an average of the samples will be considered to be characteristic; this typical representative is made up of *all* the characteristics of all the found specimens according to the law of averages and does not exist in reality.

In this example the find is not used to reproduce a single specimen, but rather to 'reproduce' an unknown species, an entire group. So in Ex. 1.16 and Ex. 1.17 we take one part of an ensemble as a representative of the whole – a non-linguistic kind of *pars pro toto* (Latin 'taking a part for the whole'). ♦

We now enter into the description of a whole class of models serving the purposes of a community of supporters of the respective theories. We shall outline these examples *sine ira et studio*, impartially, because they will certainly be discussed controversially by the proponents and opponents of these theories, and we are not a party in these controversies.

Ex. 1.18 *Anima* and *Animus*.

Jung's conception of the syzygy *anima* and *animus* (e.g. Jung 1951), a male's respectively female's perception of female respectively male essentials, are models of each of the other sex. They are archetypes[1], rather vague dispositions constituting the psychological set-up we are unaware of. They are inherited, they are part of a person's genotype, and they are projected onto the members of his community. Jung thus comes to the conclusion *mulier animam non habet* 'a woman does not have an *anima*'. This is plausible as long as the term anima is used in Jung's way, but it sounds rather strange when *anima* is interpreted as 'soul', the original meaning of the Latin word. ♦

Within the following item we compile a whole number of examples, partly esoteric, partly not quite compatible with orthodox scientific points of view:

Ex. 1.19 The Asian philosophy of *Yin* and *Yang* is very similar to Jung's conception as mentioned above, and the well-known symbol representing this idea (Fig. 1.6) is a further example of an icon (cf. Ex. 1.14). It shows that both ideas, the dark part *Yin* and the light part *Yang*, are counterparts, mutually adding up to the universal circle, and on the other hand, *Yin* bears parts of *Yang*, and *vice versa*.

This philosophy associates terms like 'female, dark, passive, humid, Moon, water, the colour black, even numbers and chemical elements like Potassium' with the idea of *Yin*, and 'male, light, active, dry, Sun, fire, the colour red, odd numbers and chemical elements like Sodium' with *Yang*. These associations seem unusual and disconcerting, and again there must be some kind of a *consensus* to agree on them. Maybe, some of these associations of syzygies like female-male and pas-

---

[1] The term *archetype* is used by Jung with the meaning mentioned in this example. There are still other meanings of the word. It may be used as a synonym for prototype (cf. Ex. 1.2). Black (1962) calls an *archetype* '... a systematic repertoire of ideas by means of which a given thinker describes, by *analogical extension*, some domain to which those ideas do not immediately and literally apply ...'.

sive-active or humid-dry will be accepted even by somebody not following the ideas of that philosophy. On the other hand, laymen may have difficulties when they try to associate even numbers or Potassium with 'female' and odd numbers or Sodium with 'male'. No mathematician will ever recognize different sexes in the natural numbers he is used to work with[1], but a supporter of the philosophy of *Yin* and *Yang* will think it quite 'natural' that even numbers are female and odd ones male. Jung, part of whose works deals with mysticism and alchemy, seems to have been in favour of these arguments. There is a very detailed line of argumentation in his books why, for instance, the number four is female whereas the number three is male. There are even more subtle arguments for the 'Systematic Symbolism of Numbers' (e.g. Bischoff 1982), which may be traced back to the early Pythagoreans and the mysticism of the *Kabbalah*. We admit we can somehow follow this line of argumentation, but we are incapable of accepting it as convincing. We cannot consent to the conclusions the mystics and alchemists drew from the construction of natural numbers.

Fig. 1.6. Yin and Yang

Another example, strongly related to the philosophy of *Yin* and *Yang*, is the 'theory of the (biological) *transmutations* of elements'. The French biologist Kervran developed a theory of biological transmutations (e.g. Kervran 1972). In short, this theory postulates the mutations of chemical elements under the influence of low energetic biological processes. Kervran promulgates that these transmutations are due to the exchange of oxygen or hydrogen nuclei between the mutating elements. According to him, biological reactions will, for instance, convert Potassium to Calcium by adding hydrogen to the respective nucleus $K_{39}+H_1 \rightleftharpoons Ca_{40}$; this implies

[1] They say natural numbers have been created by God and the rest of mathematics is the work of man; but the mathematicians do not mean it the way creationalists do, they just want to stress the fact that these numbers are the *natural* foundation of mathematics.

that – following this thesis – biological transmutations are tantamount to low energy nuclear reactions, this is not in concordance with basic physical principles. Kervran responds to any criticism based on this argument that physical laws are applicable to physical processes only, but not to biological reactions. An orthodox physicist, on the other hand, will argue that chemical reactions are also subject to physical laws, and biological processes like those of metabolism are based on chemical reactions, consequently they also underlie the generic physical principles, too. Kervran gives a lot of evidence for the *validity* of his theory. He observed that hens excreted more limestone than they had taken in with their food, that watercress seeds did not contain certain minerals he later detected in the young plants, and he concluded that this increase in substances must have been caused by biological transmutations.

There is an even more interesting article by Ohsawa (1971), who wrote a 'book condensation' of Kervran's publication and added some experiments he had carried out himself. He had 'transformed' carbon into iron using a low-cost apparatus assembled from some pieces of wire, a copper plate and a variable 110-230 V transformer. The theory of biological transmutations seems to be strongly related to the principles of the philosophy of *Yin* and *Yang*. Ohsawa, for instance, interprets the above-mentioned 'transmutation' of carbon into iron as the union of *Yin*-oxygen with *Yang*-carbon using (low) biological energy.

This theory of biological transmutations resembles that of *cold fusion*. In 1989 S. Pons and M. Fleischmann, both at the University of Utah, announced to have achieved nuclear fusion at room temperature in some sort of a jam-jar. Further attempts to reproduce these results failed, and Pons and Fleischmann admitted some misinterpretations of their experiments. Nevertheless, though incompatible with basic physical principles, the idea of cold fusion – transforming a low energetic input into a high energetic output – has haunted people all over the world ever since. Among these there are a number of reputable scientists, and even institutions working in the fields of high technologies have obviously been spending enormous sums of money on the attempt to find the *Golden Fleece of cold fusion*.

Even more irritating is the example of creationism competing with Darwin's theory of the evolution of species. Creational science has been accepted in Kansas, USA, as a theory with equal rights to evolutionism. Creationists strictly cling to the Genesis (the first book – Moses 1 – in the Old Testament) and *literally* apply it to natural sciences. Accordingly they believe that the age of the universe is not more than a couple of 10,000 years. This, of course, leads to countless difficulties, for example, in explaining the fact that dinosaurs lived some hundreds of millions of years ago whereas the maximum age of creatures is only several thousand years according to creationism. Nevertheless, this theory is an acknowledged scientific model taught in the State of Kansas, USA. When channel-hopping, we recently popped into a lecture[1] given on Trinity Broadcasting Network (*TBN Europe*); obviously it was part of a whole series of lectures. The speaker called up the names of well-known scientists as witnesses to creationism. This list reads like an excerpt of 'Who is Who in Sciences' containing names like N. Kopernikus, C. Babbage, J.P. Joule and, finally, the name of a NASA manager. Some of the cited confessions were just something like 'I believe in a creator ...', which does not necessarily imply the belief in creationism. Further-reaching propositions like the calculation of a maximum age of nature ascribed to Kopernikus may be explained by the age he lived in.

We must say that we cannot accept the arguments brought forward by the promoters of those theories. This may well be due to our scientific background, but we do not think the evidence in favour of biological transmutations, cold fusion or creationism is convincing. But we admit, the day when biological transmutations and cold fusion will be reproducible and when irrefutable evidence is produced supporting creationism all classical natural sciences will be doomed. ♦

All these examples show that there must always be a *consensus* about the acceptance of the respective models. We admit that some of the cases we have studied are on the borderline of science (or do not seem to be 'scientific' at all), but this line is hard to define. The same goes for many fields of science, and the transitions between the different branches are often fluid. The complex field of psychology is a good example to demonstrate this. Let us start off with psycho-physics, which is considered an 'exact' science: its hypotheses can be tested out by physical experiments, and we accept them if the results of research bear close scientific examination, otherwise they must be rejected. Within the branches of humanistic psychology, like, for instance, psychosomatic medicine and psycho-analysis, the testing out of a theory is more difficult, sometimes it is rather questionable, and the evaluation of the tests is often based on assumptions. So we have to be ready to accept a theory as *well-grounded*, even if it is only supported by the results of a few observations. This may eventually lead to inaccurate and in fact unscientific methods, and it explains why philosophies like *Yin* and *Yang* or similar mystical cults must necessarily depend on the *consensus* among their followers. The crucial

---

[1] This talk was broadcast on Jan. 21st, 2002, 6 p.m. CET.

question arising now is where do we leave the grounds of exact sciences and where do we enter the domain of para-sciences[1], when are our arguments still strictly scientific and when do we begin to strain our credulity to its utmost limits by believing in fantastic theories. Now we all know that there are groupings within larger communities who accept certain propositions as *evident* and therefore find themselves in opposition to the rest of the scientific world. The more these people ignore the objective rules of scientific working methods, the more they need the consensus of those who share their point of view and accept their not verified ideas as 'credible' and 'true'.

This borderline between exact sciences and the respective para-domains runs through most scientific fields. There are the spheres of orthodox medicine and those of alternative medicine, also called complementary medicine, like homeopathy, acupuncture, the cures of Aryuveda – deeply rooted in Eastern philosophies –, faith healing, herbalism or the dubious theories of magnetism to mention only some. As we have seen in Ex. 1.19, there is also a kind of alternative chemistry. The there described theory of biological transmutations is incompatible with the orthodox laws of chemistry and therefore condemned by conservative chemists, who categorize these ideas as alchemy[2]. Furthermore, para-psychological effects dealing with the occult and the physically inexplicable, like telekinesis (psychokinesis) and telepathy are widely discussed objects of 'research', even at some reputable institutions.

Obviously, the consensus among the followers of each of these schools of thought depends on some sort of blind belief in the 'validity' of the basic principles of the respective theory and the conclusions drawn from them. This agreement is strictly limited to the members of these schools of thought, and there is nothing like a peaceful coexistence between the fields of para-sciences and their corresponding orthodox counterparts. No convinced creationist will bother to take any

---

[1] We use the term 'para-science' as an analogue of para-psychology and para-physics; the term is derived from the Greek *para*, 'at the side of' or 'beyond'; thus, when entering the field of para-science we leave the domain of the corresponding 'original, i.e. orthodox' field of science.

[2] Alchemy (Arabic *alkimiya*: 'chemistry') was originally the art of transmuting metals; later alchemical ideas were applied to pharmacy (e.g. by P.T. Paracelsus) and the principles of spagyrics (Greek *span* and *ageirein*: 'to extract' and 'to collect') were developed; today the supporters of the theory of transmutation believe that in nature elements freely change back and forth into one another.

notice of the orthodox biologist's theory of evolution, to say nothing of acknowledging it, and, of course, *vice versa.*

The 'validity' of a theory is determined by the verifiable results and the recognition gained from experiments meeting certain 'quality criteria'. Then, and only then, this belief in the theory in question is rational and can be quantified by a 'degree of rational belief'. But unfortunately, the tendency to believe in irrational theories grows stronger, the less they are part of a system of rational, mathematical arguments. Mathematical reasoning will be accepted rather[1] unanimously. Nevertheless, though the impossibility, e.g. of the trisection of an angle or of squaring a circle with a straight edge and a compass has been proven, there have still been several attempts to make the impossible come true. Lengthy elaborations describe 'solutions' of these problems, and sometimes it is not easy to find out what is wrong with them. Fortunately, these 'constructions' are no longer as numerous as they were in the past; hobby mathematicians seem to have lost interest and have finally accepted the fact that their efforts are futile. Within physics the situation still seems to be strained. Again and again we learn about a new and this time 'successful' construction of a *perpetuum mobile*. Even renowned scientists are tempted to outwit nature and the laws of physics when they try to realize cold fusion[2] or to experiment in the field of biological transmutations. The reason may well be the degree of confirmation of propositions. A mathematical proof is everlasting and does not permit any exception to the proven statements – as long as the proof is accepted. Yet, there are certain schools of thought, as we have mentioned above, who deny the validity of mathematical proofs widely accepted by the rest of the mathematical community. On the other hand, the validity of physical principles depends on *consensus* just as propositions do in all the other scientific fields, and we think we can trust to them if they are confirmed by the observation of phenomena which are in concordance with these principles. Nevertheless, we might be wrong, and these principles might turn out to be invalid, and then, one day, we

---

[1] We have to qualify this statement by adding 'rather' because of the well-known debate about the foundations of mathematics, where intuitionists (cf. e.g. Troelstra 1977) call for constructive reasoning and refute the validity of certain chains of mathematical argumentation and basic mathematical assumptions leading to proofs of some famous propositions accepted by the rest of the mathematical community.

[2] Of course, these experiments may lead to the detection of yet unknown chemical reactions, but the belief in the possibility of nuclear fusion carried out at room temperature is based on uncertainty whether we can depend on basic physical principles or not.

might be able to construct a *perpetuum mobile*, perform a cold fusion or prove the possibility of biological transmutations.

We are far from disparaging the efforts of *para*-scientists who struggle with phenomena inexplicable by orthodox scientific methods, and it is hard to determine the borderline between exact science and *para*-science. A scientist always wants to transcend the boundaries of his field of science, he dreams of realizing the apparently impossible, because this would be a breakthrough causing a *scientific revolution*. Thus, the Pythagoreans transcended the borderline of then accepted scientific recognition when they postulated the spheroidicity of the earth which up to then had been assumed to be shaped rather like a disc or a cylinder. Kopernikus, likewise, left the ground of the then established Ptolemaean geocentric conception of the world when he took up again the assumption of a heliocentric universe, a model that was originally evolved by Aristarchos from Samos (~310-230 BC). There are further physical examples, like the introduction of probabilistic considerations into physics leading to thermodynamics or quantum mechanics, each of them giving rise to the development of new physical models. Strangely enough these 'outsiders' who research within the boundaries of 'normal' sciences and are often jeered at by orthodox scientists again and again make the great discoveries and break new grounds in science.

These shifts from traditional and at their time established scientific points of view to radically new theories have influenced non-physical sciences as well. For quite a long time standard geometrical ideas, for instance, have been exclusively Euclidean, and the Euclidean point of view still seems to be the most adequate to the human capacity to think in three dimensions. Still, non-Euclidean geometries and topological models have been developed using different systems of axioms which lead to a multitude of geometric *manifolds*[1] differing from the Euclidean spaces of the respective dimensions.

Non-mathematical models transcending established scientific theories of their time like, for instance, the ideas of *New Age* (e.g. Pribram 1982) are sometimes strongly connected with esotericism and occultism. The acceptance of those 'theories' depends on the period of time when they are published, and they are influenced by prevailing main-streams, the so-called *Zeitgeist*, the spirit of the times. We stress the fact that we do not support most of those ideas, but we do not want to belittle them, either. Esotericism (Greek *esoteros*: 'the inner circle') is a science intended for and intelligible only to those initiated. It contrasts with exoteric

---

[1] A *manifold* is a *topological space* which is locally equivalent to the Euclidean space.

(Greek *exoteros*: 'the outer circle') doctrines intended for and intelligible to outsiders, to the general – that means lay – public. According to these definitions any kind of science is *literally speaking* and *ipso facto* esoteric, it is intelligible only to those familiar with the foundations of the science in question, with its particular way of reasoning and its technical terminology. Likewise, occultism (Latin *occultus*: 'hidden') deals with the supernatural beyond the range of ordinary knowledge. Following this definition any scientist is an occultist trying to reveal the occult, still hidden and unknown – but both terms 'esoteric' and 'occult' are nowadays used with a different and not at all scientific meaning of the words.

We thus agree with Snow (1959) when he talks about 'two cultures' cut off from each other[1]; but the problem seems to be that there are already many different cultures and subcultures, exclusive and not compatible. Each member of each of these (sub-) cultures is an esotericist, and most of them are not willing to communicate with any of their neighbouring communities. Thus, the development of sciences depends on the unanimity among the members of each of these scientific communities, and whenever communication among these various branches is necessary there are serious problems. Whatever seems to be well-founded and 'logical' within one branch of science may be unacceptable for the members of a different community. Physically sound conceptions like fast-breeders or stem cell research may be politically inopportune. In some countries the turnout at elections tends to some 50% (or even less) of the electorate, so there may be a minority of laymen deciding on extremely difficult scientific problems with far-reaching consequences, 'evaluating' the relevant scientific models – by voting.

This quite arbitrary list of examples demonstrates the large-scale range of the term *model*. Though there are many different meanings, we see that there is one common underlying idea if we interpret a model as an *image* of a certain entity subject to a mapping. These entities may be of quite differing nature, the process of modelling may serve various purposes, and the relation between the mapped entities and those produced by the imaging processes may differ according to these purposes. We are only interested in *scientific* models and therefore single out these cases of interest. As we have shown, the borderline of science is not easily determined, and the transition from exact science to para-science and superstition is fluid. One way of classifying models may be to study the *purposes* they serve and the various subject/model-relations established by them more closely.

---

[1] These are the cultures of natural sciences and humanities.

Before we enter into these details we shall try to give a broader definition of the term model by a brief *etymological* analysis of the word.

1.2 Etyma

An *etymon* (Greek *etumos*: 'true'; plural *etyma*) is a word forming the linguistic root of terms derived from it; it is the original form of these derivations. When studying the origin of the word 'modelling', we will see that there are various terms describing the process of creating a model, the image of something 'real' or regarded as being 'real'.

Etymological roots of the word *model* lead back to the Latin noun *modus* and its diminutive *modulus*, the *module*. Originally a *modulus* was the radius of the base circle of a rotationally symmetric column. This radius was subdivided into thirty parts – *partes*, *minutes* (from the Latin adjective *minutus*: 'small') – of equal length. The *modulus* was used to define the further dimensions of the building the column belonged to. These dimensions were related to the *modulus* or to the *partes* in a particular way, thus ensuring the proportions of the building. Modern architects likewise use the *modulus* or *mudulor* as the length of a certain part of a building and equally define the dimensions of the rest of the building by certain multiples of it. Le Corbusier's *modulor* (cf. Le Corbusier 1983) is based on the proportions of the *Golden Section* of geometry applied to the measurements of the human body.

In mechanical engineering the term module is used to describe the proportions of cogwheels. It is also a term used to represent characteristic sections of flowing waters. In addition, a module is a closed subsystem in a technical system (e.g. a computer). The mathematicians talk about modules as relations between logarithms of a number relative to different bases; certain algebraic structures are called modules, too (cf. e.g. Northcott 1962). The *modulus* thus is a means to measure certain quantities and to define relations between them.

A *modus* is something measured: it describes the way something is or is done, and it is the root of the English word *mode*. Thus the *modus operandi* is the particular way in which persons perform a task or the way a thing operates; the *modus vivendi* is the way of living or of coping and arranging in a temporary compromise. The English words *modality* and *mood* as well as the French noun *mode* are derived from the Latin *modus*. The German noun *Mode* and the Italian *moda* are equivalent to the English word *fashion*, which is derived from the Latin verb *facere*, 'to make'.

Modulation – a word also derived from *modus* and *modulus* – is the change in a measure, the variation of physical parameters, e.g. the frequency of radio waves. The shift from one key to another in music is a *modulation*, too. A *modulator* is a device used for *modulating*.

The *mould* (or *mold*) as a special sort of model (cf. Ex. 1.2 and Ex. 1.4) as well as the verb derived from it can be traced back to the Latin roots *modus* and *modulus*. The German language distinguishes between *Model* ('mould') as the special case and *Modell* ('model') as the generic term.

Though modelling was well-known and performed in the ancient world, there seems to have been no general term covering all the aspects of a model we have compiled. Modelling was also regarded as an imaging process in those days. The original of those processes (Greek *prototypos*: 'original', 'prototype') was mapped into something 'similar' (Latin *simulacrum*). So there were two different terms to describe what we call a model: the *prototypos* for the original and the *simulacrum*[1] for the image of modelling (Irmscher 1987). The merits of this differentiation are quite obvious: it takes the direction of the mapping into account leading from the prototype to be copied to its copy, the simulacrum. Various modern publications classify models under these differentiating aspects, though there are slight differences in terminology. Apostel (1961), for instance, mentions the relation between *model* (simulacrum) and *prototype*. Fig. 1.7 shows this relation between the various types of models: prototype and simulacrum according to the ancient terminology.

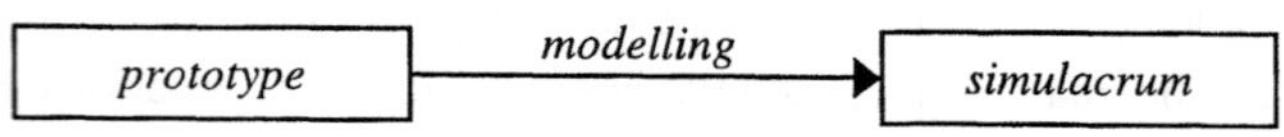

Fig. 1.7. Objectives of modelling

There is another etymological connection of the word *model* with the (old) Italian noun *modello* which reflects our ideas connected with 'modelling'. *Modello* was a last, a three-dimensional image of the human foot used as a prototype for shoemaking (the modern Italian word for 'last' is *forma*: 'shape') just as it also denoted

[1] The *simulacrum* is somehow analogous, it is similar in performance; the term 'simulator' is derived from the same root (Latin *similis*: 'similar').

a dressmaker's dummy, an image – simulacrum – as well as a *prototype*, viz. a *simulacrum* of the client's body as well as a *prototype* for the garments to be tailored. Today the noun *modello* is the Italian expression for a 'model' equivalent to modern English usage.

As we have seen, the term *model* is used to describe both sides of a mapping: the original of the mapping *process* as its *prototype* and its 'similar' image as its *simulacrum*. But we have also described examples of models serving both purposes: they are a *simulacrum* as well as a *prototype*. Unfortunately, the colloquial usage of these terms is rather vague. We return to the examples dealing with modelling cars in series mentioned in Ex. 1.6. In the terminology of automobile producers a 'prototype' is the first *sample* of a *series* of cars in working order; it serves as the basis for the production line #0. We have introduced the words 'sample' and 'series' only for better understanding of the sentence, they may both be replaced by the term *model*. The 'sample' is the *prototype*-model, and the series is, e.g. the *Tin-Lizzy*-model, *FORD*'s *T*-model of Ex. 1.6. The 'real' *prototype* as the 'original' *idea* of the series is a design study (Ex. 1.1), a technical conception or the analysis of the market resulting in 'just an idea'.

We have seen that the noun *model* is overused, indeed. Obviously there is a lack of differentiating words reflecting each particular shade of this term. Only when we talk about *modelling* there seems to be unambiguity. Modelling is a *process* mapping one model into another, no matter which type of model – *prototype* or *simulacrum* – is involved, serving as an original or as an image. Starting from this point of view we can easily make out many more examples of models, like translations from one language into another (cf. Staal 1961), words[1] and phrases as representations of certain ideas, metaphors[2], allegories (like Plato's 'Allegory of the Cave') and parables, fingerprints, traumas and phantom limb pains, to mention only some. Literally 'there are models everywhere', analogous to Bertalanffy's (1968) proposition 'there are systems everywhere'. It is reasonable to borrow this statement from Bertalanffy if we regard modelling from the viewpoint of general systems theory, and we shall return to this aspect when we describe a model as a system representing a system and the process of modelling as a system of its own.

---

[1] For the interpretation of words as 'models' cf. for instance Ubbink (1961), where he mentions Plato's concern that 'a word corresponds to the object it designates notwithstanding that it is not identical with the object'.

[2] The connection between language and modelling, especially the interpretation of metaphors as models has been outlined in great detail by Black (1962).

We shall now concentrate on the imaging process of modelling and study it more closely. But in order to distinguish between the different meanings of the term *model*, we first want to analyse the various purposes a model may serve in greater detail in contrast to Sect. 1.1 where we just touched upon them in passing. The specific purpose is one of the major criteria of classifying models, though there are further characteristics of imaging processes as we shall expound in Chap. 4.

1.3 Purposes

We are now going to examine the examples in Sect. 1.1 more closely with respect to the *purposes* they are supposed to serve. Some of them have already been mentioned there, but we want to find out the means necessary to achieve them as the purpose of a model will be one of the criteria used for its classification in Chap. 4.

The three-dimensional models of Ex. 1.1 may be characterized as *scale*-models (cf. e.g. Black 1962). Each of them represents some sort of *abstraction*, which implies that each of them lacks a certain amount of the features of its original. This loss of *information* about features of the entity mapped by modelling has been decided on by the modeller under *utilitarian* aspects. The toy car for small children is a model reduced to the uttermost. Its sole purpose is to give a rough impression of a car; this basic requirement is fulfilled if it resembles an automobile in a vague, but still significant way. Obviously, an infant associates a car with four wheels fixed in pairs to two axes connected with each other, while an in-line arrangement will not trigger off the same associations. In that case any further detail of a car is superfluous and thus omitted. For easier operation a handle may be fixed to this combination of four wheels, and it will not spoil the overall impression. This model serves the purpose of arousing associations.

The iconic models of Ex. 1.14 are constructed with the same aim. The pictograms are reduced to some characteristic features of the entities they represent. As we know by now the ability of interpreting those abstractions correctly depends on the personal situation of the observer. We cannot interpret the number symbol **9** as an equivalent to the number 'nine' if we have not learned it and *agreed* on accepting the symbol **9** to represent the number 'nine' and be *equivalent* to it. This symbol is without any meaning to those educated in other cultural groups using different numerals. Likewise, we must have seen an escalator before we can interpret the respective pictogram of Fig. 1.5 correctly. A small child with no knowledge of cars will not accept a four-wheeled representation of an automobile as a toy car.

Maybe, the child will regard it as a toy, because it is highly manoeuvrable, but it will have no further associations with it. This is also true for an experienced adult who will hardly be able to interpret the pictogram of Fig. 1.5 abstracting bleaching if he has not learned the meaning of this symbol. The pictogram of the escalator – somehow resembling the real object – may still look strange to a child of two or three even if it has used it before. Thus the acceptance of abstractions depends on a *consensus* within the community using them; this may be achieved either by associations (toy car) or depend entirely on agreement and definition (pictogram of the washing instruction).

The collector's model of a car – reproducing almost every detail of the outer appearance of the original, but abstracting from any interior fittings like the power system – serves the same purpose as the toy car, but on a different level. By concentrating on the technical details of the design, only reducing them in scale, these models associate the impression of a certain type of automobile, for example the *Tin-Lizzy*-model of Ex. 1.6. If any detail characteristic of the appearance of the car should be missing or be out of place or be not reproduced true to scale, the collector will not accept the model as *authentic*. His demands for the authenticity of the model are entirely restricted to the outer appearance.

The model of a car to be studied in a wind-tunnel is an abstraction, too, as it only reproduces the shape of the original and some parameters of the material used for the bodywork. Mostly, it is not only a reproduction on a reduced scale, but also an idealization. Certain features of the original are exaggerated, others may only be suggested. The appearance of the model sometimes differs considerably from the looks of the original. These models are made to learn about the road performance of the projected car and to find out about the costing. They are studied in order to prevent mistakes when designing the series, mistakes that may lead to a reduction in maximum speed, to wastes of fuel or to loss of control over the car. Model aeroplanes – if they are not only used as toys – may serve the same purpose. It is too expensive and too dangerous to test real planes at the risk of damage. So certain characteristics of their performance are tested on cheaper models and without the risk of lives.

Architectural models, lasts, a dressmaker's dummy and dummies used for crash tests by the automobile industry are similar examples. They are idealized representations of their originals – sometimes near to caricatures of, e.g. real feet or bodies – and are to prevent errors when building houses or planning estates, when making a shoe or tailoring a dress. The crash-test dummies are *substitutes* for corpses carrying certain human features. They offer a 'cheap' opportunity to study the impact of severe accidents on the human organs whereas the original human being as an experimental subject is too 'dear'.

Under the aspects of *abstraction* and *idealization* the fancy 'monster-car' models, garden gnomes and dolls are abstract and idealized representations as well. As abstractions they are to arouse associations with real cars, miners and babies. The exaggerated idealization of a representation is sometimes near to a caricature, like the gnomes, and may stimulate the observer. He gets a kick from the martial look of a fancy car, and the exaggerated 'sweetness' in the face of a doll evokes maternal feelings in the girl playing with it.

The *prototype*-models of Ex. 1.3 are developed in order to set certain standards and to keep them. They facilitate cheap production of mass articles of equal quality. They serve standardization and in the case of ideals (possibly represented by ideal persons) they may be a means of indoctrination helping to strengthen certain ideologies. The choice of an artist's model and the way it is reproduced in his works of art may establish a certain style of painting.

The models of Ex. 1.4 serve the same purpose. They are also prototypes, but some of them possess different technical features – like being hollow forms. The purpose of all these models is the inexpensive, fast and effective reproduction of standards preset in the models. The images of these *prototype*-models are the *series*-models of Ex. 1.6, unified series of standardized copies of the prototypes.

In Ex. 1.5 the model garments are created in order to have influence on trends of fashion and art. Very often they are exaggerated, idealizing the couturier's ideas. If these models happen to set a fashionable trend, they are idealizations *prior* to a development, they initiate it. They are thus different from the idealizing models we know from the previous discussions; those idealizations are *posterior* with reference to something already existing. These somehow trend-setting models foreseeing the future development in aesthetics are not subject to scientific considerations. But we may find similar examples of scientific models, roughly sketched approaches to new theories intended to explain observed phenomena or to solve problems arising in a field of research.

The models of Ex. 1.7-1.10 are all closely related, but they serve different purposes. Some of them are theories more elaborated than the rough sketches mentioned above. The purpose of these *theory*-models is to *describe* or to *explain* certain phenomena in the field of research of the respective branch of science. As soon as we are aware of a phenomenon we want to understand its nature. Theories are attempts to solve this problem and to make the underlying phenomena comprehensible.

To describe a phenomenon means to reflect its nature, to give an account of what it looks like. An explanation of a phenomenon shows the way it works, what effects it causes. The problem-solving capability of a theory may be limited, because the imaging process used to produce the model may disregard parts of its original. On the other hand, modelling may lead astray. A theory may be overburdened like the aforementioned condition 'Easter and Whit Sunday are to fall on the same day', because as soon as we scrutinize this statement by recalling the definitions of the ecclesiastical year, we realize that there is no day in the year representing Easter and Whit Sunday. Mathematically speaking we are talking about the empty set, and any proposition with regard to this set is true; it is thus meaningless.

Scientific models may be inadequate attempts to explain the nature of a phenomenon. Let us assume, for instance, that we observe a significant difference between the water-levels of a river at a certain place during the day and at night, say, the level at night exceeds the level during the day considerably by some 50%. Well, we might say, this is easily explained by a higher rate of evaporation in the daytime. In that case our theory is that lower water levels during the day are caused by a higher rate of evaporation, and this seems to be logical. We may test out this theory by making an experiment: we determine the rate of evaporation in the daytime and at night at the very location and compute the different water-levels of the river. Let us assume the computed data are far from those observed in nature. We will conclude that our model does not really explain the observed phenomenon. Now we may develop more sophisticated theories like the rays of the sun 'pressing down' the water surface in the daytime. Probably there will be no test confirming these theories. For further investigation we may go upstream and detect a sewage plant emptying its tanks at night, and we may think, *this* is the explanation of varying water-levels. Testing this new theory, we may find out that the nightly discharge can only partially be blamed for the rising water-level. Obviously, our current theory is better than the preceding one, but it is still not a 'full' explanation of the phenomenon in question. Five more miles up the river there is a plant consuming water in the daytime, another 'explanation' of the observation. Our model now consists of higher levels of evaporation, extraction of water during the day and sewage discharge at night. Say, this adds up to some 80% of the phenomenon. Well, there is still a problem, the *phenomenon* of a difference between observed water-levels and those computed, yet inexplicable by all the constructed theories.

To sum up, one can say that theories are means to describe and explain phenomena and the observations of phenomena; they *mediate* between some sort of 'reality' and science, but their explanations may be unsatisfactory. Models may as well be means of *testing* the *adequacy* of theories yielding a *justification* for them, like testing the influence of evaporation mentioned above. Furthermore, a theory may be accepted as a – partial – explanation of an observed phenomenon if there is no further evidence. If, for instance, we cannot go upstream, we rely on the evaporation model, though we will doubt whether it is really adequate.

A theory may as well lead to the prediction of the occurrence of phenomena not yet observed; in that case the first observation of these phenomena is evidence for the predictive qualities of the theory in question. Physical theories may thus be supported by phenomena observed in nature or produced by experiments. Likewise, the detection of 'missing links' is evidence in favour of Darwin's theory of evolution.

On the other hand, an experiment carried out in order to justify a theory may be inappropriate. The theory of the Emperor Friedrich II who believed in a proto-language, has got something striking. We know that animals – dolphins, apes and dogs, for instance – may convey information acoustically. The first protozytes communicated by chemical means only, but as soon as the organs needed for talking and hearing had been developed, they were used for communication. So, we may be sure that Eve *talked* to Adam when she persuaded him to eat the apple. So possibly, the theory of Friedrich II is not wrong, but his experiment carried out in order to justify it was inadequate. Nowadays, tests necessary to find out would certainly be different from those performed by Friedrich II in the 13th century; maybe, one day there will be a more definite confirmation or refutation of the proto-language-theory.

The justification of a theory and the degree up to which you can rely on it depend on the available evidence and on the latest developments in research. This implies, of course, that a model must constantly be reworked according to the results of the latest research. So, the giant's head constructed by Pulcher (Ex. 1.16) was regarded as an appropriate reconstruction of the creature whose tooth had been detected in Asia Minor until modern palaeontologists finally discovered the species of dinosaurs and mammoths. Likewise, Euclidean axioms were regarded as the only means to define geometry until mathematicians brought forth non-Euclidean geometries in the 19th century (cf. Sect. 1.1).

Quite a few theories are very complex. In order to make them more comprehensible, they themselves are made subject to modelling. Thus, the theory of valencies or atomistic theories stimulated the development of models of the atom with the intention to *visualize* complex theories of the structure of matter and thus *explain* them. They are idealizations and deformed representations of the underlying concepts. The electron cloud of the atom, for instance, may be represented by an ensemble of spheres each the size of a table-tennis ball representing an electron and its electric charge. The complex structure of Riemann's surfaces and their branch lines have been illustrated to generations of students of complex analysis by models knitted or crocheted by generations of professors' wives. These models resemble piles of pancakes joined together by a common radial seam (Fig. 1.8). Of course, the structure of a 'real' Riemannian surface looks quite different and is much more complicated than that of any of these caricaturing idealizations. Nevertheless, the model of Fig. 1.8 demonstrates the idea of identification of the leaves of the surface along the branch lines. It helps to understand the internal structure of the surface subject to the model.

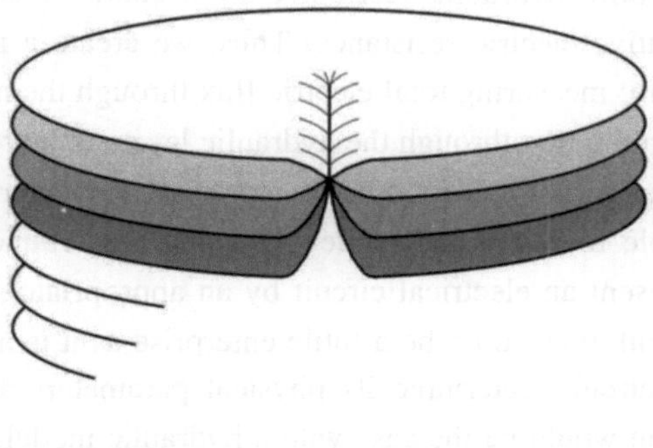

Fig. 1.8. Illustration of a special Riemannian surface

If we characterize these models with the help of the traditional terms which we mentioned, we must call them *simulacra*, being somehow similar to the objects of the various modelling processes. Sometimes the models may be even more sophisticated and *replace* (e.g. physically) an observed process by another process of the same type in an analogous manner. Some physical processes, for instance, are represented by *laws of transportation*. These laws describe the transport of entities, e.g. of particles, under the influence of a difference in potential. Their general representation is

$$\vec{q} \propto -\vec{\nabla}(\Phi), \tag{1.1}$$

with $\vec{q}$, the number of entities passing through a unit of square measure per time-unit. According to (1.1), $\vec{q}$ is proportional to $\vec{\nabla}(\Phi)$, the gradient of some potential function Φ and in the opposite direction. There are various interpretations of this generic proportion, and we thus gain a couple of physical laws. We shall just mention three examples to make that point of view clear: Fick's law of diffusion deals with the transport of particles under the influence of varying solute concentrations, and Ohm's well-known law describes the functional dependence of the transported electrical charges on the gradient of the electrical field; Darcy's formula, finally, reflects the relation between the flow of water in a porous medium and the decline of piezometric heads.

So we may by analogy examine one physical phenomenon by *replacing* the physical law governing it by another one *similar* to it. If, for instance, the object of research is a structured layer of sediments, we may represent the various layers possessing specific hydraulic resistance by a chain of electric conductors with equivalent relative electric resistance. Thus, we create a model of the given hydraulic structure; measuring total electric flux through the model, we infer a corresponding hydraulic flux through the hydraulic layers to be researched.

If an electric circuit can be used to represent a structure of sediments, it must also be possible to apply this model the other way round: that means we may, likewise, represent an electrical circuit by an appropriate structure of porous media. But we think this would be a futile enterprise as it is much easier to construct the electrical circuit, determine its physical parameters and carry out the flux-experiment than would be the case with a hydraulic model. So, obviously, there is some kind of symmetry between the two physical systems mapped onto each other: the hydraulic and the electric system. Nevertheless, sometimes only one direction of mapping is reasonable and bears a problem-solving capacity. It leads from the original, *less tractable* system to a model we can handle more easily.

Heuristic (Greek *heuriskein*: 'to find') procedures are somehow linked with these models. They are to induce new scientific methods and discoveries. A heuristic method being part of an educational system is supposed to stimulate the pupils to find out things themselves. In those cases the discovery of something new is achieved by the transfer of insights gained from the heuristic model to another somehow similar problem. If, for instance, we have to solve systems of linear equations, we have to study the theories of general n-dimensional vector-spaces and n-dimensional linear *manifolds*. In order to make the axiomatic approach to these theories more plausible, we may start off with studying two-dimensional vectors on a sheet of paper, for instance the way they are added. The only objective of this heuristic geometrical study is to develop a greater understanding of the abstract axioms defining the case of general dimensions by visualizing them in the plane.

In Ex. 1.11 we mentioned the mathematical representations of structures and processes in, e.g. social and natural sciences. These models are means of *simulation*; they represent existing problems by mathematical *calculi*, thus making the mathematical and numerical treatment of these problems possible. The mathematical calculus is an *approximation* of the process it represents. These approximations by mathematical models are also idealizations, and they are based on idealizing assumptions. We may, for instance, assume the decrease in stock to be linear when we design an inventory model. Of course, this will not be a correct representation of the real development in stock, it only illustrates the dependence of stock on time. Assuming a linear relation between stock and time means to approximate this relation by a continuous function, whereas it will often be discontinuous, e.g. when the stored goods are counted in pieces.

As a further example we mention the use of continuous probability distributions as a means to describe stochastic phenomena. We may talk about *normally distributed* parameters, say, the weight of a new-born child. Of course, there is nothing like a normal distribution of this weight, because normal distribution allows babies weighing 2,000 tons and even infants having negative weight. These values are not very realistic, but they are not entirely outside the range of the Normal Distribution. Nevertheless, within certain limits the assumption of normally distributed weight of the mentioned persons may be a good approximation of the actual distribution of weight, and it may thus be a good basis for further calculations.

These probabilistic models are *simulacra* in the way described above, but the means of representing the given problems is rather theoretic. These models consist of a mathematical calculus and of assumptions of the probability distributions of their parameters. Probability theory and statistics provide examples of hierarchical modelling. In a first phase of modelling we may, for instance, describe observed phenomena by probability distributions. We may then find out that it will be extremely cumbersome to deal with these distributions and therefore decide to replace them by more manageable ones. Thus, statistical experiments are described by idealizing discrete probabilities which may be approximated by continuous distributions to be calculated more easily.

In the examples of counting (Ex. 1.10 and Ex. 1.14) intellectual processes are represented by physical analogies. When we count and perform simple arithmetical calculations, like addition and subtraction, using our fingers, we represent natural numbers – interpreted as cardinal or ordinal numbers – by our fingers. These representations are quite obvious, and they are limited to small numbers. In this way cardinal numbers are represented by sets of outstretched fingers possessing equal cardinality. The ordinal numbers are represented by the order we extend the fingers one after the other. If the mechanical representation is an abacus, the powers of a given basis are represented by beads in separate parts of the frame. The interpretation of the position of each single bead within the instrument is due to consensus within the group of its users.

Computers represent mathematical entities by physical phenomena either in an analogous or in a discrete way. The discrete representation is achieved in a binary number system by interpreting the existence of a certain physical state as the digit 'one' and its non-existence as 'zero', the other digit of the number-system. Continuously varying physical states are used to represent mathematical entities in analogous computers.

Modelling a physical phenomenon by a set of data gained by measurements as described in Ex. 1.13 is an example of a data-model used to characterize the phenomenon by the values of certain of its physical parameters. Assigning identifying marks to an object somehow serves the same purpose. The licence plate fixed to an automobile is an unambiguous identifier – or at least it should be available only once – if it is not a fake. Yet this quality is restricted to certain 'user-groups', there may well be the same plate-numbers issued in different countries. A distinguishing mark of a person like a birthmark may help to single its owner out of a group; most probably this mark exists only once, other possible birthmarks may just look similar – though two identical marks cannot be definitely excluded. The blood group of a person may be a piece of circumstantial evidence that the person belongs to a certain ethnic group, but it will certainly not single out the person. Further values of physical parameters are even less significant, like the weight or the height of a person. Nevertheless, a set of these parameters and their combination will help to characterize a person: overweight in combination with high blood-pressure and certain habits, like excessive smoking, drinking and consumption of saturated fatty acids, are indications of the risk of cardiac diseases.

These data-models used to characterize and classify physical phenomena must be evaluated under the aspects of certain *quality criteria* they have to satisfy. There is still the problem to decide to what extent these models can be trusted and to determine their degree of adequacy when characterizing the objects of modelling. These problems will be analysed in subsequent chapters. For the moment we give the example of evaluating the significance of increased values of cholesterol for the danger of life-threatening illnesses. The common practice of limiting values as a counter-measure does not seem appropriate any longer. Obviously, the cholesterols have to be subdivided into various classes bearing different risks. Furthermore, there seem to be genotypic dispositions for certain levels of cholesterol not necessarily significant for higher or lower risks. Developments in research give rise to the supposition that cholesterol levels rising with growing age may even prolong life-expectancy.

Another example of the significance of data models is the existence of contradicting reports on the dangers of smoking or drinking while the *French paradox* seems to prove that a moderate consumption of alcoholics reduces the risk of heart attacks. People do feel uncertain when they hear that some of these 'reports on scientific experiments' have been paid for by the producers of alcoholics or tobacco. We realize that *objectivity* is of prime importance when enumerating the quality criteria these data models have to meet.

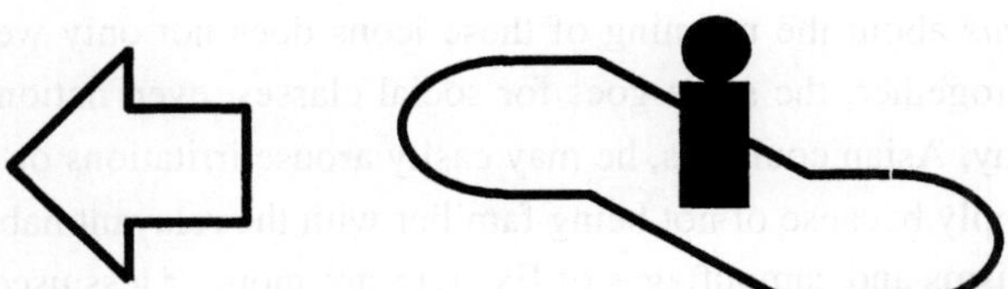

Fig. 1.9. Combination of pictograms

The symbols and icons of Ex. 1.14 are means to visualize complex situations. They condense a long-winded explanation into one sign or into a combination of a small number of signs. The meanings of these signs can be interpreted at a glance, whereas a written explanation is a complex construction of letters that form into words, which then build sentences which may even be linked, thus finally bringing forth a message which must be grasped. If for instance, we combine the two icons of Fig. 1.5 to form the pictogram in Fig. 1.9, we will learn from it that we have to turn left if we are looking for an escalator. The meaning of the pictogram is equivalent to that of the sentence 'turn left for escalator', which is a shorter representation of 'if you are looking for an escalator, you should turn left'. It may stand for the more correct, but long-winded statement 'if you want to move from one floor of this building to another one, turn left, there is a moveable means to do so'. Yet, sometimes icons may cause more bewilderment than give elucidation, and we must confess that we would prefer written warnings to some of the symbols sewn into textiles (e.g. that on the right side of Fig. 1.5).

Of course, the correct interpretation of the pictogram of Fig. 1.9 depends on the *consensus* that the left part means 'to the left' whereas the right half is the representation of an escalator. These representations are *abstractions*, and they *idealize* their objects.

There are further examples of 'icons' appealing to our different senses. Certain sounds uttered by a person signal special feelings. A groaning may be an acoustical 'icon' of pain or of well-being, for instance, depending on the intonation of the sound and the special situation. There are certain code-words triggering off reactions not comprehensible to those not belonging to the initiated. On a party we may meet a group of people roaring with laughter as soon as they hear such a code-word as only they get the hidden message while others are left outside feeling awkward. We also communicate with our eyes. There are optical signals and gestures informing us and serving the purpose of an 'icon'.

The *consensus* about the meaning of these icons does not only weld the members of groups together; the same goes for social classes, even nations. If a European goes to, say, Asian countries, he may easily arouse irritations or get into serious trouble, simply because of not being familiar with the relevant habits.

The euphemisms and camouflages of Ex. 1.15 are more or less used in the same way as we use the icons of Ex. 1.14. They are code-words with a special meaning. But their purpose is not to describe and make obvious what is meant by the various entities subject to these models. They are used with the intention to hide the negative meaning of an expression. While pictograms transform verbal communication – in the first place appealing to your mind – into non-verbal communication – in the first place appealing to your senses –, euphemisms and camouflages remain in the realm of verbal communication and make use of the ambiguity of words. But they may easily result in misunderstandings; so if we look for a model displaying clothes and thus contact with a person firming under the profession of a 'model' in the telephone directory, maybe, we get the wrong connection.

The reconstructions of Ex. 1.16 visualize entities only partly known. They are approximations of the unknown 'reality' of the past – like the giant's head in Tiberius' times – or of 'reality' of today – like a prosthesis. They abstract from essential features of the original and they idealize the original; in the latter example the prosthesis may look quite different from a real limb (for example an artificial leg) or organ (e.g. an iron lung), just simulating its performance.

The act of reconstruction may lead astray as we have seen in the example of the 'giant's tooth'. Policemen reconstructing a crime may, likewise, be on the wrong track. Reconstructions are meant to explain the origin of some finds. These models are to *induce* answers to questions like 'why ...' and 'what ...'. The example of a specimen described in Ex. 1.17 is something similar. Its purpose is the reconstruction of an unknown – i.e. of an extinct or not yet fully discovered – species of organisms. It is entirely based on the evidence given by the one found sample only – or a small number of specimens. This process is mirrored by surrealism, a school of arts, where *objets trouvés* (finds) – arbitrarily collected objects – serve the 'reconstruction' of a 'fictitious – surreal – reality' according to the artist's imagination.

The anima- animus- and *Yin-* and *Yang-* models of Ex. 1.18 and Ex. 1.19 are theories, and they serve the purpose of explanation. They try to explain the peculiarities of the natures of certain entities and their differing characters. But as Jung's anima-animus-model is restricted to the exploration of the human psyche, the philosophy of *Yin* and *Yang* is more comprehensive. It enlists a series of simulacra – of similar nature – as complementary pairs adding up to a unity. These models signal the borderline of science, and their basic statements do not stand up to close examination.

2 Systems

In Chap. 1 we demonstrated the ample range of the meaning of the term *model*. In spite of the diversity of those examples they all share some characteristics; for instance, they are all linked to a modelling process. In some of those examples the models are *prototypes* and subject to this process, in others this mapping produces the model as a *simulacrum*.

However, whether subject to or resulting from a modelling process, *prototypes* as well as *simulacra* can be characterized under structural aspects. We shall try to do so by using the methods and the terminology of *systems analysis*. The *system* is the generic term, and when in Chap. 1 we come to the conclusion, 'there are models everywhere', we have just borrowed this idea from Bertalanffy (1968), who headed the introduction to his monograph on *General System Theory* with the proposition 'Systems Everywhere'.

2.1 Definitions

A *system* is a set made up of several parts to form a whole (from the Greek *syn-*: 'together' and *histanai*: 'to set'). These *parts* (Latin *pars*: 'part' or 'portion') are interlinked and form a complex structure organizing the system. So being *structured* (from the Latin *construere*: 'to build') means that the parts constituting the system are interconnected and that there are certain relations between them (cf. Fig. 2.1); being *complex* (Latin *complectere*: 'to embrace') means that the structure is composed of a multitude of parts related by rather complicated rules. This complex structure *organizes* (Greek *organon*: 'tool') the system so that it serves a certain purpose.

The parts of a system may be systems themselves. In that case the system consists of a set of *subsystems*. When we say that the parts of a system *may be* subsystems, we want to emphasize that we look at the structure of a system from the *atomistic* point of view. There is no infinite progression of subdividing the constituents of a system. These subdivisions end when we have reached the level of the atoms of the system; we call them the *elements* (Latin *elementum*: 'basic substance') of the generic system.

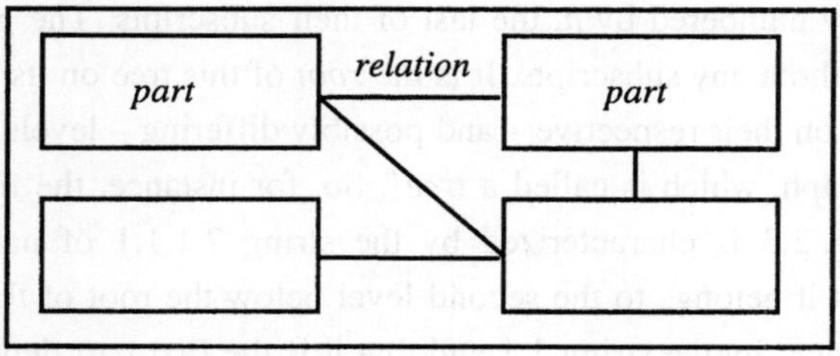

Fig. 2.1. Structural aspects of a *system*

Though the elements of a system may be further subdivided – physically, chemically or under other aspects –, they are elementary with respect to the inner logic of the system and the 'generic purpose'[1] it is to serve. They are atoms determined by the system itself.

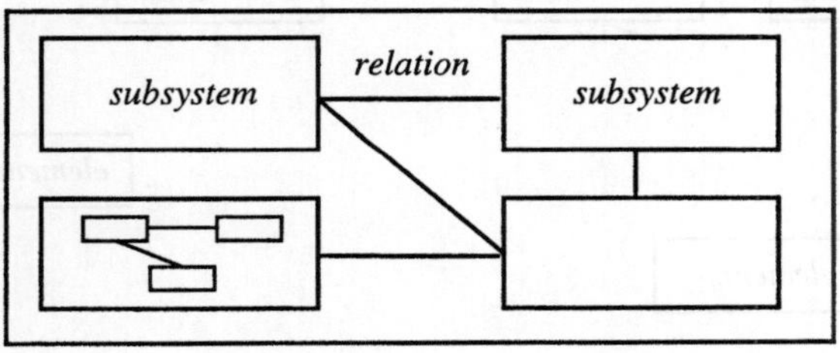

Fig. 2.2. Fine-structure of a system

This fine structure of a system leads to a *hierarchy* (Greek *hieros*: 'sacred', and *archein*: 'to rule') in subsystems and the *sub-subsystems* constituting them

[1] The term 'purpose' is somewhat too expressive. Actually, the system is defined under *teleological* aspects which are not necessarily 'purposes'. We shall discuss these considerations of teleology and finality in Subsect. 2.2.6 of this chapter.

(cf. Fig. 2.2). The hierarchical structure of a system can be represented graphically by the tree of Fig. 2.3, where the position of a subsystem is represented by a sequence of subscripts *i*, *j*, ..., *n* of natural numbers. The hierarchical level of a part of the system is marked by the subscript *i*; the following string *j*, ... represents the chain of subscripts of the hierarchical *parent* of this part. Parts with a common parent are serially numbered by *n*, the last of their subscripts. The generic system is represented without any subscripts. It is the *root* of this tree on its top level. The various *elements* on their respective – and possibly differing – levels are the *leaves* of this special graph, which is called a *tree*[1]. So, for instance, the first part of the third row of Fig. 2.3 is characterized by the string 2,1,1,1 of natural numbers which means that it belongs to the second level below the root of the tree, that its parent is represented by the string 1,1 and that it is the first part depending on this parental sub-system.

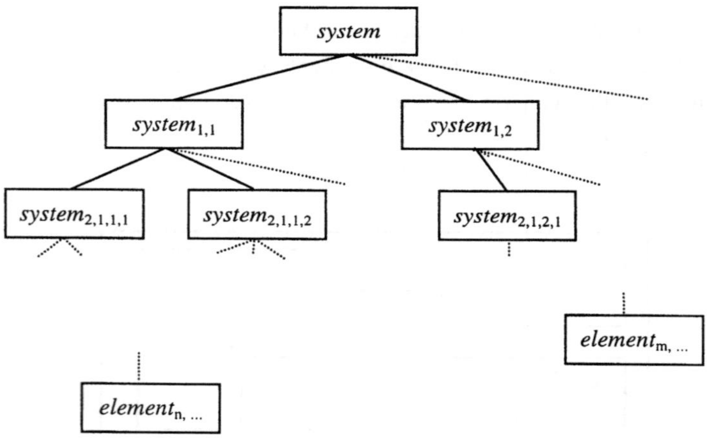

Fig. 2.3. Hierarchical structure of a system

The constituents of a system may be material or immaterial. A generic system may be purely material or purely incorporeal (e.g. dialectic), but it may also contain parts of both types. The objects of modelling processes are systems as well as the

---

[1] Any graph without non-trivial circuits – closed sequences of edges – is called a tree; this means that whenever we follow a path leading back to its starting point it is just a mere going for- and backwards the same connecting edges of the graph.

images they produce. Thus, according to our examplesin Chap. 1, automobiles as well as philosophical theories are systems. The image of an individual is a system, and the human being itself is a system, too. We said that the classification of systems under material and dialectic aspects is not disjunctive, that there are various degrees of overlapping of material and immaterial systems. A physical experiment, for example, done to test a given physical theory consists of *hardware* and of *software*; that means there is the material construction on the one hand and the ideas underlying its set-up, the considerations controlling its performance, on the other hand. A pictogram must be perceptible by the senses, so it is partly material, on the other hand, the *consensus* enabling the interpretation of the pictogram is part of its *software* and thus immaterial.

There are further ways of classifying systems; those of greatest relevance to our considerations are compiled in the following section. They will allow to distinguish between different conceptions underlying the term *system* and thus induce a differentiation of the various meanings of the word *model* we introduced in Chap. 1.

2.2 Characterizing Systems

Systems may be characterized from various points of view, for instance, under dynamic, probabilistic and other aspects. Just like the aforementioned classification of systems – as material or immaterial – these characterizations are not disjunctive either and, in addition, they may be transferred onto the term *model*. So, they will help to concentrate on those cases relevant to modelling in sciences.

2.2.1 Degrees of Abstraction

As we have already seen in Sect. 2.1, a system may be material or immaterial, this means dialectic, abstract and theoretical. Besides, the transition from one of these criteria to its complement is fluid. This is an important point when we characterize a system according to its degree of abstraction. Obviously, there is no purely material system not bearing incorporated abstract and theoretical aspects. A sledgehammer, for instance, a tool to serve the special purpose of breaking stone is something very concrete and may – metaphorically speaking – also be used 'to crack nuts' or 'to kill flies'; at the same time it also represents a 'philosophy'. It is the product and the result of a long-lasting evolution of tools. Its development is due to growing insight into the laws of power transmission. The shape of its head has been influenced by experience, gained from experiments, carried out with various hammer-heads, and the proportions of the hammer itself are the result of ergonomic considerations.

On the other hand, a theory is dialectic and seems to be an entirely immaterial system. But, in general, a theory is worked out to explain certain phenomena which occur within material systems. When we have developed a theory, we want to explain it to a body of scientists. We may use didactic aids to illustrate our lecture. These may be concrete examples or visual aids, like the knitted Riemannian surfaces or the models of the atom using table-tennis balls. Furthermore, a theory must be tested. We want to know if it 'works', if it is 'good'. So we have to carry out experiments in order to find out the quality of a theory; and these experiments involve material systems. To experiment means to manipulate material systems, and it involves observations and measurements. Yet in spite of this, there exist entirely dialectic experiments: we call such an experiment a *Gedankenexperiment* (cf. Sect. 1.1), an experiment never carried out, we just pretend it was done, and we discuss possible results. The reason for doing *Gedankenexperimente* is that they either cannot be carried out or that we do not want to do so. The former may be the case when we discuss scenarios occurring in the distant future or having occurred in the past. The latter may be the case when we are mainly interested in the aftermath of natural disasters or in other catastrophes like explosions in nuclear power plants, for instance.

Most mathematical theories are entirely dialectic, just like their models and the examples given to explain them. Of course, any mathematician can explain and understand the structure of a Riemann surface without the visual aid of a knitted model. On the other hand, some mathematical theories were once developed in order to explain observed material phenomena. In this way the theory of the natural numbers was induced by counting material objects. The representations of these objects were often corporeal themselves: the fingers of the person counting, the beads of an abacus or the marks on a beer-mat (cf. Ex. 1.10). And even elementary arithmetic was modelled on manipulations of these material representations of numbers. But soon this interrelation between the mathematical theory and the materialized phenomena of counting or, for instance, adding was relegated to the background. In a process of abstraction and hierarchical development of systems, having derived each system from the preceding ones, today the theory of numbers, for instance, has nowadays gained a degree of abstraction not suggesting the smallest relation to anything material any longer.

2.2.2 Degrees of Complexity

As we have already stated, a system can be defined as a *complex* structure. The next step to be taken is to concentrate on the *degree of complexity* of a system and thus find a new distinguishing mark.

We are aware that this criterion is far from being objective. There are systems looking rather simple to one person and very complex to another one. The motor of a lawn-mower will be a simple – i.e. not very complex – system to the mechanic repairing it. If a layman takes it apart, he will most probably not be able to reassemble it properly afterwards.

So, what we mean by the degree of complexity as a differentiating aspect is the number of parts in a system and of the relations between them as they appear to the observer. This criterion is rather a means to compare different systems under the aspect of their *extent* than being an absolute measure. We will, for example, agree that a nuclear power plant is a *more* complex system than a ball-point pen.

2.2.3 Degrees of Determinacy

We may consider a system and its parameters to be determinate and call it *deterministic*. This means that the values of the parameters defining the system are regarded as being known to the observer at any moment of time; he will thus be able always to determine the state this system is in exactly. Or we may characterize the system as being *probabilistic*. In that case the processes within the parts of the system are stochastic, and the values of the parameters defining this system are gained by estimation.

In general a deterministic system can be analysed more easily. So to start with we shall try to characterize a system as deterministic. Of course, there is nothing *really* deterministic outside pure mathematics. So we replace the probabilistic aspects of reality by the assumption of determinacy of the system in question. This is achieved by abstracting from its probabilistic characteristics and by approximating them by determinate terms and by idealizing them accordingly.

The parameter values observed at random can be characterized by their various stochastic momenta which are approximately described by means and deviations. Likewise, the probabilities determining stochastic processes and their random variables can be idealized. For instance, we may assume that they are continuous with the consequence that this will only be a vague approximation of unknown discrete distributions (cf. Sect. 1.3 and Ex. 2.3).

The degree of determinacy may vary. If there is a system consisting of parts characterized by certain parameters and of functions defined between these parts, i.e. of processes transforming sets of parameter values into other sets of parameter values, the degree of determinacy may cover these criteria in various ways. Some or all the parameters may be determinate (or probabilistic) whereas the processes are stochastic (or determinate). On the other hand, the parameter sets may be determinate or probabilistic as well as the governing processes.

Ex. 2.1 As an example we picture a waiting-line. This queuing system may be illustrated by the diagram of Fig. 2.4 with a flow of customers arriving in a queue *I* at a service facility, say, counter, waiting for a servicing process *P* at the counter. The customers released from this process *P* make up the output *O* of this system.

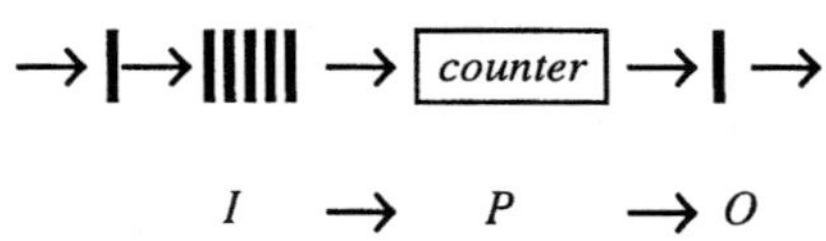

Fig. 2.4. Waiting-line model

In fact, the system of Fig. 2.4 contains two separate processes: there is the way the entities waiting for service join the queue, and there is the procedure of serving them. Both processes may either be determinate or random, i.e. the span between two entries to the queue and the time consumed by the service may possess fixed values, or they may both be random variables. But there may also be any combination of these processes, one being determinate, the other one probabilistic. In addition, a waiting-line-system may show a further structure complicating its analysis considerably, there may, for instance, be parallel waiting-lines and/or parallel service facilities. ♦

When regarding determinate parameters more closely, we see that their values are not really as determinate as assumed. If, for instance, a queuing system is described as determinate with regard to its service-time, like a production line adding a certain part to a construction every 20 seconds, we will find out by measuring that each single execution of this process will differ – as little as may be – from the value assumed as being determinate. Likewise, we may measure the length of a body; but however exactly we try to determine this parameter we will always find it only in the vicinity of a fixed value. If we were able to inspect the surface of the body in question on the level of its molecular structure, we would find out that it was a choppy arrangement of particles without any determinate dimensions at all. Anyhow, for the sake of manageability we will assume a certain real number to represent the length of this body.

So, we conclude that – complete – determinacy is an idealization and an illusion. But most probabilistic concepts are nothing but approximations of unknown probabilities themselves. Being idealizations and abstractions they replace 'real' probability distributions by purely mathematical conceptions.

2.2.4 Degrees of Self-Containment

The degree of self-containment characterizes a system by the way it is embedded into its environment and by the way it communicates with it. A *closed* system is not significantly related to any parts of its surroundings; there are only relations between the various subsystems of the system. A laboratory or a spacecraft can be regarded as closed systems.

Open systems, however, communicate with other external systems. The subsystems of an open system are related to subsystems of its environment. A plant or a service industry are examples of open systems.

When examining closed systems more thoroughly, we shall find out that there is nothing like a *really* closed system, absolutely isolated from its environment. Somehow every system is influenced by its surroundings. For instance, the team working in a laboratory will leave it after some time and, maybe, return to it later. Furthermore, there is an external supply of power, water etc. to the laboratory. The spacecraft loses energy to its environment or gains energy from it using solar generators. But these interactions between a system and its surroundings may be of minor importance to the problems we want to discuss and can thus be ignored. In an act of abstraction and idealization we consider these systems as 'closed'.

Some authors use different terms. Bertalanffy et al. (1977), for instance, call a system *isolated* if there is neither any input into it nor any output from it; they call a system *closed* if there is only an exchange of energy between the system and its environment; an *open* system – according to these definitions – finally absorbs matter from its environment or emits matter into it. As we use any kind of communication between a system and any parts of its surroundings as a distinctive feature, energy and matter are subsumed under this criterion. According to these definitions our *closed* systems are 'isolated', whereas our *open* systems communicate with their environments by which means whatever and are thus 'closed' or 'open' according to Bertalanffy et al. (1977).

2.2.5 Degrees of Adaptability

A *non-adaptive* system cannot adapt to changing environmental conditions, an *adaptive* system can. A handsaw, for instance, will not adapt to changing materials. If it is made to cut wood, it will not adapt to pieces of metal enclosed in a plank and will get stuck or blunt or lose its teeth when it hits a nail. On the other hand, self-tinting sunglasses will adapt to varying degrees of sunlight.

Some physical processes may be interpreted as adaptations to the environment, though we think they are really non-adaptive. If, for instance, we put an ice-cubicle into a heated oven, it will melt, and if a glass of water is placed into a freezer, the water will freeze. But melting the piece of ice consumes energy, thus cools down its environment and the chances of 'surviving' as a frozen body are improving. Water reacts to freezing by emitting energy, thus warming up its environment and improving its own chances of staying liquid. These phenomena seem to indicate that these systems are adaptive, but the reactions of the ice-cubicle and of the water in the glass are definitely out of the range of any purposiveness and just unintentionally help to persist in the original state of a solid or liquid body; we shall study this teleological component of the purposiveness of some systems in Subsect. 2.2.6 when we deal with the aspect of finality and teleology.

The adaptability reaches a maximum degree in the *learnability* of a system. A learnable system will improve its performance with the help of growing experience gained by experimenting. Just like animals can be trained to react to certain signals in a specific way, a computer programme can show some signs of learnability, too: a programme correcting misspellings may, for instance, automatically update its thesaurus according to the habits of a particular user.

2.2.6 Degrees of Time-Dependence

A *static* system is 'rather' time-independent. Let $t_0<t_1<...<t_i<...$ be a sequence of consecutive temporal points of observation and let $S(t_i)$ be the state of the system corresponding to the moment t_i when this state $S(t_i)$ is observed; we shall refer to t as the *observation time*, and we shall call t_i the *points* of observation time t. In a *static* system $S(t_i)$ will be constant and independent of time t.

A system is *dynamic* if its state $S(t)$ is a function of the observation time t. In general, we shall observe $S(t_j)\neq S(t_k)$ for $j\neq k$ for a dynamic system. But $S(t_j)=S(t_k)$ for $j\neq k$ is also possible; this will be the case if $|t_j-t_k|$, the difference between observation times t_j and t_k, is small. On the other hand, equal states $S(t_j)=S(t_k)$ may occur in a dynamic system *by chance* even for large values of $|t_j-t_k|$. Thus, a dynamic system may appear to be static within shorter intervals of time or just by chance if we accidentally observe the same states of the system at various points of time within a longer period of observation. We may, for instance, look at ten photographs taken of an athlete working with an ergometer. The shots may show ten identical representations of the person. When we learn that the span between the various photographs was $^1/_{100,000}$ of a second each, we can easily explain the phenomenon of a 'static' athlete with the help of the minimal time difference between the shots granting not much of a chance to notice a change in the athlete's positions. But if we assume there were ten photographs, also showing the athlete in an identical position as before, but the span between two successive shots had been 10 minutes each, then we would jump to the conclusion that the probability of shooting this series of pictures must have been pretty low or we may think that the 'athlete' was not a living person, but a dummy (a 'model') posing and thus representing a static system.

If we observe a static system, say, a building at consecutive points $t_0<t_1<...<t_i<...$ of observation time, we expect to observe $S(t_j)=S(t_k)$ for $j\neq k$. Yet, we may observe $S(t_j)\neq S(t_k)$ for $j\neq k$ whenever $|t_j-t_k|$, the difference between the points t_j and t_k of observation time, is large; this is due to the process of ageing. We know from experience that buildings look quite different after some twenty years. They look even worse if they have been damaged by the increasing acts of vandalism: nowadays buildings sometimes do not have a lifetime of more than twenty years. And even if the span $|t_j-t_k|$ is extremely brief there will be a difference between $S(t_j)$ and $S(t_k)$ in a static system. Some molecules of the material used for the building will have changed their positions or their structure. This is the effect of Heracliton's *pantha rei*, 'everything is in a constant state of flux', but we will never be able to observe these changes in the states of a system exactly if the points of observation time are too close together.

Consequently, there is nothing like a really 'static' system; being static is an abstracting, idealizing and approximating description of a system. In reality every system is dynamic. We may agree to call a system 'static' if its state is changing extremely slowly in the course of time. But under the aspects of modelling we are not particularly interested in these static systems. One of the reasons why dynamic systems are our main concern is the fact that we often develop models as a means to make predictions. These prognoses concern the future states of a system known from its past and present. For a static system this problem does not exist, because the future states of static systems will be – more or less – the same as their present states or those of the past.

As we have stated that dynamic systems are our main concern, we shall devote a separate section to their discussion. But there are two aspects we want to draw attention to already here: these are the aspects of the stability and the finality of systems. They are closely related to the dynamism of a system, but they are already inherent in its structure.

Aspects of Stability

A system is called *stable* if it persists in its original state. Whenever there is a slight disturbance shifting it into a state different from the original, a stable system will first try to resist and then return to its initial position as soon as possible. Thus a pendulum or a spring in a stable initial position will return to it after any slight disturbance – if they are allowed to; this is the example of a damped oscillator *finally* returning to its stable position. But if, for example, a spring is overstrained, it may break or lose its elasticity and remain in the imposed state of extreme elongation or being in pieces. By the way, when a stable system is watched closely, there will always be a slight oscillation of the system about its point of rest, this is due to the regulative processes leading it to this final position. There may even be physical phenomena preventing a stable system from coming to rest, like Coriolis' forces leading to the swinging of Foucault's pendulum.

An *unstable* system, on the other hand, will respond to any disturbance with a spontaneous shifting from its original state, thus possibly gaining a then stable position. A body kept in balance will fall as soon as its equilibrium is disturbed, a house of cards will collapse at the slightest breath of wind.

Though this aspect of stability will only become apparent in the course of time, when a system reacts to external influences, it is already determined by the internal (static) structure of the system. Further distinctive traits concerning the stability of a system are based on the finalistic aspects discussed in the following section.

Aspects of Finality and Teleology

Like the aspects of stability those of finality will only become apparent in the course of time when a system is allowed to develop dynamically, but they are already inherent in the structure of a system prior to any time consuming process.

Finality is a *teleological* aspect (Greek *teleos* and *logos*: 'far off' or 'end' and 'word') considering the degree of 'purposiveness' of a system. We start with purely causal chains of states used as explanations of phenomena; they are *for-*

mally teleological (Stegmüller 1969), but not goal-directed at all. If we state, for instance, 'it is raining and consequently the pavement of the street is getting wet', the fact of a rain-shower is a sufficient explanation of the phenomenon of a wet pavement. This means a process P_0 – it begins to rain – occurring at time t_0 – explains a phenomenon P_1 – the pavement is wet – observed at time $t_1 > t_0$. But we do not assume that the rain started 'intentionally in order to' wet the pavement; the explanation of a wet pavement is entirely reduced to the temporal sequence: first it began to rain at t_0 and then the pavement was wet at $t_1 > t_0$.

Material teleology (Stegmüller 1969), on the other hand, deals with '... a scale; at the upper end of the scale may be put such actions as are most deliberately purposive, the goal and means having been pondered, developed in imagination, and deliberately chosen among various alternative possibilities, before overt action began ... at the lower end ... would be such impulsive actions as ... snatching ... (a) child from imminent danger' (McDougall 1923). Braitswaith (1968) distinguishes between merely goal-directed and goal-intended behaviour. Stegmüller (1969) subdivides material teleology into two branches: into *genuine* and *ostensible* teleological behaviour. Merely goal-directed – ostensibly – teleological processes have also been subsumed under the term teleonomy. We prefer to work with Stegmüller's terminology, because it covers the various aspects with the single term 'teleology'. Thus genuine material teleology requires a determined intention to attain the goal. Ostensible teleology, however, describes those processes directed to attain a goal and serving a generic purpose, yet without any conscious planning. Examples of this kind of ostensibly teleological processes are, for instance, evolutionary processes.

If we return to the example of a wet street, but assume that this was caused by using a street sprinkler, this state cannot be explained by temporal succession alone: the sprinkler was turned on and consequently the pavement became wet. In that case the phenomenon of the wet pavement was intended, the sprinkler was used in order to wet the pavement; it is a case of a goal-intended – that means genuinely teleological – action resulting in the wet pavement. On the other hand, when studying certain surface-structures, we find striking similarities between those created by nature and those invented by engineers when they try to copy the performance of natural surfaces (e.g. in bionics). Modern coatings thus simulate the surface-structure of lotus leaves and are extremely dirt-repellent. Swimsuits and the surfaces of submarines and aeroplanes imitate the skin-structure of aquatic animals like sharks and dolphins. They thus offer less frictional resistance than conventional surfaces. In both cases of natural and man-made surfaces we will state "they are formed this way 'in order to' achieve optimum performance". But the latter are the result of a deliberate and goal-intended construction, of genuinely teleological processes, while the natural surfaces are the result of an evolution *directed* towards the goal of optimizing natural processes as a question of survival. These developments are goal-directed, *ostensibly* teleological, they lack intention (if we leave theological aspects out of consideration), and they are thus part of teleonomy.

This shows that teleology pursues an aim: it deals with *finalism* and the *finality* of systems. This finalistic perspective offers another possibility of characterizing dynamic systems, and we will therefore discuss the ideas of *equi-* and *multifinality*.

A system is *equifinal* if there are different ways (processes, starting from the same or from different initial conditions) leading to one and the same final state of the system.

Ex. 2.2 Fig. 2.5 shows a receptacle R drawn in longitudinal section, its volume is, say, 2 m^3. Above the receptacle there is a tap T with practically unlimited flow capacity, i.e. the flow Q_T through T may be adjusted to any finite value. At half the normal height of R there is an outlet O permitting a considerably lower maximum rate of flow $Q_{O,max}$ than $Q_{T,max}$, the maximum flow rate of T which tends to infinity.

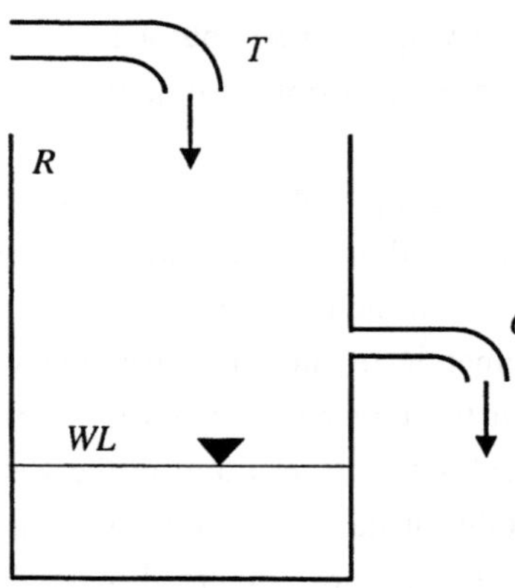

Fig. 2.5. Model of an *equifinal* dynamic system (Ex. 2.2)

Let us assume the tap is opened and 2 m^3 of water run into *R*. However we adjust the flow capacity Q_T of *T*, the system will finally reach a state, where *R* is filled to the half and contains 1 m^3 of water. But there are various ways to reach this final state with a water-level WL_1 corresponding to 1 m^3. They depend on the flow capacity selected for *T*. We have illustrated some of the possibilities in Fig. 2.6.

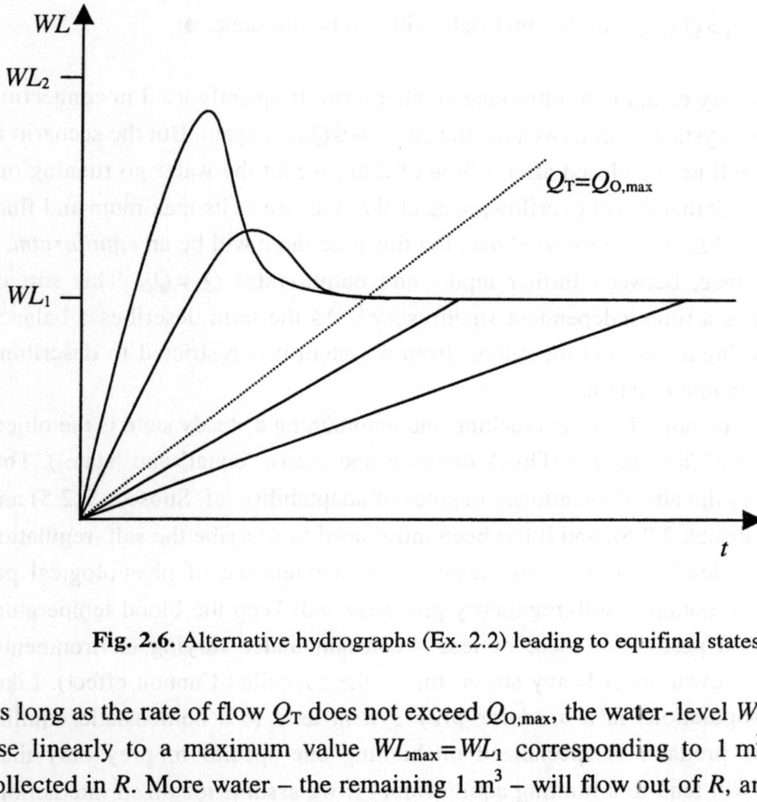

Fig. 2.6. Alternative hydrographs (Ex. 2.2) leading to equifinal states

As long as the rate of flow Q_T does not exceed $Q_{O,max}$, the water-level *WL* in *R* will rise linearly to a maximum value $WL_{max} = WL_1$ corresponding to 1 m^3 of water collected in *R*. More water – the remaining 1 m^3 – will flow out of *R*, and *WL* will remain unchanged at its maximum and final level $WL_{max} = WL_1$. In those cases with $Q_T \leq Q_{O,max}$ the hydrographs of *WL* are all piecewise linear (they form a broken line) with just differing gradients of the ascending line.

As soon as we select $Q_T > Q_{O,max}$, the situation is different. There will still be a linear rise of *WL* up to WL_1, but then the water-level *WL* will overshoot this value (up to a maximum coming close to 100% as Q_T tends to infinity) and reach a maximum value WL_{max} between the values WL_1 and WL_2 corresponding to 1 m^3 and 2 m^3 collected in *R* respectively: $WL_1 < WL_{max} < WL_2$. Then *WL* will fall, and eventually it will reach its final value WL_1, the same final state as in the case of

$Q_T \leq Q_{O,max}$. The sections of the hydrograph of WL defined by $WL > WL_1$ will not be piecewise linear as they describe superpositions of flows directed into and out of R, the latter governed by varying pressure heads.

In both cases the same final state $WL = WL_1$ will be reached even if we start from a different initial condition, if, for instance, R is not empty when T is opened, but contains already a certain amount of water. In that case there is the risk that R may overflow if $Q_T > Q_{O,max}$, but the final state will still be the same. ♦

We use this very example to introduce another term, frequently used in connection with dynamic systems. Let us assume the case $Q_T \leq Q_{O,max}$ again. But the scenario is now that T will not be closed after a flow of 2 m^3, we let the water go running on. There is no risk that R will overflow, instead WL will rise to its maximum and final level $WL_{max} = WL_1$ as mentioned above, but this time there will be an *equilibrium*, a state of balance, between further input- and output-rates $Q_T = Q_O$. This sort of equilibrium is a time-independent *steady state*[1]. As the term describes a balance between the input into and the output from a system it is restricted to describing the state of an *open* system.

The phenomenon of systems seeking and maintaining a steady state is the object of the theory of *homeostasis* (Greek *homoios* and *stasis*: 'equal' and 'state'). This term involves the above-mentioned degrees of adaptability (cf. Subsect. 2.2.5) and stability (Subsect. 2.2.6), and it has been introduced to describe the self-regulation of organisms leading to a rather steady-state-maintenance of physiological parameters. For example, self-regulatory processes will keep the blood temperature and the blood pressure – more or less – constant under varying environmental conditions or even under heavy stress (this is the so-called Cannon effect). Likewise, the populations in a *predator-prey-system* tend to a homeostatic equilibrium. If the predator is specialized on hunting one species of prey only their population will tend to a limiting equilibrium– if we assume idealized interactions between predator and prey. This is a question of supply and demand reaching a steady state. The approach to an equilibrium will be reached in cycles, either with lowering amplitudes or rather equicyclical (this is the case, e.g. with hog-market-cycles, we shall study the phenomenon in greater detail in Subsect. 2.3.2.1; cf. Ex. 1.7, too). Homeostasis is not a static equilibrium, but rather an oscillation between limiting maximum and minimum values of a parameter. The term homeo-

---

[1] This term has been introduced by Bertalanffy; he called it *Fließgleichgewicht* which is something like 'flowing equilibrium' in English.

stasis is now applied to general systems theory and describes any regulatory process causing a system to persist in stable (and perhaps steady) state, for example, a heating-cooling-system adjusted to a certain temperature, say, 25 °C and oscillating between a minimum value, say, 24 °C when the heater is turned on and a maximum value, e.g. 26 °C when it is turned off again or the cooler is turned on.

Obviously homeostasis stops any kind of progress as it is directed towards the preservation of the state of a system. Consequently, there must be completely different processes forcing the evolution of systems. The term *heterostasis* (Greek *heteros* and *stasis*: 'other' and 'states') is used to characterize these agents. According to Cannon (e.g. Cannon 1932) heterostasis leads to the transition from one homeostatic equilibrium to another one under the influence of a stress imposed on the system (Cannon reflex). Other authors (e.g. Antonovsky 1987) regard heterostasis as the basic principle underlying the development of an organism under the influence of the stress imposed upon it by its environment and determining its life from birth to death. We may, for example, spot heterostatic predator-prey-systems, like fairy rings, certain parasites or tumours developing and destroying the hosts who nourish them, thus causing their own ends.

We shall now discuss the idea of *multifinality*. Multifinal systems start from (almost) the same initial conditions, but they lead to final states that are far apart from each other. We would expect a system to end in the same final state whenever it starts from the same initial starting point, and, likewise, two systems leaving the same starting point should come to the same destination. But, as we have seen in the discussion of time-dependence, there is the effect of *pantha rei*, and thus a system cannot really be twice in the same initial state; it is impossible to have exactly identical initial conditions to start from at two different moments. Therefore, no two different systems will ever be started from the same position: though they may appear to be indistinguishable, they will certainly be different. It will even be impossible to determine the exact position a system is in; this uncertainty is due to the probabilistic indeterminacy of measurements mentioned above (cf. Subsect. 2.2.3). As an example we mention the effects of a medical treatment. A vaccination, for instance, may be successful, or it may show no effects at all; in the worst case it may have severe side-effects and cause harm to the life of the patient. The effect of the vaccination is not predictable, because it depends on the constitution of the patient at the moment he is vaccinated, which is different with each patient and cannot be found out in a sufficiently accurate way.

Ex. 2.3 We give another example taking the underlying probabilistic ideas into account. Fig. 2.7 shows a sketch of Galton's Board. This is a wooden board with a triangular array of nails (in our sketch represented by the black triangles). Balls are dropped out of the 'same' initial position; but their paths through the board are influenced by the successive contacts with the nails; so the balls will finally be collected in different slots underneath the nails. ♦

If the construction of the board in Ex. 2.3 is adjusted in such a way that whenever a ball touches a nail, the probabilities of passing this nail on the right or on the left side are 'constant', i.e. possess the same value for every nail, then this experiment is binomially distributed. If a sufficiently large number of balls is dropped, the contour of the collection of balls finally gathered in the slots will approximate to the probability function of the binomial distribution.

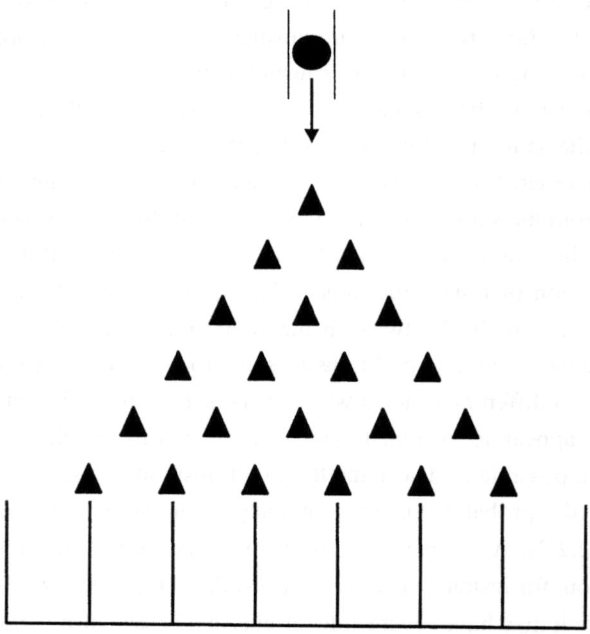

Fig. 2.7. Galton's Board as an example of a *multifinal* system

With rising numbers of the rows of a Galton's Board – and consequently of the slots underneath its bottom row – the outline of the collection of the balls in the slots will eventually approximate the graph of the density function of the Normal Distribution. Most of the balls will drop into the slots in the middle if the probabilities of a ball passing the nails on the right or on the left side are 'the same'. The number of balls collected in a slot will decrease with the position of the slots being further to the left or to the right. The distribution of the balls over the slots thus reflects the degree of multifinality of the system. A ball will fall into the slots on the extreme left or right if it is always directed to the same direction, i.e. either always to the left or always to the right whenever it hits a nail. This event – the balls always being directed in the same way – is very unlikely to happen, but it is not completely impossible. So, if the number of balls dropped into the board is sufficiently large, some of them will finally end in those extreme positions. This phenomenon is called the *butterfly effect*, and we shall return to it in Subsect. 2.3.1.1 where we want to study various effects possibly influencing dynamic systems.

Of course, probabilistic systems are not necessarily multifinal. We may, for instance, choose the input rate Q_T of Ex. 2.2 at random; nevertheless, the system will still be equifinal that means with possibly a final steady state corresponding to the condition whether the tap is closed after a flow of 2 m^3 or not. Likewise, the experiment with Galton's Board may be modified to make it equifinal. We may, for instance, open the bottom of the board and collect the balls dropping into the slots in a funnel where they are finally arranged in *one* vertical column only. In that case the number – not the order – of the balls collected in this only slot of the funnel is equifinal and the same as the number of balls dropped into the board.

2.3 Dynamic Systems

2.3.1. Definitions

As we have seen in the previous Sect. 2.2, the states $S(t_i)$ and $S(t_{i+1})$ of a *dynamic* system at two consecutive points of observation time $t_i<t_{i+1}$ will generally be different: $S(t_i)\neq S(t_{i+1})$. We represent this time-dependent progress of the system in the following diagram:

$$S(t_i)\rightarrow S(t_{i+1}),\ \text{for}\ i=0, 1, 2, 3, \ldots . \qquad (2.1)$$

It is convenient to introduce the process P transforming $S(t_i)$ to $S(t_{i+1})$ into the representation of (2.1). Furthermore, we call $S(t_i)$ the ***Input*** and $S(t_{i+1})$ the ***Output*** of the respective ***Process*** P of this system. We thus transform this diagram (2.1) to

$$\textbf{\textit{Input}}\rightarrow\textbf{\textit{Process}}\rightarrow\textbf{\textit{Output}}; \tag{2.2}$$

we shall refer to the chain (2.2) as $I\rightarrow P\rightarrow O$, and we call it an *I-P-O-diagram*.

So, by successively combining I-P-O-diagrams, we construct a chain of consecutive progresses of the states of a dynamic system. If we distinguish the various constituents of this chain by their respective subscripts i, this chain can be represented graphically more precisely by the following sequence

$$I_0\rightarrow P_0\rightarrow O_0\equiv I_1\rightarrow P_1\rightarrow O_1\equiv I_2\rightarrow \ldots\ O_{i-1}\equiv I_i\rightarrow P_i\rightarrow O_i\equiv I_{i+1}\rightarrow \ldots\ . \tag{2.3}$$

Here the output O_i of the i-th process P_i is identical with the input I_{i+1} of its successor P_{i+1}.

This analysis of a dynamic system may be applied to any of its subsystems, and they can be represented by similar sequences of consecutive I-P-O-diagrams. Due to the further relations connecting the various subsystems of a system, an output generated within one of these subsystems is at the same time the input of a process incorporated into another subsystem. The most general representation of an I-P-O-diagram will thus be described by the network of (2.4).

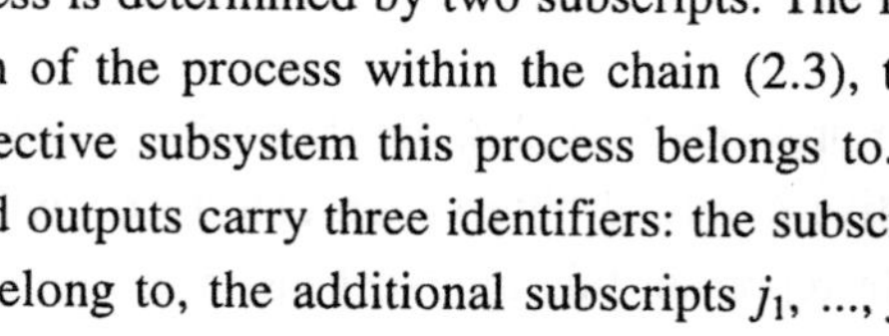

(2.4)

In (2.4) each process is determined by two subscripts. The first one i indicates the respective position of the process within the chain (2.3), the second subscript s identifies the respective subsystem this process belongs to. Consequently the respective inputs and outputs carry three identifiers: the subscripts i and s determine the process they belong to, the additional subscripts $j_1, \ldots, j_m$ refer to the various subsystems they come from as outputs, and $k_1, \ldots, k_n$ determine those subsystems they serve as inputs. The representations of (2.3) and (2.4) may be compressed into

$$I_{i,s} \rightarrow P_{i,s} \rightarrow O_{i,s} \tag{2.5}$$

which is a diagram analogous to (2.3); here $I_{i,s}$ is the sum of the m components $I_{i,s,j1}$, ..., $I_{i,s,jm}$ of the input $I_{i,s}$ to $P_{i,s}$. Likewise, the n components $O_{i,s,k1}$, ..., $O_{i,s,kn}$ of the respective output are summed up in $O_{i,s}$.

A system is thus characterized by its structure, and by the various transforming processes representing its parts. The structure of a system is determined by its subdivision into subsystems and their interdependences. Each subsystem displays a functional relation between its output and its input. This relation is determined by the process representing the respective subsystem. Each process may be broken down into a more detailed structure by successively subdividing it into its parts.

2.3.2 Some Special Effects

A dynamic and open system Σ is subject to internal interactions between its components; on the other hand, there are also interactions between Σ and its environment. Some of these interactions create quite special effects. They may be initiated by the system itself due to internal interactions and put down to the structure of the system, or they are caused by the environment of the system and are thus due to external interactions.

This distinction is not very accurate, because the 'effects caused by internal interactions' may be influenced by the environment of the system as well. And *vice versa*: those effects which we have just described as being 'caused by external interactions' may just as well reveal characteristics of the performance of the system. Accordingly these relations between 'cause and effect' describe only the most important components of the set of the respective phenomena. Furthermore, our distinction between effects caused by internal and external interactions respectively is *ipso facto* theoretical as we are dealing with open systems whose boundaries either include or exclude the crucial interactions and involved parts. Thus, any external interaction may be transformed into an internal one – and *vice versa* – by a redefinition of the boundaries of the respective system. But somehow the various effects seem to depend in a characteristic way on varying degrees of either internal structural characteristics or of external stress exerted on them.

2.3.2.1 Effects Caused by Internal Interactions

The internal structure of a system Σ and its features according to the characterization given in Chap. 2.2 will lead to some peculiarities in the performance of the system. And here steady state and multifinality must be mentioned above all as they are the characteristics contrasting most. Steady state and homeostasis are caused by *feedback*, multifinality occurs in probabilistic systems and leads to the so-called *butterfly effect*.

Feedback

Feedback describes the structure of those systems Σ evaluating part of their output O in order to control their transforming process P. So, part of the output O from Σ is taken off and input into the system again. This structure may be represented by the diagram of Fig. 2.8.

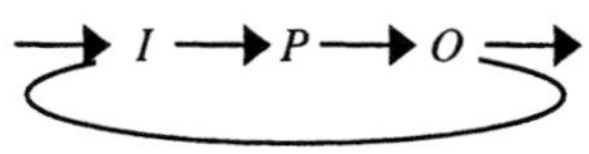

Fig. 2.8. Feedback Systems

The fed back information adjusts P in order to establish steady state conditions; feedback, therefore, is an essential part of self-regulating systems. The control unit of a central-heating, for instance, must constantly compare the temperatures of the water leaving the system and returning to it in order to achieve constant temperatures in a building. It will generate a steady state between the energy produced by the system and the energy leaving the building. Of course, the difference between the temperatures θ_O of the water leaving the system and θ_I of the water returning to it also depends on external parameters, like the air temperature θ_a outside the building. The heating system will thus adjust to this external parameter, and it will be self-regulating for any constant value of θ_a. If this system is controlled by a thermostat, the produced room temperature θ_r will be input into the system again where it is constantly compared with a preset value θ_p. The control unit will be activated whenever $\theta_p - \theta_r > \Delta$, a defined maximum difference. θ_r will oscillate within an interval of temperatures containing θ_p as described in Subsect. 2.2.6.

Of course, a steady state may develop for certain input-conditions even if the respective output is not fed back into the system and evaluated. In those cases the building up of steady state is entirely due to the internal structure of the system Σ and to its input I, as is the case in Ex. 2.2.

The Butterfly Effect

The butterfly effect is caused by probabilistic interactions between the components of a multifinal system, like, for instance, in Galton's Board (cf. Ex. 2.3). The term was introduced by E. Lorenz in a 1972 lecture, whose title was 'Predictability: does the Flap of a Butterfly's Wings in Brazil set off a Tornado in Texas?'; this effect is the essence of chaos theory and of multifinality as described in Subsect. 2.2.6.

As we have seen in Ex. 2.3, it is rather improbable that the final position of a ball dropped into a Galton's Board with many slots at its bottom will be on the extreme left or right, but it is not impossible. The larger the number of slots the lower the probability that the ball will finally land in one of these extreme positions. The probability of this happening – when the number of the slots approaches infinity – can be compared with the probability of a butterfly flapping its wings in Brazil and thus causing a tornado in Texas.

Naturally, we take up a strictly determinate position of argumentation. So we may believe that two balls dropped from exactly the same position and dropping under exactly the same environmental influences (gravitational, atmospheric and so on) will finally end in the same slot of the board. If we follow this line of argumentation, we still have to admit that it is (almost, cf. Subsect. 7.3) impossible to have two balls starting from the same initial position on the board. Or, if we carry out the experiment under predictive aspects, i.e. we want to predict the final position of a ball dropping into a board, we shall find out that this is (almost) impossible and, likewise, it is (almost) impossible to determine the exact initial position the ball started from when we know the outcome of this experiment. So the strictly deterministic approach to the movements of a ball within the board is inappropriate, because it is erratic and is thus related to probabilistic arguments concerning the initial and the final positions of the balls.

There are further criteria helping to characterize the development of a dynamic system in the course of time. They are mainly due to the internal characteristics of the system, but sometimes it may be quite difficult to determine whether this time-dependent behaviour is governed by external or internal influences and often both of them will be overlapping. We shall describe some of these effects in the following sections.

Cyclic Development

When we talk about the cyclic development of a system, we mean that some or all of its output parameters will oscillate between maximum and minimum values. This oscillation may occur quite unsystematically and without any regularities. Or there may be a constant alternation between ups and downs leading to a periodic, cyclic time-dependence of these parameter values (e.g. cf. the example of the homeostatic central-heating-system described in Subsect. 2.2.6).

A well-known cycle is that of the hog-market; we mentioned it in Subsect. 2.2.6 where we described homeostatic systems. If hog prices are down, the breeders will reduce the production of pork and invest in other rural products. After some time this will lead to a fall in hog supply and consequently cause the rise in pork prices, at which the breeders will feel inspired to invest into the hog production again, thus, after some time, causing a surplus of pork with eventually falling prices, closing the cycle. The observation of these cycles leads to the result that their periodicity is some seven years. Strangely enough, the number of students of certain sciences follow these rules of the hog market development. Only those studying exotic subjects seem to show steady – but low – numbers, but within the fields of economics, technology and other studies chosen by masses of students their number cyclically – with some delay – follows the demands of the labour market.

Oscillations of the output of a system between two values close to an equilibrium were described in Subsect. 2.2.6 where we studied homeostatic dynamic systems. By self-regulation they swing to and fro between maximum and minimum values; the difference between these values may be very small, but it still exists (cf. Ex. 1.7, too).

The Joseph Effect

The *Joseph* effect refers to the (momentarily) unretarded development of the parameters of a system, which may eventually – in a cyclic process with no defined frequency – change to the opposite direction. This technical term was introduced by Mandelbrot, who observed this effect when watching economic parameters (e.g. Mandelbrot 1997), but also distinguished it when he analysed the development of environmental data (e.g. Mandelbrot 1968). The name of this effect was derived from the interpretation of Joseph's dream described in *Genesis* 41: '*... there come seven years of great plenty throughout all the land of Egypt. And there shall arise after them seven years of Famime ...*'. This effect is connected with the criterion of stability described in Subsect. 2.2.6. A stable system returning to its neutral initial position follows a steady trend towards this initial (and at the same time final) state. The actual position of the system may oscillate with a decreasing amplitude tending to zero, but never disappearing.

The Noah Effect

Another effect observed by Mandelbrot is the *Noah* effect. You speak of this effect when you observe the sudden and immediate marks of external stresses on the reactions of a system. Mandelbrot noticed this effect when he observed the growth of financial parameters (e.g. Mandelbrot 1997), but it also influences the development of environmental processes (e.g. Mandelbrot 1968). The name of the effect is derived from *Genesis* 7: '*... and the waters prevailed exceedingly over the world* ...'. The *Noah* effect is subject to external stress upon the system, but, on the other hand, it is also influenced by the internal structure of the system.

Again, we try to portray this effect in terms of stability. Under this aspect the *Noah* effect describes the reactions of extremely instable systems to disturbances (like the house of cards we mentioned in Subsect. 2.2.6). The (temporary) collapse of stock market prices in the aftermath of the *WTC* mass murder of Sept. 11th, 2001 may serve as another example. Obviously, computerized broking led to extremely sensitive and instable systems responding to external stress in chain reactions. But while the house of cards collapses to the stable system of a heap of cards, stock prices were just moved from one instable position to another one.

Sometimes we observe sequences of *Noah* and *Joseph* effects depending on each other. Thus a *Noah*-like collapse of stock market prices may be followed by a further temporary decrease in these prices in a cyclic *Joseph*-process with varying frequencies and eventually diminishing amplitudes. This may well be due to a general uncertainty amongst the participants in the market, yet it may as well be caused by political or economic consequences of the event having triggered off the *Noah*ian collapse.

2.3.2.2 Effects Caused by External Interactions

Of course, any open system interacts with its environment and reacts to information input into it from external sources. For example, we have seen in the preceding Subsect. 2.3.2.1 that central heating systems adapt to external conditions and that they are self-regulating in constant interactions with changing external environmental parameters. Some of these cases are quite important as they lead to effects frequently occurring in practice – the impact of external parameters may be rather predominant –, and it may be worthwhile to study these systems separately; others – like the *placebo* effect – are of minor importance to our considerations and will therefore be disregarded.

The Pygmalion Effect

The *Pygmalion*[1] effect is the unintentional shaping of a system according to the perceptions of its observer, caused by the mere fact that the system is observed. A surgeon, having developed a new surgical technique, may be inclined to give a better after-care to those patients treated with these techniques than to those having been operated in the conventional way. This may lead to a better rate of healing among the former in comparison to the latter. Of course there is still the possibility that there is no significant difference between the rates of healing at all, but anyway the better after-care will improve the chances of healing for those patients treated with the new method.

A special case of *Pygmalion* effects, sometimes observed in educational sciences, is the *Rosenthal* effect (e.g. Rosenthal and Jacobson 1971) which has a critical look at the more successful careers of pupils thought to be superior and explains this by the fact that they are given more attention by their teachers.

With the help of these effects we can explain the interactions between an observer of a system and the system itself, and they lead to *self-fulfilling prophecies*[2]. Whenever, for instance, stock market products are evaluated by investors using more or less the same pricing procedure, predicted prices and market prices will – generally – converge very soon. Option pricing, for instance, is commonly based on the discrete algorithm of Cox, Ross and Rubinstein (e.g. Cox et al. 1979) or on its continuous limit, the well-known formula of Black and Scholes (1973). In both cases the evaluation procedure is the same for all members of the market, and it will cause the actual prices of the derivatives and those predicted by the respective calculi to converge eventually.

---

[1] *Pygmalion* was a Greek sculptor, who created the statue of a maiden, *Galatea*, obviously matching his *anima* (cf. Ex. 1.18) so that he fell in love with it; this statue was later brought to life by the Greek goddess *Aphrodite*.

[2] A similar, but opposing phenomenon is the effect of self-contradicting prophecies, when, for instance, the warning 'your conduct will lead to a dreadful end' shakes up the person thus addressed and causes him to mend his ways.

The Hawthorne Effect

There are other examples of systems being driven by expectations of their observers – sometimes these expectations are mere assumptions set off by the systems themselves. As an example we shall study the so-called *Hawthorne* effect. This effect is called after an experiment made at *Hawthorne*'s, an American company. Here they studied the influence of light on the effectiveness of the employees' performance. The evaluation of the collected data showed no significant relation between performance and light.

This effect stands for the interaction between the observer of a system and the system itself and is well-known to any consultant: whenever he is to analyse the performance of employees or groups of employees, forming subsystems of the generic system of an organization, he will often be considered to be a 'jobkiller'. Consequently, the observed system will make efforts to adapt to the idealizing perceptions of the client, and the studied performance will – generally speaking – be more effective than the everyday results.

The Procrustes Effect

The *Procrustes*[1] effect is in a way related to the *Pygmalion* effect. Here a system is also shaped according to the perceptions of its observer, but it is done more deliberately. The influence on the system may be exerted unwittingly, but there is still some purpose of pushing it into the direction of a certain performance predetermined by the observer.

If, for instance, statistical data are analysed, then certain measured parameter values may be neglected as being insignificant outliers. Deliberately or not, the selection of these outliers will decide on the results of the analysis, thus violating the dictates of objectivity. Just like *Procrustes*, the evaluating observer manipulates the results of observation to fit a certain scheme – and his expectations. We admit that this may be done quite unintentionally. Furthermore, the transition between the *Pygmalion* and the *Procrustes* effect is fluid, and the borderline between intention – inherent in the *Procrustes* effect – and mere negligence is difficult to draw and finally leads to the question of deception and honesty in science. Sometimes the *Procrustes* effect may be regarded as a somehow exaggerated *Pygmalion* effect. It renders self-fulfilling prophecies, too, but they are attained the 'tough

[1] *Procrustes* was a giant in Greek mythology, who seized travellers, fixed them to his bedstead and either stretched them or chopped off their feet in order to make them fit it.

way', leaving no room to the observed system *not* to fit the observer's expectations, the way the *Pygmalion* effect does.

The Halo Effect

A halo (Greek *halos*: 'threshing floor') is a circle of light round a luminous body – like the sun or the moon – and outlines its contours. Accordingly, an observed system causes a halo effect when one of its features predominantly covers others of its characteristics and thus impedes their investigation by the observer. He will just recognize the surpassing element of the system and come to wrong conclusions regarding it as a whole.

2.4 Systems Analysis

We regard modelling as an imaging process: mapping one system – subject to the mapping – into another one – the model. Thus scrutinizing *modelling* is tantamount to analysing a system, and we shall continue with the description of some basic definitions and methods of systems analysis. With respect to modelling the aspects of systems analysis are limited to the input I and to the output O of the imaging *process* yielding a model. So our considerations will be rather restricted, not taking the dynamism of the process P transforming I into O into account. The mappings, imaging the objects of modelling by the models themselves, constitute the dynamism of modelling; they will be described in Chap. 3.

A model is the *projection* of an I-P-O-chain and is an I-P-O-chain itself. Modelling is an imaging process. It maps the original system – subject to the mapping – into another system, its image. So first of all, there is the problem of comprehending the system to be modelled. Furthermore, we have to interpret the knowledge correctly we gain from modelling. These problems will be the central topics of this section.

First we are going to describe the mapping of the original problem subject to modelling by means of *systems analysis*. Let us call the original system Σ_I, as it is the *input* of modelling. We represent it by the chain $I_I \rightarrow P_I \rightarrow O_I$. We have to analyse this system Σ_I, and the problem is how to map it appropriately into its model. Thus the system Σ_I *is* the input of a problem-solving process of mapping. We refer to Σ_I with respect to this process by I_P: I characterizing it as the *input* into a process and the subscript P referring to the *problem-solving* capabilities of this *process*.

Accordingly the problem-solving procedure of modelling itself will be called P_P. This modelling procedure P_P generates an output O_P which is the model of Σ_I, and it is an *I-P-O*-chain of its own; we call this chain $I_O \rightarrow P_O \rightarrow O_O$, and it constitutes the system Σ_O. The mapping process regarded as an *I-P-O*-chain itself is represented by Fig. 2.9.

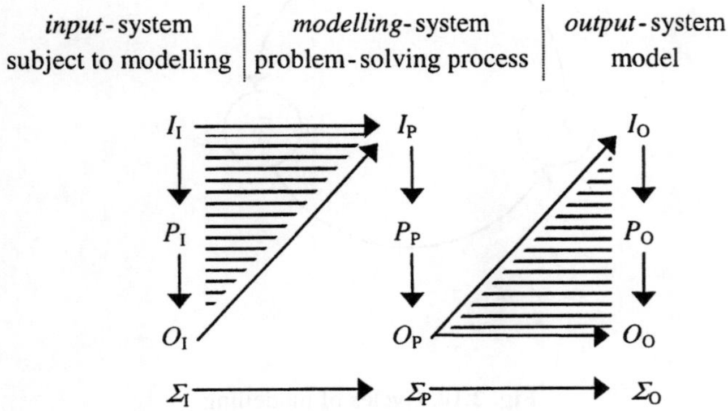

Fig. 2.9. Modelling systems

We thus interpret modelling as an *I-P-O*-chain of its own, which is represented by the horizontal bottom chain of Fig. 2.9. If Σ_P is the system $I_P \rightarrow P_P \rightarrow O_P$, we conclude that modelling is also a dynamic system $\Sigma_I \rightarrow \Sigma_P \rightarrow \Sigma_O$ and that each component of this modelling chain is a system regarded by itself. The input into this modelling chain is the original system Σ_I to be modelled (the *prototype*), the process is the problem-solving mapping system Σ_P, and the constructed model constitutes the output-system Σ_O (the *simulacrum*) with respect to Σ_P.

The precision of the reproduction of the input system Σ_I reflected by the output Σ_O of the mapping will be of varying quality. Therefore we have to *evaluate* the *quality* of a modelling process under consideration. Evaluation is a *process* of its own, and it is performed in a separate procedure of comparing the performance of the system Σ_O developed by modelling with the original system Σ_I subject to the modelling process.

In general the evaluation of a model will show major discrepancies between the original system and its image. To a certain degree these discrepancies are due to the abstractions and idealizations typical of modelling, and we have to accept them as one of the characteristics of a model. On the other hand, they may give rise to a closer analysis of the original system Σ_I input into the process of modelling, and this scrutiny may lead to a more sophisticated model. The diagram of Fig. 2.9 thus comes to full circle as shown in the cyclic representation of Fig. 2.10 below.

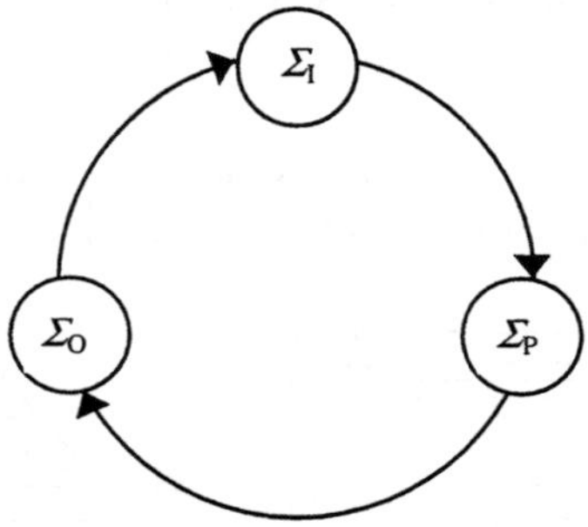

Fig. 2.10. Cycles of modelling

According to Fig. 2.10 the development of a model is a cyclic – possibly never-ending – process. When the developed model Σ_O is compared with the original system Σ_I, this comparison may reveal that the model does not represent its original appropriately. In that case the system Σ_I input into modelling must be analysed again and even more closely than before. On the other hand, the evaluation of a model may suggest that there are quite a few major discrepancies between the performance of the model and of its original Σ_I, but that, nevertheless, the model may be used for a start.

Ex. 2.4 The discussion of the following problem may help to illustrate the assertion that a model does not necessarily represent its original appropriately. Let us assume we are to construct a tin with minimum surface. This is a well-known problem in elementary calculus and may be phrased like this: 'The cost of materials of the production of a tin is proportional to the surface of the tin; minimize the cost'. If you have a careful look at the wording of this sentence, you will realize that it actually does not reflect the problem appropriately as the minimum value of the surface of a body is zero. The 'solution' of the problem would consequently be not to produce any tins at all in order to consume a minimum amount of material. Naturally, we must reject this model as absurd. We may try to formulate the problem anew in order to get a cost-effective production: 'Minimize the surface of a tin, 500 ml in volume'. This reformulation does no longer allow the previous no-can-no-cost-solution of the problem, and we conclude that under these conditions a spherical tin with a radius of some 4.92 cm will be optimal. This model is much better than the previous one, because it does not shut down the factory producing tins. But spherical tins will cause problems when you try to affix labels to them, when you want to store them, or when you try to open them using a conventional tin-opener. So what we really want is 'a cylindrical tin with a certain cubic capacity (e.g. 500 ml) and a minimum surface'. The problem worded this way will lead to the well-known optimum cylinder with diameter and height of equal size of some 8.60 cm. ♦

Sometimes we are even ready to accept a model, though we are well aware of the fact that it is only a poor representation of its original. We would like to explain this by an example taken from the field of the development of (e.g. commercial) computer systems. It may seem absurd, but the 'perfectionist' (analyst or programmer) might become a real burden in this branch. He will always try to develop the *perfect* system meeting all demands. But as he has to convert progress and daily born new ideas into his project, he will never succeed in developing a system which can be translated into action. His *less perfect*, but *pragmatically* working colleague will try to define interim stages and an adequate schedule to be carried out, once he has begun to develop his system. He will store up any technical progress or new idea for further improvements of his product. Thus, after some time he will finish off his project with a version of a system, not perfect and perhaps not up-to-date, but working. Then he will start all over again (cf. Fig. 2.10) making use of his store of further-reaching ideas. This iterative process of systems development will lead to an improved system (*release*). We owe (though imperfect) models we can work with to this pragmatic, cyclic approach to a given problem. The result of modelling under these aspects will only be an imperfect and momentary solution of the original problems, but gradually we can improve the developed system Σ_O and may eventually achieve a model coming as close to reality as we wish.

The problem-solving system Σ_P of Fig. 2.9 contains a complex process P_P mapping the original system Σ_I into the resulting model Σ_O. This process is multi-layered and includes many facets of intuition and creativity. These components of modelling are out of reach of any textbook. They are subject to experience and skill and cannot be taught, they can only be developed in practice. Furthermore, the methods of model-building will vary with the fields of application, and the quality criteria a model has to come up to also depend on the area of application. For instance, environmental models and modelling in social sciences may depend on data sets of different extent and the relevance of the quality criteria may vary with the field of application.

Nevertheless, all these varying points of view on modelling have something in common: the strategies of systems development and those of evaluating the developed systems. We shall try to compile the most important of these steps constituting the creation of a model; they include aspects of systems analysis as well as of systems design. The evaluation of models will be dealt with separately.

The term *systems analysis* is frequently used to cover two distinct aspects. These are the analysis of a given system to be modelled and the design and development of a system to be produced by modelling. Both components mesh together, as systems development is based on analysis. Thus parts of the analysis will later be incorporated into the design. The quality of the design of systems is evaluated by means of analysis of the model Σ_O and its counterpart Σ_I in reality.

2.4.1 Analysis

The first phase of systems analysis is the definition of the problem to be considered and of the model to be constructed. As we have seen in the preceding example, the vague definition of a problem may lead to solutions without concrete value. At the very beginning of dealing with a project the scope of the model to be developed and the purpose it has to serve must be made perfectly clear.

Quite often the job of a systems analyst (modeller) is defined by a woolly 'let us develop a system ...' and its purpose is just put down as '... in order to model ...'. A modelling-project based on these vague premises is doomed to fail for various reasons.

First of all, there is the risk of misunderstanding between the client and the person accepting the order due to different levels of insight into the problematic nature of modelling. Very often the client will be a layman in the respective field of research. His expectations concerning the qualities of a model may go beyond the possibilities an expert sees under scientific aspects. Another reason for a misunderstanding between layman and expert is the problem of communication: scientists use a technical terminology incomprehensible to an outsider and thus open to an inappropriate interpretation. So, before anything definite is settled, a lay client should be informed about the significance of models or any kind of systems development. He should know about the character of a model – that it is only an image – and about the loss of information inherent in the object of modelling due to the mapping process producing the model (cf. Chap. 3). Due to this loss a model will only be an imperfect reproduction of the system to be considered and will mirror only parts of it.

Even if both parties agree that a model is a mere image of a given system, there may still be diverging ideas about the cost-benefit ratio. There should be an agreement about the available resources: budget, manpower, the time-span granted for development etc. Without doubt, there is some sort of correlation between the quality of a model and the expenditure provided for its development. The involved parties should therefore decide whether they want the development of a system to be done 'fast and dirty' at low costs or if they want an extremely sophisticated and dependable system. The development of the latter may turn out to be a never-ending and consequently expensive procedure.

The boundaries of the system to be analysed have to be defined, and if research is restricted to a subsystem, the respective boundaries within the generic system must be determined. In both cases the various interfaces between the system or subsystem in question and its neighbours have to be established in order to meet the demands of this system as well as those of its neighbours. These interfaces comprise a lot of information. First of all, they describe the range of the subsystems and mark out the borderlines separating them. Furthermore, they show the way information is exchanged between neighbouring subsystems, data-formats, periodicities of data-exchanges, the hierarchical structures reigning the relevant set of subsystems and the priorities derived from them.

If there is an agreement about all those details basic to the development of the system in question, it should be put down. There must be a written declaration of consent to the definition of the project, otherwise the inevitably arising discrepancies between the expectations the clients of the design have and 'reality' – the system developed by following the dictates of low cost, of low manpower and of meeting narrow deadlines – will lead to irritations.

When these questions concerning the run-up of a project are settled and the decisions are properly documented, only then the design of the system enters its second phase: the analysis of the original *object-system* Σ_1, the system subject to modelling.

Here the used methods depend a lot on the respective field of research. Within social sciences we are easily confronted with the problem of the *objectivity* of ascertained data. The efficiency expert, for instance, will frequently be regarded as a *job-killer* with the consequence that each participant in a process he wants to analyse will overemphasize his own importance and try to belittle the influence of his colleagues – his neighbouring subsystems – in order to demonstrate his own indispensability. Similarly, a psychologist studying some abnormal behaviour will have problems in getting dependable answers to his questions in a survey. The way the input-system is *transformed* into the modelled output can also be effected by *non-objectivity*. This problem may be due to deliberate misinterpretation of the object-system leading to reports on request. As we think the 'models' thus gained cannot be regarded as scientifically sound, we would rather omit them from our further considerations. Nevertheless, you come upon them more often than you would expect to do. But of course, *non-objectivity* may also be caused unintentionally by the modelling scientist. If he is the member of a certain school, he may be biased and interpret observations he made in a way supporting his theories and prejudices – even if they do not.

Modellers working in the fields of natural sciences are in danger of unconsciously making these mistakes. They may also give biased misinterpretations of facts, but we think in their field of research this will be an exception, because research in natural sciences is less influenced by schools of thought contradicting each other than, for example, research in social sciences. Furthermore, the proofs and refutations of theories in natural sciences are much more stringent and can be checked more conclusively. The highest approval or disapproval of a theory or model will be gained in mathematics, because of the epistemic content of mathematical propositions (cf. Sect. 6.1, too).

When a project is started, there will be a first, rather rough analysis of the given system. In this phase – called the *feasibility study* – the object-system Σ_I and the demands made on the system to be developed are studied from the point of view of the feasibility of the realization of the project. This aspect will cover the range of *technical* feasibility. The feasibility study tries to answer the question whether there are sufficient technical capabilities to realize the modelling project in question. During this phase of systems analysis we may still come to the conclusion that it is impossible – given the actual state of technology – to realize the project. The project then fails because of technical conditions.

Another aspect of the feasibility of a project may be the *Cost/Benefit–Analysis* (*CBA*). The results of this analysis may be that the realization of a project is technically feasible, but not worthwhile under economic aspects. We would like to point out that *CBA* – when carried through improperly – can be misleading. In order to explain this, we split up the benefit B into two components $B=B_E+B_I$. The first summand B_E is the *external* benefit, i.e. the benefit the planned project may get under market conditions. In general B_E will lead to a cash flow. The second summand B_I of the benefit B is its *internal* component. It reflects the growing experience of the technologies involved in the project. Part of this experience is represented by the highly qualified staff and is thus raising the cost of the development. Due to the rising market values of experts it is a question of money to bind these persons to the organization realizing this project or similar developments. The internal benefit will pay off during the subsequent cycles of the development of the project. There may well be negative *external* benefits caused by the first few cycles in the lifetime of a project resulting in massive financial losses. On the other hand, the insight into the relevant problems may grow considerably, and it may help to pile up an enormous amount of know-how, thus raising B_I and the future external benefit B_E.

So when deciding in favour of a project, we must be aware that there may be hard times until an appropriate return on investment will be obtained. In addition, cost of systems development will rise excessively, because the 'value' of the qualified personnel increases during the lifetime of a project. So it might turn out to be a question of economic survival if you cannot keep the system on the market for long. This can be proved by quite a few examples of outstanding technical systems, which could not be put onto the market or maintain their holds there. We do not want to drag up the legends about stockings resistant to ladders or tyres immune to punctures and never losing their tread. But there certainly were remarkable products like the video-recording-system *VIDEO 2000* or the rotary piston engines accepted by the public, but suddenly vanishing, because an overwhelming and financially better situated competition based on other technologies – sometimes of a lower standard – swept them off the market.

Some demands on technical systems are time-dependent. Certainly, the first nuclear power plants would not pass today's standards of security, neither would the space technology of the 1960s. Another obstacle to releazing technical projects can be political interests. So, for example, some Western European countries have decided to give up the technology of nuclear energy under the pressure of political trends. They are ready to shut down their high-tech nuclear power stations, but at the same time they are prepared to cover their energy requirement from Eastern European reactors reflecting a lower standard of security. They even accept to live in the potential fallout of these stations. Political or legal considerations may obstruct the organization of everyday life under aspects appearing reasonable at first sight. Examples are the refusals to accept electronic signatures or to introduce electronically encoded identity cards in some countries. These considerations also impede the installation of automatically charging road tolls or carrying out national censuses. Many of these projects fail, because they may possibly infringe the individual's rights by storing computerized data.

All these restrictions must already be observed in the run-up to systems development. When an agreement on an exact definition of the project under consideration and its objectives is reached, and when the feasibility of its development is determined under each relevant aspect, may it be technical, economic or legal, then finally a closer analysis of the given system can be started.

As our main interest lies in the field of open dynamic systems, the analysis of these systems will be process-orientated. It will cover the relations within the system as well as those established between the system and its environment. We shall start by determining these relations between the system and its environment. This is achieved by detecting the input and output relations between the system and the relevant parts of its surroundings. In that case the system itself will be treated as a *black-box*, i.e. we do not regard its internal structure as relevant. The situation may be represented by the diagram of Fig. 2.11.

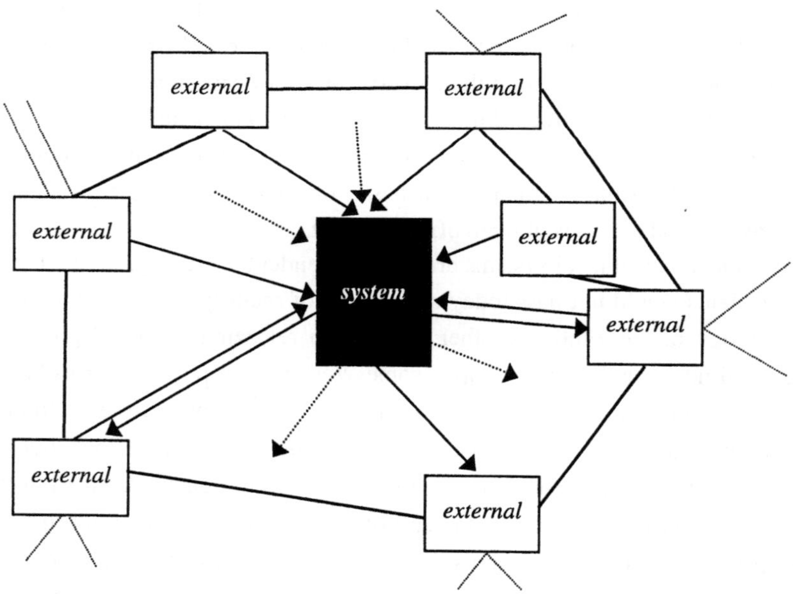

Fig. 2.11. The generic system seen as a *black-box*

Here the generic system is the *black-box* in the centre of the diagram. Its internal structure is invisible and the system is related to parts of its environment only. These external structures are referred to as *external* in the diagram of Fig. 2.11. We realize that the environment of the system, its complement, constitutes a system of its own. The structure of the external system is reflected by the lines joining the various external parts of Fig. 2.11. The arrows representing inputs and outputs into or from the black-box-system signify the relations between the system and its environment. There are cases of external structures emitting the input into the system and equally being addressed by its output, thus representing a *dialogue* between the system and those parts of its environment. The dotted lines and arrows indicate further less significant external structures and communications.

Analysis so far has been entirely focused on detecting the relations between the system and parts of its environment. These relations are the various combinations of inputs and outputs concerning the system under consideration as mentioned above.

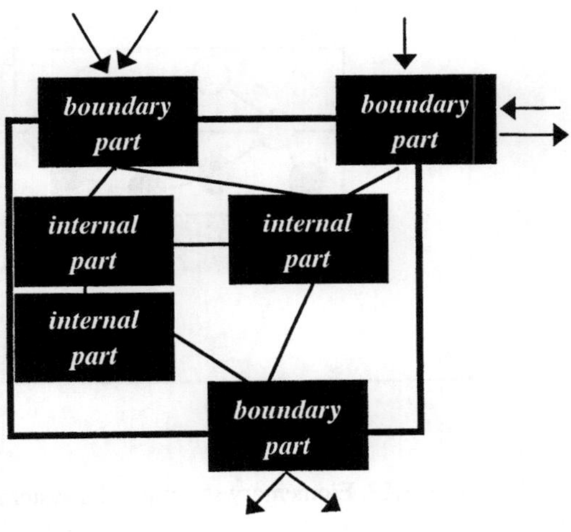

Fig. 2.12. Internal structure of the generic system

In a second step of specifying the system we shall open the black-box of Fig. 2.11. Now we are only interested in the *internal* structure of the system, and by opening the black-box representing it we will discover a further structure of interconnected black-boxes. Each of these black-box-systems of Fig. 2.12 is embedded into its own environment akin to the representation of Fig. 2.11 valid for the generic system.

Within the generic system there will be subsystems entirely internal and representing the internal functions of the system. Furthermore, there will be subsystems serving the communication between the generic system and its environment. We call the latter the *boundary*-parts of the generic system. By successively opening the black-boxes of the lowest level we are working on, we may eventually derive the fine structure of the system on the atomic level of its elements.

During this process of studying the structure of a system by steadily gaining more insight into its details, the subsystems of lower levels within the boundary-parts are incessantly sorted out. We have given a graphic representation of this process on the lowest level of the elements of the system in Fig. 2.13. The idea is that every opening of a black-box representing a boundary-part – what level so ever as long as it is above the level of the elements – will show subsystems either internal or external or being a part of the boundary.

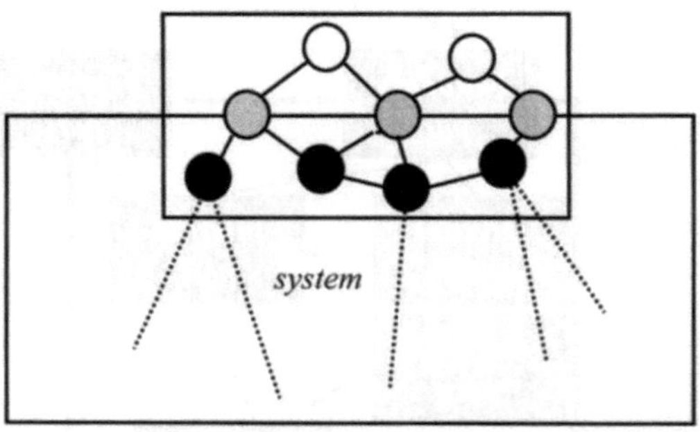

Fig. 2.13. Elementary structure of a system

In Fig. 2.13 the elements finally detected as external are represented by white circles, the internal elements of the system are the black circles of the figure, and those elements still serving boundary purposes are shaded grey. The dotted lines indicate further internal relations between parts of the system irrelevant to this diagram. So by successively opening the black-boxes down to the level of the elements, we can determine the fine structure of the system as well as of its boundaries.

Fine-structuring the system in question will lead to a hierarchy in descriptions with increasing degrees of detailed specification. The concern of systems development may be restricted to a level of a rather low degree of specification. This will be the case when we are only interested in the development and organization of global structures. The development of a network of interconnected power-supplies, for instance, does not depend on the internal structure and organization of each single power station. In that case it is sufficient to know the locations of the plants, their capacities and some more global data relevant to the study. On the other hand, if we take the cyclic aspect of systems analysis into consideration, we may prefer to go into further details of all the subsystems detected right at the beginning. In that case further studies and developments will focus on only those subsystems relevant to the particular project we work on.

Eventually, the analysis may have traced down the structure of the generic system or of one of its subsystems to a level containing only elements as its constituents. These elements are the processes P in an I-P-O-network constituted by the elements of (2.3) - (2.5). There are various ways of exploring how these processes transform the set of inputs into their outputs. These techniques depend on the respective fields of application. They include measuring, observation, interviews, conferences and black-box-analysis. Details about these subjects are beyond the scope of our investigation, and we refer to the ample supply of books dealing with the methods of collecting data and their statistical evaluation.

2.4.2 Approach to Incomplete Systems Analysis

Though the analysis of a system depends to a large extent on very special characteristics of the system under observation, there are some general aspects of the analysis of dynamic systems. They are due to the I-P-O-structure of these systems and to the fact that the knowledge of only one or two of the components I, P or O allows conclusions about the remaining unknown elements of the chain. These special facets make it possible to specify a system whenever a scrutiny of all three structural elements I, P and O is not feasible and the analysis would thus be incomplete.

The Black Box Aspect

One aspect of analysing a system Σ is to regard its process as a black box and to disregard its internal structure. In that case the output O is related to the corresponding input I into the system Σ. Any model representing this system Σ will reflect its I-O-relation without considering the process P transforming I into O or its internal structure. For the most part, this analysis will not lead to a definite determination of *P*, it just offers means to model the performance of the system rather accurately.

Ex. 2.5 As an example to illustrate the idea of black-box-analysis we describe the transformation of periodic discontinuous sawtooth signals into smooth graphs of trigonometric functions as reflected by Fig. 2.14.

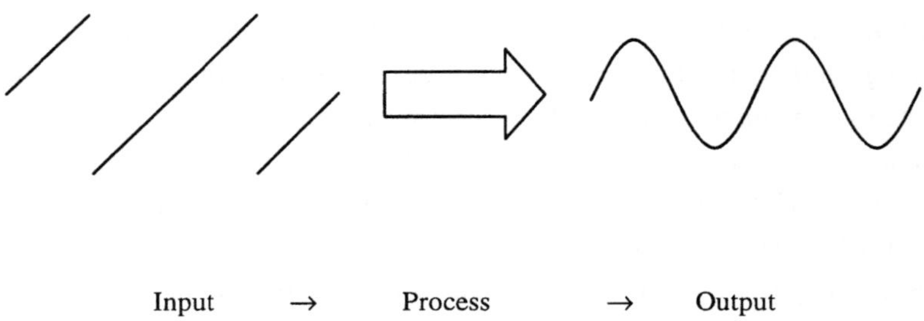

Fig. 2.14. Harmonic analysis

The sawtooth curve *I* input into the system is transformed into a resulting trigonometric curve; this is the output *O* of an unknown process *P* of the black-box-system Σ: no further details of Σ are known yet. When studying the relations between the parameters of *I* and *O*, we may, for instance, find out that the harmonic sine output possesses the same frequency as the sawtooth graph input, and its amplitude depends on the gradient of the linear input *I* in the following way: if *I* is the $2p$-periodic expansion of $f(t)=c\cdot t$ for $|t|<p$, the output *O* will be $2c\cdot\sin p^{-1}\pi t$.

Accordingly we may create a model Σ_0 of the input system $\Sigma_1=\Sigma$ by determining the frequency and amplitude of the input signal and generating a corresponding trigonometric output. Of course, this model will not represent the internal logic of Σ, but it will serve the purpose of simulating its performance.

The internal structure may in fact be an electronic device, e.g. a Wien-Bridge, sorting out the main component of a harmonic analysis of the sawtooth signal and suppressing all succeeding summands of this analysis. Generating the output signal *O* the way we have described it above – *just* generating the sine function $2c\cdot\sin p^{-1}\pi t$ whenever *I* is linear, $2p$-periodic with gradient c – is quite different from the performance of a Wien-Bridge. The latter analyses the input *I* and transforms it into the output signal O according to a predefined procedure of harmonic analysis and filtering. The Fourier-series of the $2p$-periodic expansion of $f(t)=c\cdot t$ for $|t|<p$ will be

$$f(t)=2c\sum_{i=1}^{\infty}\frac{(-1)^{i+1}}{i}sin\frac{i\pi t}{p} \tag{2.6}$$

and the *filtering procedure* of the Wien-Bridge will yield the observed output $2c\cdot\sin p^{-1}\pi t$. ♦

The black-box-approach, on the other hand, just correlates I to an appropriate output signal O, but it does not derive O according to a procedure representing the process governing Σ. The situation is in some ways similar to the trisection of angles we discussed in Sect. 1.1. Of course, we are able to construct an angle equalling the third of, e.g. 90° or 180°, but we cannot construct it by *trisecting* the given angle of 90° or 180°. The black-box analysis thus allows to draw conclusions concerning the input-output relation of a system and to derive a procedure of generating *general* pairs of corresponding input-output phenomena from the observation of *special* pairs of these phenomena.

The White Box Aspect

Up to a certain point this aspect is *dual*[1] to the black box aspect. We now assume that only the process P is known transforming, e.g. the periodic linear input signal I of Ex. 2.5 into the corresponding harmonic oscillation output by the system.

Anybody acquainted with electronics will be able to analyse the way a Wien-Bridge works. So if we learn that the process P is based on a Wien-Bridge and if we know the set-up of the circuit, we may predict its performance and the output O it will produce depending on the parameters c and p of any periodic expansion of a continuous function $f(t)$ for $|t|<p$ serving as an input I.

The white-box-analysis thus derives a procedure of generating any general pair of corresponding input-output phenomena from the analysis of the transforming process P in an I-P-O-chain.

---

[1] The term is derived from the principle of *duality* used in mathematics to describe terms which are related by the exchange of certain of their essential parts.

The Direct Aspect

We may also use Ex. 2.5 in order to explain this aspect, but the scenario is different now. In this case the process P transforming input I into output O is known, for instance, the Wien-Bridge and the way it operates, i.e. all the parameter values determining it, like capacities and resistances of the various components of the electronic circuit. With this knowledge of the transforming process P we can predict any output O generated by the Wien-Bridge whenever we *observe* a certain input signal I determined by its parameters c and p. In this case I and P are the known components of the I-P-O-chain permitting a precise and unambiguous prediction of the corresponding output O.

This direct aspect analysis requires information about the process P of an I-P-O-chain and the knowledge of a specific input I into the system Σ described by this chain; it then allows to predict the output O of this system

The Inverse Aspect

The inverse aspect is *dual* to the aforementioned direct approach to the analysis of a system. In this case the transforming process P is known and a special output O created by P will be observed. The knowledge of P and O permits conclusions about the structure of the input I into the system. Thus knowing, for instance, the way a Wien-Bridge works and observing a certain trigonometric output O of this system, we infer that the input signal I is periodic with a certain frequency, but it will in general be impossible to determine the definite structure of I: in the case of our example the input I may, for instance, be any of a family of periodic signals, continuous or not.

2.4.3 Development

The phase of analysing the given object-system Σ_I – input into the modelling system Σ_P – is followed by the development of the model, the output Σ_O of Σ_P. Of course, the procedure of systems development depends on the respective field of research and cannot be described in general terms covering each single application. Nevertheless, we think there are some general intermediate stages which the various fields of systems design have in common.

To start with we subdivide the phase of systems development into the design of a rough conception[1] of the system to be developed and into the organization of

[1] This is also called *rough copy, preliminary* or *global design.*

further details[1]. The procedure is similar to that of the analysis of a given system Σ_I. We begin with the generic system to be developed and look at it as a black-box, thus only regarding its relations to its environment. By successively opening the black-boxes of subordinate levels we then *develop* the fine-structure of the model-system Σ_O. So eventually, we shall reach the level of the elements of Σ_O reflecting the desired highest degree of specification.

Again, we may decide to start from a certain level of detailed specification and to focus on only one – or a few – of the subsystems of one level if we deal with the development of subsystems: this collection of subsystems corresponds to the system Σ_O to be developed. In that case further subsystems, on the same or on different levels, will be regarded as black-boxes; they are only of interest as parts of the generic system correlating with the select subsystems relevant to Σ_O. We concentrate on the relations exclusively established between these neighbouring parts and those to be modelled, as only they are of relevance to Σ_O. The reason for a subdivision of the input-system Σ_I into subsystems is that we want to define sub-projects of systems development, because the generic system Σ_I is too complex to be treated as a whole. So we break it down into parts which may either be modelled simultaneously by different teams, or we may decide – due to limited available resources – for a consecutive development and set priorities in the process of the development of these subsystems. We thus attend to the realization of one particular subsystem or of a set of subsystems in order to get something to work on (cf. Fig. 2.15). By gradually including further subsystems into our model, we try to expand – in a cyclic successive modular development – the realization of the partial structure until the complete generic system is eventually covered. In the course of a modular step by step systems development those subsystems left out of consideration for the time being will be parts of further upgradings. No matter whether the components of a generic system are developed at the same time or consecutively, the interfaces between these required constituents of the generic system have to be taken into account in any case. It goes without saying that they will be taken care of in the case of simultaneous developments of the subsystems, as each partial project must be finished at the same time; thus interfaces will be noticed as the subsystems will not work without them. When developing consecutively with just one team, these interdependences are missed more easily as they are not of importance for the time being and the requirements of the neighbouring subsystems are not so obvious then.

[1] Other names for this phase of systems development are *final* or *detailed design*.

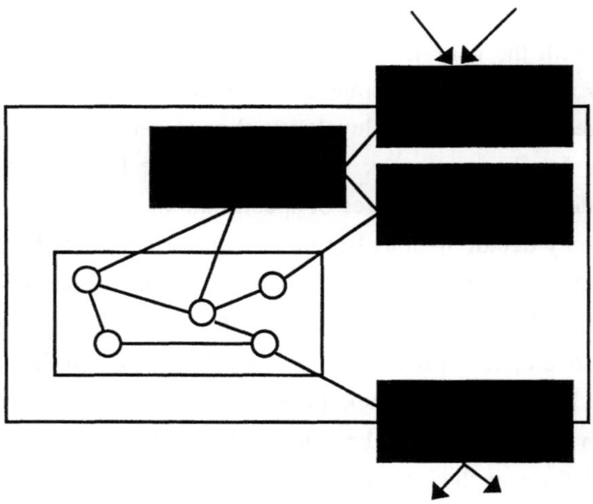

Fig. 2.15. Systems development focused on subsystems

Within this phase of devising a rough conception of the subsystem to be developed great importance is attached to the internal relations between the parts of this subsystem and to the relations between them and the rest of the external parts (respectively their elements; cf. Fig. 2.15). In the case of a cyclic, modular development and upgrading of a system the subsystems left out of consideration are those related to still coming cycles of systems development. Consequently the improvements achieved during the further progress of modelling the generic system are already determined by the preselected initial subsystems; these aspects are prior to the development of the total model.

The relations and interfaces between a modelled system Σ_O and its environment as well as those between its parts must constantly be compared with the respective relations relevant to the original system Σ_I. The relations within the model should represent their counterparts within the original system Σ_I under two aspects: first they must correctly interconnect the imaged parts representing the relations between their respective counterparts in the original system Σ_I. On the other hand, and secondly, a relation is defined by a certain flux of information. This information exchanged between the various parts of a system Σ_O, gained by modelling, must represent the relations adequately that govern the original system Σ_I. The representation of the interconnections between the parts of the original system Σ_I by modelled relations must be complete and without redundancy. They must mirror the information the parts of the original system interchange. As modelling is associated with abstraction, idealization and/or approximation, there will, naturally, be a loss of information characterizing the original system. On no account must those items lost by modelling contain relevant knowledge of the system subject to modelling; being of relevance depends on the objective of the project. The decision what part of the information in Σ_I is dispensable when modelling is made according to the utility principle. First of all we dispense of any features of Σ_I not necessarily needed to achieve the purpose the model Σ_O is expected to serve; this will, for instance, be the case when we produce models used as toys (cf. Ex. 1.1). We may also decide to dispense of certain information in order to facilitate modelling, though we may find this difficult, because we feel that just that piece of information really matters, but has to be sacrificed for the sake of simplicity. A model developed under these conditions will only be an initial *first* approximation to a final solution of the problem and it must be improved in the course of subsequent cycles of systems development. Sometimes we will decide – prior to any modelling – to do without certain details of the original system in our first attempts and to include them into the model later. If, for instance, we want to develop an inventory control system, we may decide to start with a deterministic model assuming linear decrease in stock. The resulting system will be simple, yielding rather vague predictions. Further improvements of this system may successively take more realistic considerations into account, like probabilistic demand and probabilistic costs.

Besides, we are forced to do without certain data of the original system when they are not obtainable, or we decide against them because of the cost-benefit ratio when we try to get them. It is possible that we even omit a piece of information unintentionally, because we are not aware of it, though it may be an essential part of the system in question (cf. example Ex. 1.16 of Tiberius' giant's head).

When the level of the *elements* of the system (the generic system or one of its subsystems) to be analysed is reached, these elements represent processes which cannot be broken down any further. These elementary processes governing the original system must be reflected by the processes within the model as exactly as possible, but they will be represented by other means. These representations are subject to abstraction, idealization and approximation, like, for instance, a mathematical calculus represents a physical process. Likewise, a computer-programme and an automatically managed data-base represent registers originally manually manipulated. As a further example we mention the analogous representation of physical phenomena, like the flux of water through porous media represented by electric circuits (cf. Sect. 1.3), making use of the analogous structures of Darcy's formula and Ohm's laws.

As we have pointed out, a model is a simplified representation of a system, and the loss of any characteristics of this original system due to the projection into its model is often determined under utilitarian aspects. Still, the model-system Σ_O must represent the corresponding input-system Σ_I adequately according to the purposes of the project. If this is the case, can be checked by input-output-analyses regarding the transforming processes as black-boxes. In these tests the concordance between the modelled output and the original one is evaluated under the assumption that the modelled input represents the corresponding input into the original system adequately.

Models are adjusted to the original systems – the reality they represent – so that they are unbiased and come up to the *expectations* of the modeller. There are aspects of the logic of testing models which we shall discuss in greater detail later on, and there are techniques of testing based upon statistical methods. These statistical procedures are not the subject of our exposition, because they are dealt with in a large number of reference books, but we will mention some of these books in Chap. 11 where we give suggestions for further reading.

Testing a model is a bottom-up procedure. It starts on the level of the elements of the model-system Σ_O. When each element of a subsystem has passed its tests, the whole subsystem is regarded as one element of a generic subsystem and tested itself. We thus proceed from tests of the elements on the lowest hierarchical level of a system to tests of its subsystems and, finally, to tests of the complete model-system Σ_O.

Some of these tests are based on sets of *test data*[1] representing original data in a simplified form or special data developed to test the system. These test data are input into Σ_O, and the respective output generated by the model is compared with expected results computed by alternative methods (cf. Sect. 9.3). Concordance between data produced by the model and those computed is a piece of evidence for the assumption that the modelled *procedure* is adequate. Nevertheless, it does not prove that the model is adequate; on the contrary, we should perhaps be more modest and state merely that concordance between the computed data and those modelled does not prove the *inadequacy* of the constructed model.

In a second series of tests data representing an input into the original system Σ_I are fed to the model Σ_O. The corresponding output generated by the model is com-

---

[1] These are data developed to test a system or a system component.

pared with the respective data measured in Σ_I. Concordance between modelled data and those observed will again be in favour of the model. But this again does not prove that the model is 'correct'. These results of the tests only support the hypothesis that the designed model represents reality appropriately, and we may conclude that they are 'a piece of circumstantial evidence' that this model is 'not incorrect', even rather dependable[1]. Concordance has only been established for some data sets, and this does not imply further acceptable results of the model-system Σ_O in future applications. In this sequence of tests – when at its very beginning data observed in reality are compared with those gained by modelling – we cannot expect concordance to be achieved at once, but the model-system must be tuned and adjusted to provide an output we can accept as an adequate reflection of our observations. This phase of tuning and adjusting a model is also called the *calibration* and *recalibration* of Σ_O (cf. Subsect. 5.1.2, too); interestingly enough the etymological roots of the word *calibration* refer to the etyma we already described in Sect. 1.2. One of the word stems is the Arabian *qalib*, which means 'form' or 'model', another one is the Greek *kalopodion*, which is a 'last'. As we have seen in Sect. 1.2 the 'last' corresponds to the Italian *modello* and *forma*; thus 'to calibrate' originally was a mere synonym of 'to model', but nowadays it is used to describe the process of adapting a developed qualitative model quantitatively to the values of the parameters measured in the original system.

The system Σ_O thus tested and calibrated meets the necessary requirements on the basis of available evidence with the mentioned restrictions. There are different demands on the *degree* of concordance between data measured in the original system Σ_I and those produced by Σ_O. If, for instance, the developed system Σ_O represents a commercial cashing up procedure, the clients will expect a *complete* correspondence between figures calculated in reality and those computed by the system Σ_O. On the other hand, it may be impossible to calibrate a system Σ_O to such an extent that there is absolutely no difference between its output and data measured in reality. For instance, any model designed for predictions, like weather forecasts, stock market pricing etc., will generally deliver an output differing from the actual future developments. In those cases we shall be quite content if there is at least *some degree of agreement* between reality and the performance of the model. But how much do we permit the model to deviate from reality? Obviously, the acceptance of a model as a system representing its object-system Σ_I adequately

---

[1] We avoid terms like 'valid' or 'reliable', as they will be used with a different meaning (cf. Sect. 5.2.).

enough is based on *consensus* among those working in the relevant field of research; a model thus only represents the state of the art of modelling available within this field, which may sometimes be of a rather low standard (cf. Chap. 8).

2.4.4 Documentation and Introduction

As soon as a model-system Σ_0 is tested and calibrated to the client's satisfaction it can be released and come into regular operation, which – at the beginning – must be accompanied by an extensive training of the users.

The training of the users and the operation of a system depend on an appropriate documentation. We already mentioned the importance of documentation in connection with the definition of a projected system. Of course, documentation should accompany each step of systems development. Detailed records are an effective means to avoid misunderstandings between the client and the producer of the system as well as among the members of the team developing it or among project teams working simultaneously on various subsystems which have interfaces in common. Furthermore, a comprehensive documentation is needed for the maintenance of the system. Programmers, for example, tend to develop a programme code without any further documentation. To them the code is self-documenting as they are up to their ears in their programmes. If asked to keep proper records of their programmes, they will promise to hand them in later. But as time passes by and under the stress of everyday work this promise is soon forgotten. One day the programmer may have left the company when a change in his computer code is due, and someone who has no idea of the programme must do it, a Herculean task that can only be appreciated by someone who had to update a computer-programme written by somebody else without any further documentation at hand than its source code.

Yet, the documentation used for training and operating a system serves a different purpose. It does not contain the internal details of the system Σ_0 and of each phase of its development, at least not all of them. It rather describes the functions of Σ_0 and demonstrates how to handle its components in order to achieve certain reactions. The way this documentation should be written depends on the users of the systems; it must be adapted to their comprehension. We all have seen instructions for the operation of technical systems – video recorders, cameras etc. – more confusing than enlightening. But, obviously, it has been widely recognized that this is a problem, and most producers of technical systems run teams specialized in writing instructions. The members of these teams are no technicians in the first place, but rather men of letters possessing a certain technical knowledge. When training the operation of more complex technical systems, e.g. the new model of an aeroplane, or of power stations etc., the situation is different, as the users are qualified: they have been trained on the systems and they possess the necessary technical know-how. Nevertheless, the better the documentation the more it helps even the expert to understand and control the system in question.

2.4.5 Operation and Operational Evaluation

When the developed system has been introduced to its users and when they have been trained properly, the system can be operated regularly. The maintenance of the system will be taken on by special service teams, who escort its operation. These teams offer advisory services and further training if required; they look after the relevance of the documentation and update it if necessary. They may need some support themselves whenever extremely unexpected situations occur, but in general the developer of the system will be free for a new beginning. This may be the design of a totally different system, but it may also be the development of an improved release of the former system following the mentioned cycle of systems development.

As soon as a system is operated regularly, problems may occur right at the beginning. In general these problems can be solved in no time; they may lead to minor revisions of the design of the system which will then run again to the satisfaction of its users.

Every day a system works properly is evidence supporting the hypothesis that the system is 'good'. Each breakdown of a system is incompatible with this claim and casts doubt on it. If a system has been running properly for a very long time and then suddenly breaks down, this may be a sign of ageing or it may be evidence that an absolutely unexpected situation has occurred. We remember an explosion in a refinery in the Rhine-Main-Area in Germany. It happened in the early 1970s. Of course, the refinery was maintained properly and supervised by the authorities. But one day, something unexpected happened. Even now no-one really knows what started it, but there was a heavy explosion. Unforeseen circumstances had caused the coincidence of a spark and a leakage in the same place and at the same moment. Among the left-overs of the main entrance of the plant there was a table announcing proudly, 'We have been working without accidents for XXXX days'. We do not remember the correct number of days 'they had been working without any accidents', but it was in the range of some thousand.

Another and better known example will be that of the *CONCORDE*, the supersonic commercial aircraft developed in the 1960s and the 1970s. It served its purpose – to transport people across the Atlantic in minimum time – well. There had not been any serious accidents with the *CONCORDE*, and it was regarded as a safe – though not at all economic – means to cross the Atlantic Ocean. The minor accidents or faulty functions of this aeroplane were examined thoroughly and led to more reliable series of the plane. But suddenly, in July 2000, there was that disastrous crash in Paris. This catastrophe was caused by an improbably unfortunate constellation of circumstances. Of course, flying this plane was stopped immediately, and the authorities scrutinized the wreck and the surroundings to find the cause of the accident and weak points in the system. Nevertheless, the crash did not *retrospectively* declare the *CONCORDE*-system a failure; on the contrary, we have learned from it that the particular situation that had occurred had asked too much of the system, and about one year later the engineers obviously had the particular problems under control. Now the *CONCORDE* is flying again.

Likewise, the refinery in Germany had worked perfectly over the very long period of XXXX days. The staff running this plant had reason to be proud of this performance. Nobody could be expected to anticipate the exceptional event of the explosion, so nobody can be blamed for it except those who may have caused it by irresponsible actions, but certainly not the technicians and the staff. Those examples show the awkward situation of those who will be called to account. It is impossible to foresee and thus prevent any technical breakdown or any misuse despite a careful maintenance and obviously clear instructions. We have to be prepared for unfortunate constellations of circumstances of any kind, above all human errors and thoughtlessness. Some so-called accidents in the past make this clear. And even if the cases dealt with are sometimes absurd and beyond common sense, it is worthwhile to report one of them: a lady decided to bathe her lap-dog and then dry it in the microwave oven for some twenty minutes. Who is to be blamed when the pet was well-done afterwards? This lady knew from experience that some 1,000 g of meat would be cooked within twenty minutes. Can we expect the manufacturers of microwave ovens to anticipate any possible way of misusing this appliance? If we do, their instructions will grow massive and unreadable, because they will contain an infinite number of warnings. Anyhow, there will always be someone inventing a new form of misuse nobody knew before. So the only safe way of warning would certainly be 'do not use this utensil at all'. This also explains all those unreadable, frustrating and even deterring directions we have to study whenever we are ill and must take a medicine. These instruction leaflets – constantly getting longer – are a counter-measure of the pharmaceutical industry to protect themselves from excessive claims for compensation which might ruin any company.

If a system never passes its prime tests or if a theoretic model cannot be calibrated and adapted to reality, the development was incorrect, and these systems will be rejected. Yet, having passed those tests is an argument in favour of the respective system, though it still has to prove its qualities in daily operation. *CONCORDE* was regarded as a very safe means of transport before July 2000, and its dependability has obviously been enhanced by the meticulous inspections and the improvements carried out during the period following the accident; so it is even safer today. But we will always keep that fatal crash in the backs of our minds and somehow feel that there is still a risk involved in flying with the *CONCORDE* – like flying with any aeroplane. One, day there may be a new unfortunate combination of conditions, a situation the engineers failed to foresee or discarded as improbable. Then the system must be re-checked again.

On the other hand, we sometimes decide to operate a system even if there is the risk of failure. This may be the case if we hope to gain a deeper insight into a problem area and to learn by experimenting with it. Thus the failure of our system may give rise to the development of improved and safer models. When, for instance, C. Barnard transplanted the first heart in 1967, there was not much of a chance for his patient, L. Washkansky, to survive the operation. In fact, he died 18 days after the transplantation. But Barnard's team had learned a lot about immune reactions and operation techniques, and they could improve their conception. A second transplant was carried out in 1968, and the patient, P. Blaiberg, survived for almost two years. Today, hearts are transplanted in large numbers all over the world, and the risk of failure has been reduced to a tolerable minimum. This example stands for a multitude of medical and pharmaceutical developments having claimed many victims when they were first tested.

Growing insight into scientific problems may lead to the decision to reject a system we accepted in the past, because we thought it was tried and tested with a positive result. *Heroin*, for instance, was originally developed as a remedy. It was used in paediatrics to soothe asthmatic attacks, an application nobody would suggest today when the possession, trade and consume of the narcotic *heroin* is strictly forbidden in most countries. There are less drastic examples of drugs taken from the market, because one suddenly realized that they have severe side-effects and that they are more of a poison than a medicine, e.g. *thalidomide*, which was prescribed as a well tolerated sedative in the 1960s, but turned out to cause foetal deformities if taken by a mother in early pregnancy. The understanding of the pharmaceutical effects of *cannabis* seems to lead to a contrary reaction. This narcotic – less hard than *heroin* – seems to be an ideal analgesic, and it will probably not take long until *cannabinol* (*THC*), its active part, will be legalized as a remedy. Even *thalidomide* enjoys a renaissance; it is used as a cure for leprosy, and only recently there has been evidence of healing effects of *thalidomide* on cancer.

To sum up we can say that *systems* are well-defined organized collections of parts serving a particular generic purpose. We have learned about their internal structures, and we have seen that under classifying aspects those of an open and dynamic system seem to be most important for our further investigation.

Testing a developed system plays a major part in determining its qualities; these will have to satisfy certain quality criteria to be studied in Chap. 5.

We have introduced the procedure of systems design as an *I*-*P*-*O*-chain of systems themselves constituting its elements as $\Sigma_I \rightarrow \Sigma_P \rightarrow \Sigma_O$, and we have discussed the structures of the input- and of the output-systems. The central part of our further discussion will be the system Σ_P mapping the original input system Σ_I into its image Σ_O. This corresponds to what we realized in Chap. 1 where we saw that there is one common idea inherent in all the examples when a model is interpreted as an *image* of a certain entity subject to modelling. Thus, the following chapter will be devoted to the discussion of mappings.

3 Mappings

3.1 The Imaging Process

The examples compiled in Sect. 1.1 show that models are connected with an imaging process, which either results in the model as an image (*simulacrum*) of an underlying entity or produces an image as a copy of the model in question. In that case the model is the *prototype* of its copy. These relations are depicted in Fig. 3.1 (resembling Fig. 1.7). The arrow in the diagram represents the mapping or the imaging process. This mapping produces an *image*; the object of the imaging process, the projected entity, is called the *counter*-image of the mapping.

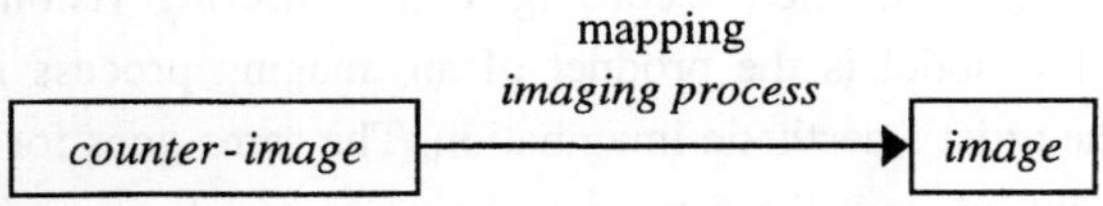

Fig. 3.1. Imaging process

As we will frequently refer to this diagram, we want to represent it in a more simplified and generalized form. The imaging process will just be *process P*. P transforms its *input* I (the counter-image) into its *output* O (the image), and the diagram of Fig. 3.1 may thus be reduced to the following $I \rightarrow P \rightarrow O$-diagram:

$$I \xrightarrow{P} O. \tag{3.1}$$

When we apply this scheme to the examples we started with, the respective models will be the output (image) of the underlying process in the cases of the three dimensional representation of an object, the philosophical theory and the mathematical representations. The counter-images of these models are 'reality', the real world, i.e. the original object, the world as it is seen by the philosopher; in general it represents the process within the underlying system Σ_I subject to modelling. The process P_P of modelling within the modelling system Σ_P leads to the model Σ_O output by mapping the 'reality' Σ_I in question.

In the cases of figures to be reproduced, models posing for artists, moulds and exemplary personalities to be imitated the situation is quite different. Here, the models themselves are the counter-images Σ_I of a 'modelling' process Σ_P reproducing them in more or less perfect copies Σ_O.

In order to demonstrate this, we will look more closely at the example of a model posing for an artist. This model is the counter-image of an imaging process resulting in a work of art, the model's image as it is seen by the artist. On the other hand, the model herself – let us assume the model is a she – is chosen by the artist according to certain aesthetic ideas which – for their part – are time-dependent and wholly subjective. As we have already mentioned, Rubens' model posing for a nude study certainly has little similarity with Picasso's, and *grief* has been represented by various *pietàs* modelled according to the differing visions of various artists. So the artist's model is the product of an imaging process itself, whose counter-image is the artist's aesthetic imagination. The same goes for all the other examples we have introduced: the figures to be reproduced are modelled on an original, the moulds used for modelling a multitude of copies are modelled themselves on an original. The exemplary personality is chosen according to some moral ideas, and each society has its own ideals to be imitated. The model garments and the *AIRBUS* model are developed in the same way: they are modelled on aesthetic ideas or they represent certain technical concepts and ideas of economic transportation.

So, in general, modelling is a multilayered process containing a whole hierarchy of *I-P-O*-diagrams The output of one of these diagrams is the input into the succeeding one. We condense this idea into the following schema:

$$I_0 \xrightarrow{P_0} O_0 = I_1 \xrightarrow{P_1} O_1 = I_2 \;\ldots\; O_{i-1} = I_i \xrightarrow{P_i} O_i \;\ldots\;. \qquad (3.2)$$

This schema (3. 2) corresponds to the chain (2.3) of Sect. 2.3 describing the states governing a dynamic system.

The last (*modello*) is an example of a hierarchical system of imaging processes. In a first step the last is modelled to represent either an individual foot or a foot of a certain size and shape in general. In the second phase of modelling shoes are produced fitting the last. Moulding a statue may serve as a further example to clarify this point: first a person is chosen to pose for the artist. In a second imaging process a figurine is modelled on the model of the first process; this reproduction will be in a different material, e.g. in wax, and it may also be on a different scale. In a third step a mould is made from the wax model and is then used to produce copies of the wax model.

The example of a measure (*modulus*), described in Chap. 1.1, also fits into the *I-P-O*-scheme. Measuring is the process of imaging certain physical phenomena, like heat, extensions or distances, into real numbers; their images are the temperatures and the lengths measured respectively.

We now want to study the structure of an imaging process more thoroughly. When modelling we accept a certain loss of information contained in the counter-image of the model, the object of modelling. As we remember, this loss of information is necessary to reduce the features of the counter-image in order to serve the purposes of a model. In Ex. 1.1, for instance, a car is modelled to serve either as a toy or to demonstrate certain functions. In those cases the features of the original car are reduced to serve those purposes: there are only the four wheels reproduced by the toy car, and there are only the shape and some material parameters when the model is supposed to be tested in a wind-tunnel.

On the other hand, the loss of information may be immanent in the process of modelling whenever phenomena or parameters of the original system are not known yet. So, of course, we will have to do without them when modelling the object-system; this is due to a lack of insight into the complexity of reality. So, to Pulcher (cf. Ex. 1.16) the only logically consistent way to model the giant's head belonging to a giant tooth was to give it a human countenance, as he knew nothing about mammoths. Likewise, modern models of dinosaurs present them with a skin colour something between a bluish grey and a greyish green, the colour thus reflecting our assumptions derived from the appearance of contemporary reptiles.

This shows that a model will either dispense with details not serving its purpose or that it has to dispense with them when they are not available. We represent the imaging process leading to models under these aspects in Fig. 3.2.

In Fig. 3.2 the *counter-image* to be modelled is represented by the horizontal line at the top of the diagram. The left part of this line shows all the information contained in the counter-image, but neglected by the modelling process. The rest of the counter-image subject to modelling is the *co*-image of the model. The horizontal line at the bottom of Fig. 3.2 stands for the model; its left part, which is the image, reflects the features of the counter-image decisive for modelling. These modelled features are derived from the *co*-image, the relevant part of the counter-image with regard to modelling.

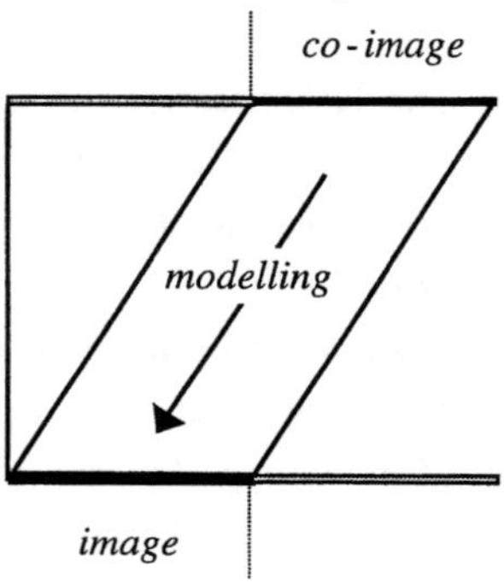

Fig. 3.2. Forwarding information by modelling

The right-hand section of the model – outside the image – represents additional characteristics of the model irrelevant to the modelling process and not representing any features of its counter-image. These additional characteristics may, for instance, include the material the model is made of: so the model of a dinosaur, exhibited in a museum, may be made of a synthetical material, not at all reflecting the tissue of its counter-image.

3.2 Structure Preservation

As modelling is tantamount to imaging, it is therefore subject to general aspects of mappings. The concepts of mapping are used in many different fields of scientific research. One approach is to describe a mapping as a *morphism* (Greek *morphe*: 'form'), a transformation or projection preserving certain structural characteristics which are part of the form of the mapped object.

A clear and unambiguous description of morphisms may be found in mathematical theories, though the term has frequently been used in other fields of sciences as well, like in philosophy and modelling. The basic idea, when defining morphisms as a means of modelling, is to consider mappings between sets of entities each one structured by a certain *relation* between its elements.

So let $S_1=\{X_1, X_2, X_3, ...\}$ be a set of elements X_i for $i=1, 2, 3, ...$ and let $S_2=\{Y_1, Y_2, Y_3, ...\}$ be a set of elements Y_j for $j=1, 2, 3, ...$. Let, furthermore, ρ be a 'relation'[1] between the elements of S_1 represented by $\rho(X_m, X_n)$, if the relation ρ holds for the elements X_m and X_n of S_1. Let X_m and X_n be mapped onto the elements $*X_m$ and $*X_n$ of S_2 and let the relation ρ between the elements of S_1 be mapped onto the relation $*\rho$ defined in S_2. The prefixed asterisks *(...) thus represent the images of the mapping from S_1 to S_2. A morphism is defined as being structure-preserving with $*(\rho(X_m,X_n))=*\rho(*X_m,*X_n)$. This means that the relation ρ is transferred from the original S_1 to its image S_2: whenever the relation ρ holds for the elements X_m and X_n of S_1, its image $*\rho$ is valid for the respective images $*X_m$ and $*X_n$ of X_m and X_n in S_2.

It may be difficult to identify the representation of these elements and of the relations between them outside of pure mathematics, but it has grown a habit to call a mapping a *morphism* whenever it preserves *some* sort of structure. We will follow this terminological agreement, but we emphasize that with our conception of modelling based on the ideas of systems theory there are two different relations governing each of the systems Σ_I and Σ_O, the respective input and output of the modelling process Σ_P. First we realize that there is a relation ρ_1 between the parts constituting a system the way it is reflected in Fig. 2.1. If X_m and X_n are two different parts of Fig. 2.1 connected by a line representing a structural interconnection between X_m and X_n, relation ρ_1 holds for X_m and X_n, and we represent this fact by $\rho_1(X_m,X_n)$. Secondly, there is another relation ρ_2 between the respective inputs and outputs of an *I-P-O*-chain. This relation holds for an input *I* and an output *O* whenever they are connected by an associated process *P*, and we represent it by $\rho_2(I,O)$. In the case of the process described by (2.4), transforming a set of inputs into a set of outputs, the respective relation ρ_2 is valid for each pair $(I_{i,s,j_u},O_{i,s,k_v})$, $u=1, ..., m$ and $v=1, ..., n$, and it is symbolically represented by $\rho_2(I_{i,s,j_u},O_{i,s,k_v})$.

We return to Fig. 3.2, and in order to give a more detailed representation, we introduce a new term borrowed from the theory of mathematical structures:

---

[1] We use the word 'relation' in the broader sense of the term including mathematical relations, e.g. between numbers like $1<3$ as well as compositions (operations), like addition $1+3$ or multiplication $1\cdot3$

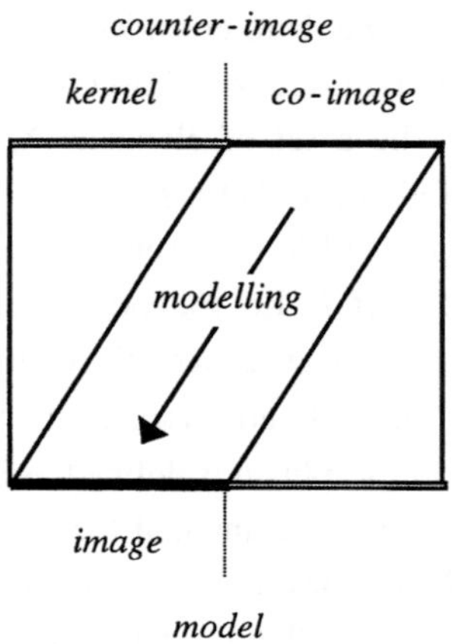

Fig. 3.3. The *kernel* of modelling

The *kernel* of a mapping is the set of all those parts of the counter-image of a model that are lost when modelling. It represents the set of information contained in the object-system Σ_1 considered as being *irrelevant* to the model or *out of reach* of modelling.

The architectural model of a building, for example, will represent the proportions and outer appearance of the building. Static and technical details – especially of the inner construction like staircases – are part of the kernel of the mapping. Yet, the model also possesses features which are not connected with any characteristics of the original building, for example, the material – cardboard – or the size of the model. We call this part of the model of Fig. 3.3, made up of the additional information defining it, but not representing any feature of the counter-image of the model, the *co-kernel* of the mapping. In our example of an architectural model the material and size of the model are part of this *co*-kernel.

If we disregard the orientation of the arrow representing the direction of modelling in Fig. 3.3, the rest of the diagram will be invariant when it is rotated through 180°. This invariance will be extended to the arrow if we make it point into both directions, the way it is shown in Fig. 3.4. In Fig. 3.4 the top row contains the *kernel* and the *co-image* of the projection; the bottom row is made up of the *image* and the *co-kernel* of the mapping. What has originally been called 'counter-image' and 'model,' mutually correspond in a dual way: we can no longer tell which of the rows actually is the 'model' and which is its 'counter-image'. If we return to our example of an architectural model, this means that the card-board house is a *model* of the concrete building and that, *vice versa*, the concrete building is a *model* of the card-board model construction as well. At first glance this seems to be puzzling, but it becomes clear when we look at the different purposes an architectural card-board house model may serve. It may be modelled on an existing building and thus be the product of an imaging process. This is what we called a *simulacrum* in Sect. 1.2. But the card-board house may be the *prototype* of a building, and in that case the building (i.e. its outer appearance) will be modelled on the card-board 'model'.

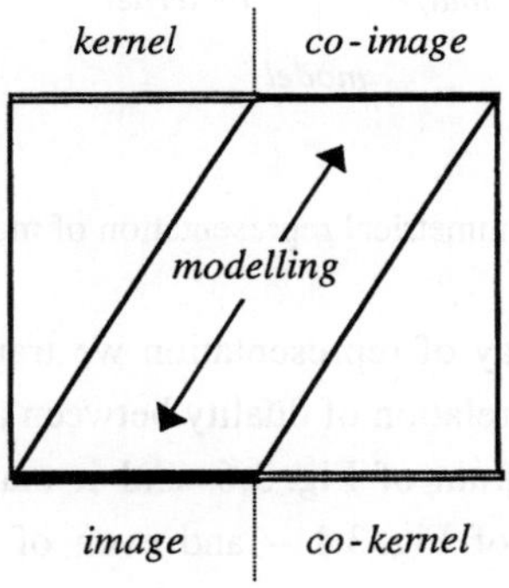

Fig 3.4. The *co-kernel* of modelling

We see that the terms 'modelling' and 'mapping' are synonyms, two words describing the same process. A *model* thus is the result of a projection and any imaging may be interpreted as a process leading to a *model*. Furthermore, there is not much difference between the top and the bottom lines of the diagram in Fig. 3.4. Both lines represent models, each of them is a model of its counterpart, the line facing it. The only difference may be the aspect of the utility the modelling process is intended to serve. We consider this and call the intended *object* of the modelling process the *co*–model. Thus we assume that the bottom row in Fig. 3.4 is still the model as it was in Fig. 3.3, but the name of the top row has been changed from *counter-image* to *co*–model in the symmetrical representation of modelling as represented by Fig 3.5.

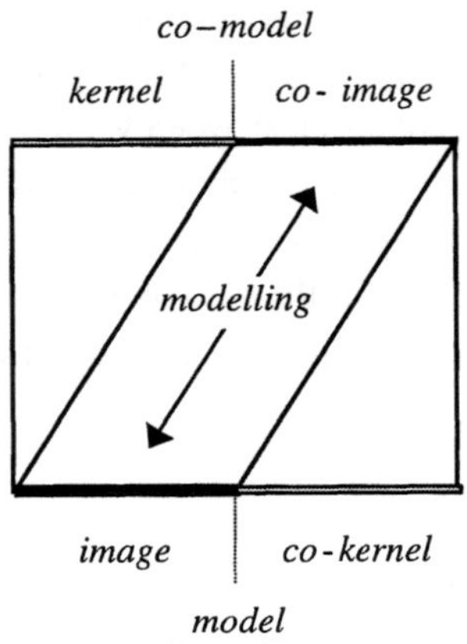

Fig. 3.5. Symmetrical representation of modelling

To find a more simplified way of representation we transform Fig. 3.5 to a horizontal diagram reflecting the relation of duality between a model and its *co*-model. This is the symmetrical diagram of Fig. 3.6, and it may be regarded as a more sophisticated representation of Fig. 3.1 – and even of Fig. 1.7 we started with. Within this diagram the prefix *co*- serves the only purpose to distinguish the initial intention of modelling a counter-image into its model to be developed.

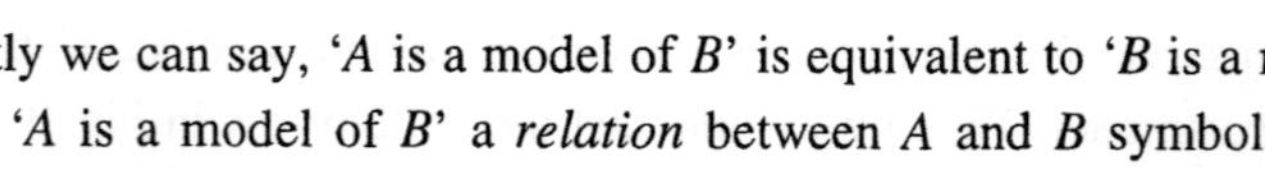

Fig. 3.6. Dual representation of modelling

Consequently we can say, 'A is a model of B' is equivalent to 'B is a model of A'. Let us call 'A is a model of B' a *relation* between A and B symbolically represented by $A=\mathrm{mod}(B)$. Our conclusion is thus described by the logical equivalence

$$A=\mathrm{mod}(B) \Leftrightarrow B=\mathrm{mod}(A). \tag{3.3}$$

In (3.3) the symbol '$\Leftrightarrow$' stands for the statement 'is logically equivalent'. We thus conclude if $A=\mathrm{mod}(B)$ then $B=\mathrm{mod}(A)$, and *vice versa*. A relation meeting condition (3.3) is called *symmetrical*. We realize that our concept of modelling leads to a symmetrical relation between models. In spite of that, the two models A and B still serve different purposes, and we distinguish them by calling one the *model* and the other one the *co-model* of the imaging process. The latter is the object of modelling and the former is its product, the model in the narrow sense of the word.

In this way we describe the process of modelling by using characteristics of mathematical relations. These relations may be defined by further properties. For instance, regarding modelling-relations, the proposition

$$X = \mathrm{mod}(X) \text{ for all } X, \tag{3.4}$$

mod is a *reflexive* relation, is true. Proposition (3.4) means that everything X is a model mod(X) of itself. This seems to be self-evident, and we may say that the modelling relation 'mod' is reflexive. But in order to avoid false conclusions, we have to study this statement (3.4) more thoroughly. 'Any X is a model of itself' stated in (3.4) does *not* mean that the symbol X on the left of this equation is *identical* with that one within brackets on the right. This is due to the fact that a model is produced by a modelling process *transforming* its input into its output. To demonstrate the consequences of this point, we return to our conception of systems analysis outlined in Sect. 2.4. Using the terminology of that section the proposition of (3.4) can be represented by the *I-P-O*-chain $X = \Sigma_I \rightarrow \Sigma_P \rightarrow \Sigma_O = X$. Within this chain X occupies two quite different positions. The imaging process represented by the arrows of this diagram consumes time and space. On the other hand, the identity $X \equiv X$ does not permit any variations in space or time. If we succeeded in doubling X perfectly, producing some sort of a 'clone', but even more perfect, i.e. being identical with X in its structure up to the last elementary particle of its constituents, even then this 'clone' would still occupy some space of its own, different from that of its original and would thus not be *really* identical with X. If, on the other hand, we only study the original on its own, we shall have to observe it and – insignificant as they may be – there will be intervals of time between two observations, and everything is due to changes in the flux of time as we have seen in Sect. 2.2, where we cited Heracliton and his *pantha rei*.

Our remark concerning the *reflexivity* of the relation 'mod' seems to be restricted to concrete physical systems. But on second thought, it is also valid in the case of an abstract, incorporeal system. An immaterial system like a theory is also time- and space-dependent. It sets off varying interpretations in the course of time, and these interpretations depend on space, where it is interpreted, by whom and what purpose the interpretation is expected to serve. This may well be the beginning of ample secondary literature accompanying the life-time of a theory. We wish to exclude purely mathematical theories from this comment, as we think that once a mathematical proposition is proved, it will no longer be the subject of any further discussions about its validity[1]. A mathematical theorem may be the fundamental principle of further scientific progress, and there may also be efforts to publish it and its proof under varying *didactic* aspects so that they may serve certain special purposes in teaching; but as soon as the proof of a mathematical proposition is conclusive, the case is settled from the purely scientific point of view.

These considerations may seem to be rather far-fetched. But there have sometimes been discussions about *true* models. We want to stress that there is nothing like that, nothing like a true model. A 'model' being really true implies the identity of Σ_I and Σ_O, which is an unattainable objective. So, the *reflexivity* $X=\text{mod}(X)$ of the relation 'mod' does *not* mean that X is a *true* model of itself. It is just a model, the modelling process does consume time and space, and $X=\text{mod}(X)$ does not imply the absolute identity of X and mod(X).

A further means to characterize relations is their *transitivity*. Regarding the mod-relation transitivity is defined by

$$A=\text{mod}(B) \text{ and } B=\text{mod}(C) \Rightarrow A=\text{mod}(C). \tag{3.5}$$

This definition implies that whenever we know that A is a model of B and B is a model of C, we can conclude that A is also a model of C. A symmetrical, reflexive and transitive relation is an *equivalence relation*. But 'mod' is *not* transitive. We want to demonstrate this fact by a simple counter-example. It will be somehow similar to the architect's card-board model, just a little different. The *Khan* was periodically weighed in gold by his supporters. This procedure probably kept him from being too strict with diets. The result of weighing him and substituting his weight by an amount of gold led to a certain pile of metal representing and thus modelling the *Khan*. If, in a second step, we had measured certain parameters of this heap of gold, like its volume or its density or its electrical conductivity, these values would have represented the heap of gold, they would have formed a model of it, but not of the *Khan*, his volume and his density would have been different and his electrical conductivity certainly, too.

---

[1] Nevertheless, there is the exception of the debate about the foundations of mathematics mentioned in Sect. 1.1.

From this we conclude that the relation 'mod' is not an equivalence relation defined to be symmetrical, reflexive and transitive. Consequently it is also incorrect to say that there will necessarily be a model-relation between *A* and *B* if both *A* and *B* are models of the same counter-model, say *C*.

3.3 Chains and Invariants

In Sect. 1.1 we developed the idea of hierarchical models. If, for instance, we take up our example of an architectural model, we may construct a hierarchy in models starting with the architect's model of his visions embodied in a card-board model. This will serve as a prototype and it will be modelled by being realized in the actual building. Students of architecture may then produce card-board models of the building again, and these card-board houses may be photographed to be models of the students' etudes. Each of these models, the architect's, the engineers', the student's and the photographer's, is a model of the original perception of the architect. In that case successive modelling is a *transitive* relation, but that need not necessarily be so, as we have seen in Sect. 3.2. Because of continuous loss of information the final photograph may show only little resemblance to the architect's visions, but some of his ideas will still be represented in the picture.

We want to look into the structure of hierarchical chains of models more carefully. We return to Fig. 3.3 and change it to the horizontal representation of Fig. 3.7 where only the image and the *co*-image of the modelling process are explicitly mentioned.

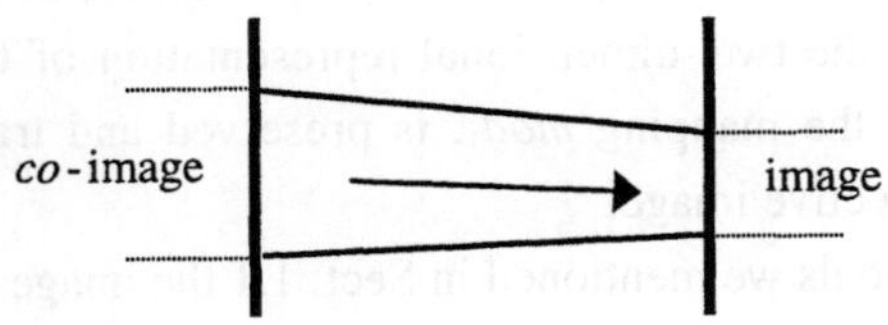

Fig. 3.7. Horizontal representation of modelling

In a second step we compose the chain of Fig. 3.8 containing two successive modelling processes, mod_1 and mod_2.

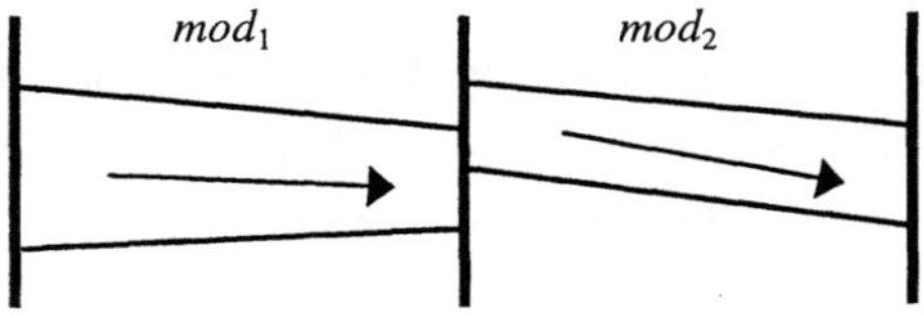

Fig. 3.8. Chain of successive modelling processes mod_1 followed by mod_2

Within this example of a hierarchical system of modelling the *co*-image of the second imaging process mod_2 comprises parts of the image as well as of the *co*-kernel of the first mapping mod_1. This is a representation of a special, but quite frequently occurring case. There may also be situations where the *co*-image of mod_2 is entirely included either in the image or in the *co*-kernel of mod_1, or it may, for example, contain all of the image of mod_1.

We return to the chain of successive architectural modelling described above. If the photography of the student's model is a coloured print and we produce a copy of it in a second modelling process mod_2, in the same format, but in black and white, mod_2 will preserve the proportions and the scale of its counter-image, but the colour (part of the image of the previous modelling process mod_1 of shooting the coloured picture) will be lost and will thus be part of the kernel of mod_2. But the material both models are made of – photographic paper as part of the *co*-kernel of mod_1 – will be (more or less) the same and will thus appear in the image of mod_2. If we modify this example, so that mod_2 transforms the coloured photography into a coloured slide, the image of mod_1 will be part of the *co*-image of mod_2. But in both cases the two-dimensional representation of the photography, part of the co-kernel of the mapping mod_1, is preserved and transported by the mapping mod_2 to its respective image.

Within each of the models we mentioned in Sect. 1.1 the image of the modelling process represents its counter-image in a different way. The examples of a last and of a statue image the foot and the posing model into an inanimate object. The shoe in the former example is to fit the original foot (or an idealized standard-foot of a certain size and shape) as well as possible. The statue is to represent the posing model as similar-looking as possible, probably in a different size, but with more or less the same proportions. The models produced by these processes are *analogous*. In the latter case the statue is even a work of art and therefore also represents the artist's very personal perceptions.

Measuring a physical phenomenon leads to a real number-representation of the degree of the phenomenon. These models are *digital*. The image of a continuously varying quantity will appear as discontinuous in a digital time-dependent representation. Fig. 3.9, for example, shows the continuous growth of a variable v_c depending on time *t*, like time *t* itself or the weight of an organism, and its discrete representation v_d.

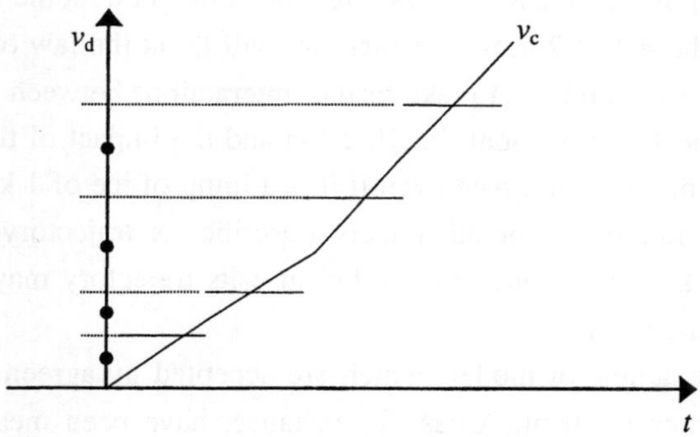

Fig. 3.9. Discontinuous models of continuous phenomena

Some physical phenomena can also be modelled analogously, like the weight of a body which may be represented by the elongation of a coil spring, or the distance between two points *A* and *B* in space which can be represented by the length of a piece of taut rope connecting *A* and *B*. These analogous models allow the comparison of degrees of the underlying phenomenon. Two distances, for instance, are of the same degree if they can be represented by the same piece of rope; accordingly distance d_1, for instance, is of a degree less than that of d_2 if the extension of the rope representing d_1 is not long enough to connect the two points separated by distance d_2.

The free fall of a body may serve as an example of a *numeric* model. The trajectory of the body can be modelled by the law of gravity representing the physical position of the body by a mathematical equation. Within this example the model is made up of a mathematical calculus representing its counter-image, the trajectory of the falling body.

When we define a modelling process '*mod*', its kernel is of great importance. By definition it is made up of all the information contained in the counter-image of the process *mod* that is not transported into its image. This information of the *prototype* lost by modelling may still be significant, so that the developed model cannot be accepted as a *simulacrum* of the original. Consequently the scientific community working with a particular model has to decide whether they accept the model as more or less accurately representing 'reality' – its counter-image – or whether they reject it as inappropriate, not being *similar*.

We may thus accept the law of free fall as an adequate description of the trajectory of a lump of ice of, say, 1 kg falling from a height of some 2 m. If the body falling from the height of 2 m is a feather, we will think the law of free fall is not adequate. In that case we have to take further interactions between the falling body and its environment into account, like friction and the impact of the movement of the air surrounding the body. And even if it is a lump of ice of 1 kg, we will think the law of free fall does not adequately describe its trajectory if the body is dropped from a height of some 30 km, because its trajectory may end above the ground if it has melted away.

As a further example of models which are accepted by agreement we mention the systems of measurement. Areas, for instance, have been measured by acres, one acre being the area of land a farmer could plough in one day (another measure was the morning, the area ploughed within one morning). This measure depends on the skill of the ploughing team, on the quality of the used equipment and on the quality of the ground to be ploughed. Furthermore, what is a day and what is a morning? So, there have been different *acres* and *mornings* depending on the locality of measurement. Further examples are *gallons* representing different volumes in America and Britain, the *ton* etc.

Even the attempts to define units of measurement less dependent on particular constellations and conditions, like the acre and the morning, have to be accepted by those working with them. Length and mass were defined by *standards*, the proto-metre and the proto-kilogramme respectively, kept in Paris. These standards are expensive designs made of platinum and expensively kept under almost invariant conditions of their environment – nearly constant temperatures, atmospheric humidity and so on. We call these environmental conditions 'almost invariant' and 'nearly constant', because we doubt there are really constant values of physical parameters, like temperature and humidity; but they are kept *sufficiently* constant, sufficient for the scientific community to agree that the proto-metre and the proto-kilogramme were constants themselves. These proto-measures were models (*simulacra*) of the scientific definitions of 1 m and 1 kg and they 'posed' as models (*prototypes*) for copies of the metre and the kilogramme. But these prototypes were also subject to ageing, and the proto-kilogramme of 1980 was not identical with what it had been in 1970. So the idea is to redefine measures by more invariant relations depending on other physical invariants. There has obviously not yet been a revised version of the definition of masses accepted by the scientific community. But the definition of the metre has been reworked and traced back to 'natural constants'; this definition now depends on the speed of light.

But what do the terms 'natural constant' and 'physical *invariant*' mean? Can we really be sure that these 'constants' do not vary in the course of time and within the space surrounding us? Will the speed of light be a constant forever and was it like that in the past (cf. Ex. 6.10)? Can we be sure that light does not expand with a different velocity somewhere else in the universe? Definitely, we cannot be sure; we just believe in the assumption that, e.g. the speed of light is an invariant, everlasting and valid everywhere – yet there seems to be evidence that this parameter may not be a constant after all (cf. Webb et al. 2001); consequently the definition of the metre depending on the speed of light may not be *constant* and consistent either.

And even if we were able to define invariant measures, we would still have to admit that modelling by measuring suppresses a lot of information included in the object of measurement. So, for instance, weighing an object does not reflect any other physical parameters of this object like, e.g. its volume.

In conclusion, we want to sum up that we have found out that modelling is a process of projecting ensembles of entities into each other. Some of the features of the counter-image of a modelling process will be invariant in this process, others belong to its kernel and are lost, i.e. not transported to the model, the image produced by the projection. Among these invariants, there are – with prime importance – certain structural aspects, which may most easily be regarded from a mathematical point of view, and for some time now modellers have already adopted the mathematical terminology as far as those invariants characterizing *morphisms* are concerned.

3.4 Morphisms

Modelling is an imaging process preserving various features of its counter-image and projecting them into its image. We have collected the mathematical aspects of these mappings under the generic term of *morphisms* in Sect. 3.2. One of the features inherent in morphisms is the inner structure of the system to be modelled. This inner structure may well be defined by different relations valid for the entities constituting the system in question. In this section we shall describe these structure-preserving mappings from an entirely mathematical point of view, because it permits unambiguous definitions of these characterizations of modelling processes.

In these examples, borrowed from mathematics, the mappings are the so-called *-morphisms* and they possess the additional feature of preserving parts of the underlying mathematical structures. If – for a start – we deal with those algebraic structures most relevant to modelling, the mappings are *isomorphisms* and *homomorphisms* defined to map these structures[1]. An isomorphism is structure-preserving and one-to-one, a homomorphism is the more general case of a mapping many-to-one; it also preserves the algebraic structure of the modelled object.

Ex. 3.1 We study the rotations of the square of Fig. 3.10 through multiples $z\cdot{}^{\pi}/_{2}$ of ${}^{\pi}/_{2}$, the factor z being an integer; this mapping thus transports one edge of the square onto another one. Rotation in a clockwise sense is regarded as positive, whereas it is negative when the square is rotated in a counter-clockwise direction.

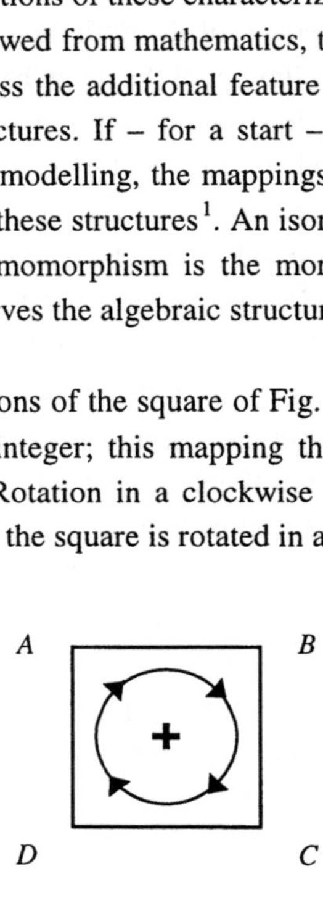

Fig. 3.10. Rotation of a square as described in Ex. 3.1.

Each of these rotations of the square will permute its vertices. If we start with the left hand vertex at the top of the square and list the vertices in their clockwise order, they form the string (*ABCD*) for the original square of Fig. 3.10. Each rotation through $z\cdot{}^{\pi}/_{2}$ will change this order cyclically; thus we get (*DABC*) for $z=1$, for $z=2$ (*CDAB*) and finally the permutation (*BCDA*) of (*ABCD*) if $z=3$; $z=4$ will lead to (*ABCD*) again and cannot be distinguished from the square corresponding

[1] There are further special cases of morphisms, like *mono-* and *endo*morphisms; for a detailed study we refer to the relevant mathematical publications, e.g. Northcott (1962).

to the rotation with z=0 we started with. Likewise, z=5 will be the same as z=1, z=-1 will be equivalent to z=3 and so on. So, there are just four distinguishable positions of the square, no matter how large the number z of multiples of $^{\pi}/_{2}$ will be through which it is rotated. Each position of the square can be characterized by one of the integers **0**, **1**, **2** or **3**; **0** will be equivalent to a rotation through $0+4\cdot z\cdot{}^{\pi}/_{2}$ where z is any integer. Likewise, **1** will be equivalent to $1+4\cdot z\cdot{}^{\pi}/_{2}$, **2** will be equivalent to $2+4\cdot z\cdot{}^{\pi}/_{2}$ and **3** to $3+4\cdot z\cdot{}^{\pi}/_{2}$. As there are only four different positions of the square, there are also only four different rotations of the square we can tell apart. We may represent each of these distinguishable rotations by either **0**, **1**, **2** or **3**[1].

We can now map the set Z of all integers z onto the rotations of the square through $z\cdot{}^{\pi}/_{2}$. Any rotation through a multiple of $2\cdot\pi$ will leave the square of Fig. 3.10 invariant, and we conclude that the kernel of this mapping is the subset of Z containing the integers $0+4\cdot z$. So mapping Z onto the four integers **0**, **1**, **2**, **3** representing the four characteristic positions of the square is *many-to-one*, any integer $0+4\cdot z$ coincides in the same image **0**, likewise, $1+4\cdot z$ in **1**, $2+4\cdot z$ in **2** and, finally, $3+4\cdot z$ in **3**. Moreover, this mapping preserves the additive structure of the integers z: if z_1 is mapped onto the rotation through $z_1\cdot{}^{\pi}/_{2}$ and if z_2 is mapped onto the rotation through $z_2\cdot{}^{\pi}/_{2}$, then z_1+z_2 will be mapped onto a rotation through $(z_1+z_2)\cdot{}^{\pi}/_{2}$, where all rotations are oriented according to the respective sign of the integer z[2]. A mapping between algebraic structures, many-to-one and *structure-preserving* as described above, is a *homomorphism.*

Likewise, we may represent the four distinguishable positions of the square by numbers **0**, **1**, **2**, **3**, as mentioned above. Whenever an integer z_1 is mapped onto the rotation $*z_1$ of the square and z_2 is mapped onto $*z_2$, their sum z_1+z_2 will be mapped onto the rotation $*z_1\oplus *z_2$ within the set of integers **0**, **1**, **2**, **3** representing these rotations. The 'addition' $\oplus$ of these numbers is *modulo* 4 and defined by the following table, where the top row and the column on the very left contain the summands of $\oplus$, the rest of the entries are the respective results of $\oplus$ in the scheme of Fig 3.11:

---

[1] The bolt type of these numbers stands for this representation governed by arithmetic laws described in Fig. 3.11.

[2] Further algebraic laws, like the commutative and the distribution law, are preserved by this mapping, too. For a detailed proof cf. e.g. Northcott (1962).

⊕	0	1	2	3
0	**0**	**1**	**2**	**3**
1	**1**	**2**	**3**	**0**
2	**2**	**3**	**0**	**1**
3	**3**	**0**	**1**	**2**

Fig. 3.11. Addition *modulo* 4

This mapping $z \rightarrow {}^*z$ is also a homomorphism and it is equivalent to the homomorphism of the integers onto the rotations of the square. The kernels and further structural elements of these mappings correspond; they are similar.

As we have seen, there are only four different rotations we can tell apart, and each of these is represented by one of the numbers **0**, **1**, **2** or **3**. The addition (carried out successively) of these distinguishable rotations follows the same rules of addition *modulo* 4 represented by the scheme of Fig. 3.11 for the numbers **0, 1, 2, 3**. So if we map **0** onto the class of rotations through $0+4\cdot z\cdot{}^{\pi}/_{2}$, **1** onto the class of rotations through $1+4\cdot z\cdot{}^{\pi}/_{2}$, **2** onto $2+4\cdot z\cdot{}^{\pi}/_{2}$ and **3** onto $3+4\cdot z\cdot{}^{\pi}/_{2}$, this mapping will preserve the algebraic structure. This mapping is *one-to-one*, which means that there is exactly one original element in the counter-image corresponding to each distinguishable element of the image; the kernel of this mapping contains **0** as its only element, corresponding to no rotation being carried out at all. A structure-preserving *one-to-one* mapping is an *isomorphism.* ♦

An important difference between iso- and homomorphisms is the *reversibility* of isomorphisms. As isomorphisms are one-to-one and structure-preserving, there is a definite correspondence between elements of their counter-images and of their images which can be inverted and are thus structure-preserving in both directions, too. Accordingly we have seen in Ex. 3.1 that we can represent, e.g. the rotation of the square through $2+4\cdot z\cdot{}^{\pi}/_{2}$ by the number **2**, and also map the number **2** one-to-one onto the class of rotations by $2+4\cdot z\cdot{}^{\pi}/_{2}$; both mappings are structure-preserving, and each of them can be inverted.

Homomorphisms, on the other hand, are many-to-one. Thus, there is no *unique* counter-image to each element of the image. And as mappings have to be *unique* by definition, i.e. assign a unique element as an image to each element of its counter-image, there is no inversion to a homomorphism. In the previous example Ex. 3.1 we have explained this idea of mapping algebraic structures in great detail, because the habit of talking about iso- and homomorphisms has frequently been adopted by modellers. The algebraic structure somehow equals what we called relation ρ_2 in Sect. 3.2. If, for instance, the integers z_1 and z_2 are summed up to the integer z, i.e., if the equation $z_1+z_2=z$ holds, we may represent this relation between z_1, z_2 and z by a ρ_2-operation as described in Sect. 3.2, like the relation $\rho_2(z_1,z_2;z)$ being valid for the integers z_1, z_2 and z, where ρ_2 is the addition, z_1 and z_2 are the summands and z is their sum. This can be represented by an I-P-O-chain, its input I is made up of the two summands z_1 and z_2, the process P is their addition and the output O is their sum $z=z_1+z_2$.

There are many more structures in mathematics apart from the algebraic ones, and there is the relation ρ_1 – mentioned in Sect. 3.2 – valid for systems, too. This relation ρ_1 describes the interconnection of the parts of a system, and it is essential when the internal structure of a system is considered. So, mathematical *topological* (Greek *topos*: 'place') structures are important features of modelled systems. These structures are explained quite differently. They are related to different mathematical entities and they are defined by a terminology of their own. If topological structures are the objects of research, they can, for instance, be mapped (one-to-one) by *homeomorphisms* (Greek *homoios*: 'like') and, generally speaking, by *continuous* mappings (many-to-one), both preserving the topology[1] defined by the structure of neighbourhoods within the topological spaces under consideration.

As isomorphisms and homeomorphisms are both mappings one-to-one, each of them can be inverted; they thus lead to greater 'similarity' between the structures to be mapped and their 'models', the structures they are mapped *onto*, i.e. there is no significant kernel or *co*-kernel – according to the structures – of these mappings. Nevertheless, there are features of the objects of the mappings getting lost, and the structures they are mapped *into* may possess additional features not corresponding to any quality of an original. This fact becomes quite clear when we regard topological mappings.

---

[1] There are also discrete topological structures (cf. e.g. Lefschetz 1956) we leave out of consideration here.

The structure of topologies is defined by the *neighbourhoods* of a point P in space. A neighbourhood of P consists of all those points in its vicinity, say, in a distance less than $\varepsilon>0$ (cf. Fig. 3.12) however small ε may be.

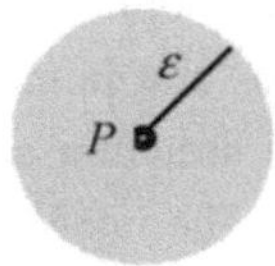

Fig. 3.12. Two-dimensional neighbourhood of a point P

In the planar case of Fig. 3.12 the neighbourhood of a point P is the circle shaded grey. In space it will be a sphere; a one-dimensional neighbourhood is just an interval containing point P. Nevertheless, there are also topological structures with neighbourhoods defined as being different from intervals, circles and spheres. We shall try to explain the fundamental ideas of topological mappings:

Ex. 3.2 We look at the following topological spaces T_1 and T_2: T_1 is the line L, T_2 is the quadrant Q of Fig. 3.13.

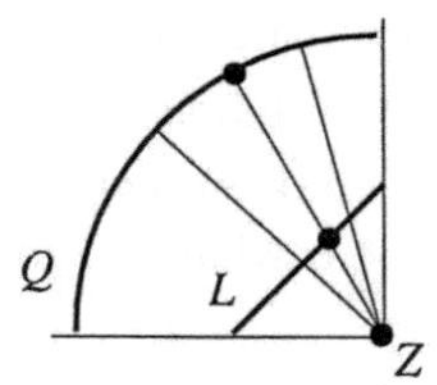

Fig. 3.13. Topologically, but not 'totally' equivalent spaces

By projection from point Z a one-to-one correspondence between the points of L and those of Q is established. Likewise, this projection renders a mutual correspondence between the neighbourhoods of the points of L and of Q. These facts are represented in Fig. 3.13 by the respective projecting lines. The one-to-one correspondence between the points of the topological spaces L and Q and the preservation of the topological structures of neighbourhoods in L and Q make the projection of Fig. 3.13 a *homeomorphism*, the topological analogue of the algebraic isomorphism: both are one-to-one and can be inverted; but whereas an *iso*morphism preserves an algebraic structure, a *homeo*morphism does not change the topology of a space. ♦

We now proceed with more general topological mappings; they are only *continuous* and correspond to homomorphisms preserving algebraic structures, but just like the homomorphisms these mappings are many-to-one and, for this reason, cannot be inverted. We give an of this type of projection:

Ex. 3.3 In Fig. 3.14 the rectangle *R* with vertices *A*, *B*, *C* and *D* is projected vertically into the horizontal line *L*. *R* is the object of this mapping, the line between the points *X* and *Y* is its image. This mapping is continuous as it projects the neighbourhood of a point *P* of *R* into the neighbourhood of the image **P* of *P* in the horizontal line *L*.

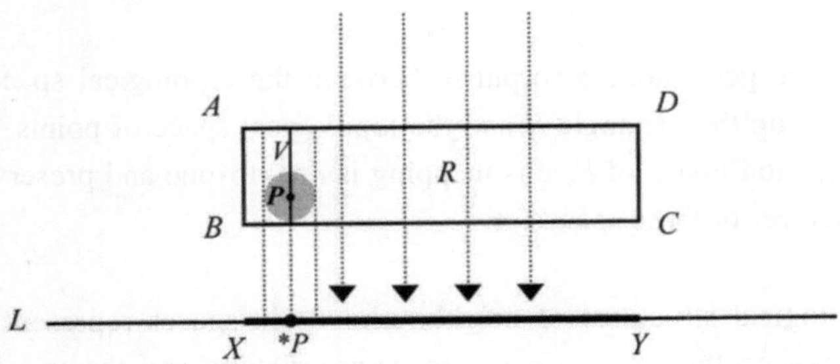

Fig. 3.14. Continuous mapping of a rectangle onto a line between two points

But this mapping is not one-to-one, it cannot be inverted, its object and its image differ in dimensions, and there is a whole vertical line *V* in *R* containing *P* and being projected onto the single point **P* of *L*. The line *X*-*Y* between the two points *X* and *Y* of *L* represents the basis *B*-*C* of the rectangle. This projection of *R* into *L* is continuous, but not a homeomorphism, the kernel of the mapping is its second dimension, its height represented, for instance, by its edge *A*-*B*. ♦

But we may interpret the projection in a different way in order to get a one-to-one mapping between the rectangle *R* and the straight line between *X* and *Y*, part of the horizontal line *L*:

Ex. 3.4 In Fig. 3.15 we do no longer interpret the rectangle R as the union of its points, but rather as the union of parallel vertical lines V, each representing the height A-B of R. This different interpretation of the constituents of the rectangle and its image yields a one-to-one mapping between the vertical lines V filling up the rectangle and the points $*P$ of its image.

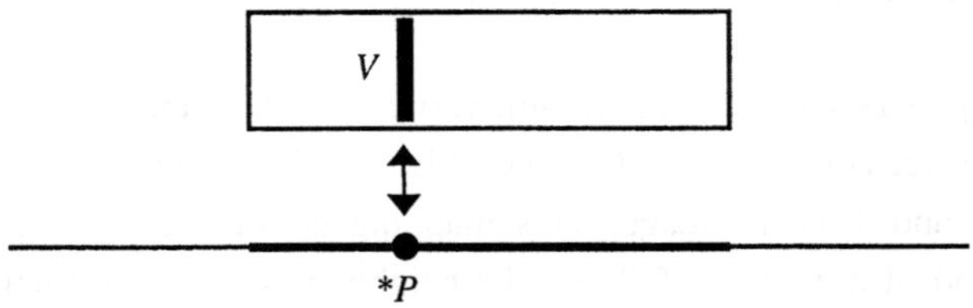

Fig. 3.15. Homeomorphism between a rectangle and a line between two points

In this way we get a homeomorphism between the topological space of vertical lines V filling up the rectangle R and the topological space of points $*P$ constituting the horizontal image of R; this mapping is one-to-one and preserves the topological structures of these spaces. ♦

These topological structures of neighbourhoods in spaces represent the internal relations between the parts of a system appropriately, and the topological mappings we have described mirror the modelling process. But even if topological mappings are homeomorphisms, some additional information can still be contained in their images and counter-images, not transported to their respective counterparts. In Ex. 3.4, there is a homeomorphism between the space of parallel straight lines V filling up a rectangle R and the one-dimensional space of the points of the interval, yet both spaces are of different dimensions. Likewise, the homeomorphic correspondence between the points of a straight line and those of the quadrant described in Ex. 3.2 does not reflect the differing curvatures of these systems. We can adapt the two spaces to include the curvature in the modelling process – if we think it is essential – by continuously deforming these spaces, i.e. by either straightening the quadrant or by bending the straight line. But both spaces will still differ in size and in their location in the plane.

We would like to discuss another example of mapping between topological structures. In that case the spaces cannot be deformed continuously in order to reflect information lost in the modelling process.

Ex. 3.5 In Fig. 3.16 there are three circles; they are homeomorphic and thus topologically equivalent from the aforementioned point of view. But they differ in the way they are embedded into the three-dimensional space surrounding them: circle B is knotted (it is the *trefoil*–knot), circle A is flat and unknotted, and circle C, while unknotted, is twisted in space.

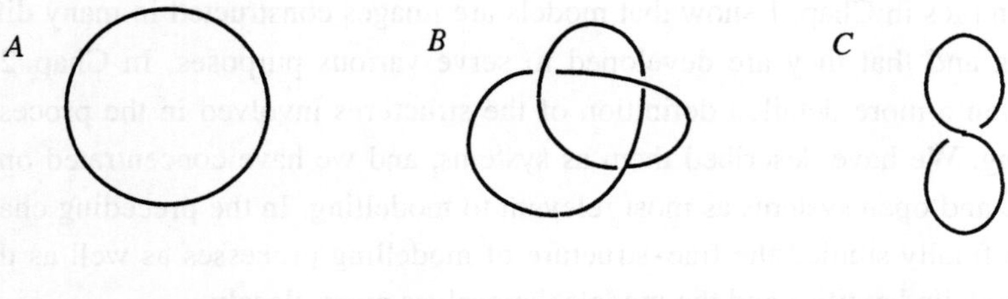

Fig. 3.16. Different embeddings of a circle into the three-dimensional space

Though there are one-to-one mappings between the points of A and B transporting the structure of point-neighbourhoods from A to B, A cannot *continuously* be deformed (knotted) to look like B, nor can B be continuously unknotted in order to look like A. The only – discontinuous – way of knotting A in order to form the trefoil-knot B is to cut it into one straight line and knot its ends to a closed line again. And still, if we regard only unknotted circles, there are many different ways to embed them into space, like for instance, circles A and C of Fig. 3.16. ♦

4 Characterizing Models

The examples in Chap. 1 show that models are images constructed in many different ways and that they are developed to serve various purposes. In Chap. 2 we have given a more detailed definition of the structures involved in the process of modelling. We have described them as systems, and we have concentrated on the dynamic and open systems as most relevant to modelling. In the preceding chapter we have finally studied the fine-structure of modelling processes as well as those of the modelled entities and the models themselves more closely.

We now want to describe means to distinguish models and select those of prime importance to a specific field of research. There are quite a few publications dealing with the problem of characterizing models. Some of them underline the fact that some models are metaphors (e.g. Black 1962) or icons (Frey 1961), while others regard modelling under the aspect of mediating: models are more comprehensible representations of complicated *co*-models which are difficult to understand (e.g. Morgan and Morrison 1999). Extending these subdivisions to the utmost, we will get quite a detailed categorization of certain types of well-known models.

Classification according to the mentioned criteria is tantamount to regarding models under the utilitarian (more precisely *teleological*) aspect compiled in Sect. 1.3 where we studied the purposes a model may serve. But this way of differentiating between the various purposes is too rough-and-ready to do justice to the enormous amount of facets in the field of modelling; it is a strong limitation, because it means to map the entire number of models onto those few purposes we can distinguish. In our terminology the class of purposes is a model of the whole of models, and its mapping has got a kernel of its own. If we consider that there are only a few classes models can be subdivided into under these aspects, we realize that this kernel must be very extensive.

An ideal classification of systems should take into account that each single model is *sui generis* – something unique – and special, but this would be too impractical a means to *classify* systems as there would be as many criteria as there are models (either created in the past, existing or to be set-up yet). So, to be realistic, we have to lump different models under certain generic aspects together, considering as many different features of each individual model as possible, not just the purpose it serves. So, we no longer ask the question, 'What purpose does this model serve?'; our question is rather 'What are the model and its *co*-model made of, what are their *contents*?'. This criterion – the contents – should be defined in order to permit a classification of systems, i.e. the concentration of different models into one characteristic class. But the criteria must be defined very accurately so that we can still distinguish between models that are basically different.

Of course 'purpose' will be amongst the different features when analysing the contents of a system, serving as a model or as its *co*-model, but its meaning is different from the purpose the model has to *serve*, because this teleological aspect is restricted to the modelling system Σ_P, whereas our considerations concerning the contents of a system are more general and also relate to the systems Σ_I and Σ_O. As soon as the contents of a model and its *co*-model are determined, we can characterize the underlying modelling process by the way these contents are distributed over the image and the kernel and over the *co*-image and the *co*-kernel of the mapping.

4.1 Contents

As we have expounded in the preceding chapters, modelling is an *imaging* process transforming an open dynamic *system* Σ_I, the *co*-model of the mapping – its input-system – into the model-system Σ_O – its output-system. As the kernel of modelling, containing the information lost by the mapping, is determined deliberately for utilitarian reasons or unwittingly because of a lack of knowledge, we shall study the *contents* of both systems Σ_I and Σ_O. These contents will be finite ordered sets – *strings* – of characteristics of the systems, and we shall denote them by $C=(c_1, c_2, \dots)$. If necessary, we shall add the respective additional subscripts I and O to determine the system under consideration. Within these strings the positions of their elements are significant and must not be muddled up.

We begin with the fact that a system Σ is just a structured collection of its parts, but by outlining further characteristics of Σ we can give it a further-reaching coherent framework. So there are, for instance, the generic 'purpose'[1] the system Σ has to serve and its internal fine-structure to mention two of them. The hierarchical order of these properties of Σ does not really matter; this means that the order of the characteristics c_1, c_2, ..., identified by their respective subscripts 1, 2, ..., is arbitrary and not hierarchical: we just have to determine the position of the various c_is within the string in order to tell them apart.

By the definition given in Chap. 2 a system has a certain *structure* S. Again, we shall indicate the position of the system in a modelling process – as a *co*-model or as a model – by the respective subscripts I and O. Accordingly S_I is the structure of the *co*-model, *input* into the mapping-process, and, likewise, S_O is the structure of the model-system constituting the respective *output*. The structure, the building plan, of system Σ is determined by the relations between its various parts: it characterizes the *static* aspects of a *dynamic* system. The structure S of a system Σ can be specified to a certain degree of detailing by successively opening the *black boxes* of Σ; we have described this process in Sect. 2.4. The most detailed representation of a system is reached on the level of its *elements*, but perhaps we want or we have to do without this *maximum* degree of information about the structure of a system. This means we are either not interested in further details, because they are irrelevant to the model we want to develop, or we simply have to cope without them if they represent relations between certain parts of the system not yet known to us.

Ex. 4.1 There are various possibilities of producing scientific models of the human body. Drawings and maps may show its individual parts under different aspects, like, for instance, the bone structure, the digestive system or the nervous system. Sometimes these drawings are rather complicated and multilayered: parts of the maps may be foldaway to allow a look at graphic representations of organs deep inside the body. The model may be three-dimensional and true to scale, an accurate sculptural representation of the human body and its parts under the various aspects mentioned above. Some of these models even allow to move parts of them in order to gain insight into hidden regions. In addition to these *hardware*-models, there are *software*-models of the human body like, for instance, textbooks on medical science. The most comprehensive of these software-models is the complete knowledge of the human body and its parts. As medicine is not an exact science this model as a theory is not consistent: there have always been incompatible

---

[1] This is the teleological aspect of the definition of systems; we refer to Sect. 2.2.

medical views, like those of complementary and orthodox medicine. Other medical theories turned out to be untenable from the scientific point of view, like *phrenology* developed by F.J. Gall in the early 19th century, a theory associating mental faculties with certain parts of the cranium, thus providing a means to analyse a person's character and define his faculties by simply studying the shape of his skull.

The specification of medical models is imperfect: there have always been gained new insights into the relations between the various subsystems of the human body, many of them have even not yet been detected or described in full detail, e.g. the relations between the activities of the glands and the development of the bone structure. Consequently, any model of the body is incomplete and provisional, but we decide to set up incomplete models, deliberately neglecting characteristic features of its *co*-model when, for instance, we construct a model of the skeleton used for instruction and do without considering relations between the bone system and other subsystems of the human body. Of course, we know that the skeleton cannot work without the support of muscles and sinews, for example, but we ignore them in order to make our model of the skeleton transparent. Just looking at a skeleton, one might imagine that the human bone system could do very well without any *intervertebrals*; but whoever has suffered from the symptoms of a *lumbago* knows that the spinal column needs those pieces of cartilage to work properly. ♦

This digression on the development of medical models demonstrates the importance of the structure of a system. Therefore we start the definition of the contents by setting $c_1=S$ and thus $C=(S, ...)$ with the respective subscripts I and O yielding $C_I=(S_I, ...)$ and $C_O=(S_O, ...)$ respectively. This way of arranging the characteristic contents of C does not – necessarily – imply that the *structure* S of a system is of *prime* importance; it just means that the structure S has been positioned under the subscript '1' and must not be moved from this place.

As we have seen, the degree of detailing a system is determined under the utilitarian (teleological) aspect of what we want to use our model for. So we shall now outline those purposes the model and its *co*-model have to serve. The second item we enter into the list of contents of a system Σ is the *generic* purpose G it is supposed to serve. We abstract – for the moment – from the internal structure of Σ and from the various purposes of its *subsystems*. The model of a skeleton mentioned in Ex. 4.1 is supposed to explain the set-up of the human bone system to the students: so its generic *purpose* is one of explanation. The *co*-model of this skeleton, the human body, serves a different *generic* purpose, if any at all; we do not want to discuss this question here; we would certainly enter into a fruitless dispute with all sorts of schools of thought if we did so! So we set $c_2 = G$ in $C = (S, G, \ldots)$. In addition to the internal *static* structure S we recognize G as a characteristic determining a system Σ.

This generic purpose G of a system depends on the way its subsystems operate, but it can be defined without considering these subsystems and their interactions. A more careful specification will consider the processes taking place in a dynamic system; they are part of its *performance*. The following example will demonstrate the meaning of this component of the contents of models.

Ex. 4.2 In Ex. 1.1 we distinguished between various models of an automobile by describing the respective purposes they are to serve. Another distinguishing aspect is the *performance* of these systems and of their parts. The performance of a real car, for instance, is that its movement is caused by the rotation of its wheels. With the toy-model of a car the causal connection is exactly the opposite: the wheels of the toy rotate, because the car is moved. Likewise, the manoeuvrability of the model and of its *co*-model is achieved by different means of performance. The real car has a perfectly devised steering system at its disposal, whereas the wheels of the toy-car are fixed and cannot be steered; the toy is moved by force and changes its direction without turning its wheels. The toy-car is powered by the child's muscles, whereas the co-model needs a complicated system of interacting parts like, e.g. an engine and gears.

If the model of a real car is used just for instruction at a driving school, the situation is different. Here, the performance of some of the subsystems of the automobile has to be represented as exactly as possible. The model should give the learner a good insight into the operations of the single parts of the original, like clutches and brakes. Quite often, this model consists of parts of a real car – the *co*-model – prepared for closer inspection, so parts of the casing may be removed for, example; other parts of the model are representations on a smaller scale. Further aspects of the performance of a car are not taken into account, because they are of no importance to the purpose of instruction, a real engine, for instance. ♦

We learn from this example that the transport of each single of the elements c_i of the set C of the contents of a given system Σ_I into its model Σ_O leads to a loss of information contained in c_i. But as different models are constructed under different utilitarian aspects, prime importance is put onto the aspect of keeping some features of all the c_is more invariant than others. The interactions between the various parts of the systems and the way they transform their respective inputs into the corresponding outputs are the *performance* of the system under consideration, and we include it into the string of contents of the system as its item P. This performance P of a system is a means to serve its generic purpose. With $c_3=P$ we characterize a system Σ by the string $C=(S, G, P, ...)$, in which P is a further essential determining it.

These examples, which we have just discussed, show that the material M used for the representation of a system is a further criterion defining its contents C. The model of the human skeleton, as an example, will be made of plaster or a synthetic material, but not of the substance bones are really made of. The toy-car may, for example, be made of wood or plastic, but not of the materials used for the production of a real automobile. A building is constructed of concrete and other materials, but it will be modelled in card-board or something similar, quite different from the materials used for the *co*-model. Thus the aspect of *material* covers the physical and technological representation of the *co*-model including its dimensions as well as those of the model. Furthermore, we have already seen that concrete physical systems may be represented by dialectical models, and *vice versa*, that dialectical mathematical theories (models) are represented either by dialectical models, too, or by material ones (e.g. the *abacus* of. Ex. 1.12). So the representation of a material system may be immaterial, but we can as well model an incorporeal system by a corporeal one. In fact, all conceivable combinations between material and immaterial systems – whether they constitute the *co*-image or the image of a modelling process – are possible. We conclude from this that the aspect of the material of a *co-model* and of its *model* characterizes the process of modelling itself. Therefore we include this criterion as a further parameter $c_4=M$ into our set C defining the various systems under consideration. The set C of contents of Σ is thus represented by the string $C=(S, G, P, M, ...)$.

Up to now, we have only studied the internal properties of a system in question, like its structure S, its generic purpose G, its performance P and its representation M which characterizes it as being incorporeal or corporeal and if necessary describes the material the system is made of. The generic purpose G, a system Σ has to serve, certainly refers to its environment, but it is only an incomplete description of a part of the interactions between Σ and its surroundings.

As our concern is the study of *open* systems, we have to consider the way E these systems are embedded into their environment. This embedding covers the way a system is spatially embedded as well as the temporal aspect and the various interactions between a system and the neighbouring parts of its complement. In order to illustrate the spatial embedding of a system, we refer to the examples Ex. 3.13 and Ex. 3.16 demonstrating different embeddings of models and their *co*-models due to different curvatures and the various ways a circle can be embedded into the three-dimensional space. An example of the temporal embedding of a dynamic system is its *being dynamic* itself, comprising the process of ageing. We could, of course, – for a certain span of time – abstract from the fact that every system Σ is embedded in space and in the flux of time and regard it as being *closed* and *static*, but in the long run the way an *open* system communicates with its surrounding systems is deciding, and we include E as another characterizing parameter into the set C of contents of Σ. This leads to $C=(S, G, P, M, E, \ldots)$. We want to give a further example supporting our view of the embedding of a model into its surroundings.

Ex. 4.3 We have already mentioned the dynamic process of the free fall of a body in Sect. 3.3. In addition, we discussed the physical processes significantly affecting a body falling in the terrestrial atmosphere there. This analysis already considers various aspects of embedding the body falling into its environment as well as the interactions between the system under consideration and its environment. We will thus give thought to the effects of friction if they are significant, or else we will ignore them; we will, likewise, consider the impact of the Brownian movement of the surrounding gas particles or disregard it; and perhaps the initial position of the body right when it is dropped is also an important characteristic of the system. ♦

We admit that some of these criteria characterizing a system are mutually dependent. Thus the items on our list of elements of the *contents* of a system are not disjunctive, and this list is not complete either. Characterizing a system by its contents is a means to do so thoroughly, and for each particular system further characteristic items must be added to the set C of its contents; these items depend on the special case and cannot be compiled as generally valid criteria. For our discussion we think that those items we have described above generally characterize a system sufficiently exactly, and we will thus assume that a system is defined by the string $C=(S, G, P, M, E, \ldots)$ leaving any insight into any further components of C – represented by the dots ... – out of consideration.

Possibly, the features of a model constituting its contents as being characteristic of the system will be selected arbitrarily and subjectively; nevertheless, this selection is a means to classify the system modelled or gained by modelling under the described aspects. Thus, classifying the modelling process is tantamount to distinguishing the classifying contents of its model as well as of its *co*-model and to comparing their respective components. The classification of systems by their contents C as described allows to characterize *any* system under consideration as exactly as necessary by adding further relevant items to our incomplete set C of defining items.

4.2 Selection

When the systems Σ_I and Σ_O constituting the respective input and output of the modelling system Σ_P have been characterized by their contents C_I and C_O, we have to decide which aspects of the components of C_I should be selected to be mapped into C_O. Of course, there are still the problems of utility and availability, but it may be useful to disregard part of the information contained in the elements of C_I in order to construct a simple model, adequately representing its *co*-model. Moreover, there may well be information contained in C_I we cannot convert into a model; this will be the case whenever we do not know this information or if we do not know how to interpret it.

To give an example of this aspect of determining the selection of components of the contents of a *co*-model we return to Ex. 4.3, the example of free fall:

Ex. 4.4 In order to describe the path-time law of a body falling within the terrestrial atmosphere correctly, we have to take several points into consideration: the law of gravity, friction, varying density of the atmosphere depending on the height over ground, the impact of the Brownian movement of the gas particles surrounding the falling body and the pressure of the rays of light affecting the trajectory, only to mention some. Of course, there are further effects influencing the falling body, like gravitational interactions with every single celestial body and the deformation of the body under consideration due, e.g. to acceleration and to changes in temperature. Furthermore, there may even be physical phenomena still unknown, but determining the trajectory of a body in free fall.

Certainly, a model considering each of these physical phenomena influencing the free fall of a body will be the most exact one, describing the trajectory most competently. On the other hand, we cannot take into account what we do not know yet. But if we refuse to model unless we know each single phenomenon affecting the object under consideration, we will never model, because we never know what physical phenomenon relevant to the model will be detected next. Furthermore, even if we know about relevant physical phenomena, like the gravitational interactions between masses, this information may still be incomplete, because we will, for instance, not be able to detect and measure the gravitational impact of each single celestial body on the experiment of free fall.

And even if we could, if we knew each single phenomenon influencing the free fall of a body, and if we were able to quantify it, we might still prefer to do without considering these parameters and only apply the *law* of free fall, if the falling body was a solid rock of some 500 kg falling from a height of some 2 m. ♦

As we have seen, the degree of specifying the information contained in C_I is determined by the utilitarian consideration of finding a reasonable and suitable model. On the other hand, the choice of information contained in C_I may unwittingly be limited by its unavailability. Information being 'unavailable' may mean that it concerns phenomena not yet detected or that we know about these phenomena, but cannot or do not want to take them into account. As an example we refer to Ex. 4.4 where we mentioned the gravitational interactions between a falling body and the matter of the universe which are left out of consideration for the sake of simplicity when developing a model. Below we describe an example where we would like to possess more information when modelling, but where we do have to content ourselves with the little, just available.

Ex. 4.5 When forecasting the weather we rely on experience and on available information. Without any records about the weather of the last days and without any further evidence we may just induce that tomorrow's weather will be the same as today's. But if we possess records of the past, we may predict the weather of the near future according to a trend we may have detected, like temperatures constantly rising during the last three days. If we have merely evidence gained from observations or from records of our ancestors, then we may make use of the Hundred Years' Calendar or believe in rural sayings and thus regard the halo of the moon as a sign of approaching bad weather or an evening glow as an indicator of a fine next day.

Whenever there are other more reliable possibilities of gathering information – like a tight grid of observation points and satellite supported observations – the expert will make out a rather accurate forecast for the next two or three days. But it needs the expert, the meteorologist, to interpret all the data collected by meteorological stations and by satellites fairly correctly[1], a layman would be lost. ♦

These examples present some of the aspects which determine the way the contents of a *co*-model are selected for modelling. First there is the principle of 'as little as possible to get the best results', which means that when modelling we content ourselves with a minimum amount of information contained in the *co*-model in order to get an adequate and dependable model, like we did when we calculated the trajectory of a solid rock of 500 kg falling from a height of 2 m, only applying the law of gravity.

Sometimes, of course, our modesty is not voluntary, but enforced on us by a minimum of available information. Then we are obliged to construct a provisional model and accept it – *for the time being* – as representing its *co*-model adequately; maybe, it will even serve as an early stage for a more complex development. But this makes clear that the quality of a model depends on the growing insight into the contents of the system Σ_I, its *co*-model. The better the quality of a model, the more exactly it represents its counter-image with regard to the purpose the modelling process is supposed to serve. The criterion of the *exactness* of modelling the physical parameters of reality can be described quantitatively by *degrees*; we will discuss it in subsequent chapters.

The qualitative aspect of modelling is determined by the way elements out of the contents C_I of its input system Σ_I are regarded as being relevant and selected to be part of the *co*-image of the mapping. We want to illustrate this idea by a system of cylinder drums, as sketched in Fig. 4.1.

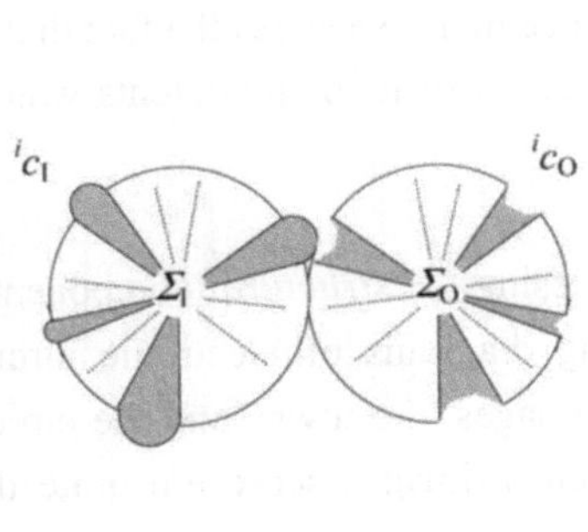

Fig. 4.1. Illustration of the process of characterizing models by selecting contents

1 And even he will make mistakes with sometimes fatal consequences, as we all know.

This diagram of Fig. 4.1 shows how those parts ${}^{i}c_{\mathrm{I}}$ of the contents C_{I} of the input-system Σ_{I} regarded as most significant for the process of modelling in question are selected and protrude from the drum representing Σ_{I}. This drum is regarded as a punch and the ${}^{i}c_{\mathrm{I}}$ leave imprints on the surface of the drum representing the model, the output-system Σ_{O}; these imprints are called ${}^{i}c_{\mathrm{O}}$ respectively. But as some of the information contained in each ${}^{i}c_{\mathrm{I}}$ is not covered by the mapping and thus gets lost, these accompanying imprints ${}^{i}c_{\mathrm{O}}$ are only rough images of their counter-images.

When we defined the contents C_{I} and C_{O} by strings $(S, G, P, M, E, \ldots)$, we stressed the fact that the order of their items is not hierarchical, but chosen arbitrarily. There is room for an extensive number of items to be added to both lists of contents C_{I} and C_{O}, on demand. Accordingly the number of combinations of items available for the definition of a particular system rises considerably. If we interpret the cylinder drums representing the process of modelling in Fig. 4.1 as the drums of a musical box, each model produces a tune of its own.

To sum up, the way items are selected from the contents of a system in order to be modelled depends on the period of time and the particular situation the modeller lives in, on his knowledge, on his insight into the problems to be modelled and on his demands on the accuracy of the output of the model, to name some of the criteria. So when modelling we deal with mappings, and when mapping we have to decide what parts of the contents of the *co*-model we regard as relevant to the modelling process; furthermore, each of these parts represented by the model may have a kernel of its own, features we do not want to model or cannot model, though they are significant parts of the original system.

The following examples show how the contents of a special system serving as input Σ_{I} or as output Σ_{O} of a modelling process Σ_{P} will be selected according to the purpose Σ_{P} has to serve. Once more we stress the fact that the modelling system Σ_{P} is a system of its own, made up of its own contents which we do not consider in detail here.

Ex. 4.6 As an example stressing the *structural* components of a system we regard maps of towns and wiring diagrams of electronic circuits. Both represent the structure of their counter-images, the towns and the circuits respectively, but the generic purposes their counter-images serve are quite different: a town is supposed to lodge people, whereas an electronic circuit may, for example, be meant to transform received electro-magnetic waves into music if it is a radio. The maps and diagrams of this model are means of orientation or they may serve as prototypes if a wiring diagram is the basis of the production of, e.g. radios or TV-sets.

The modelling system Σ_P serves the purpose of documentation, not considering the material M the houses or electronic units are made of or the way E they are embedded in space, though a street-map may reflect some topological measures like distances if it is reproduced true to scale. Likewise, the performance of the subsystems will be of relevance in the case of a wiring-map of an electronic circuit, whereas it is without relevance to a street-map; in the latter case there may just be some additional information, like, for instance, 'school' or 'church', reproduced in the map to facilitate the orientation of its user. ♦

Parts of the contents will be the same with different systems and they will only be characterized by some few or even only one differing element of their respective specification. Consequently, our examples will deal with models similar under many aspects and only differing in facets of the relevant contents of the systems involved.

Ex. 4.7 This example demonstrates the importance of the *generic purpose* G of a system to be transformed into its image: the objective of the development of an aircraft may, for instance, be that it is to serve the purpose of a fighter plane. Of course, there are further characteristics of this system like, for instance, its performance P. But in a step by step process of systems development the initial idea may just be to construct this special type of aircraft within a certain span of time without any specification of further details like its performance. Thus, at that very moment, the generic purpose G is the predominant characteristic out of the set of contents of this system. ♦

In general many of the items characterizing a system are related. In the preceding example the general purpose G of the system must be described in greater detail by specifying the performance P of the plane; often these criteria used to characterize systems are not disjunctive and overlap.

Ex. 4.8 We now want to define more demands in order to specify the *performance* of the fighter plane of Ex. 4.7 to be built. These may, for instance, be that it reaches a certain maximum speed, is still manoeuvrable at a minimum speed, can be flown at a very low height automatically avoiding any contact with buildings, trees and so on; furthermore, the plane is not to be tracked down by radar or anti-aircraft-missiles. Other characteristics regarding the realization of the aircraft, like the material M to be used for its construction, will probably not be part of these objectives, though they are important parts of the contents of the system to be constructed. So, the general purpose G the system has to serve and its performance P will be of prime importance when modelling it at this stage of the process of systems development. The result Σ_O of the modelling system Σ_P will be a design study of the future plane. ♦

In our example of the development of a technical system, we have seen that the initial objective was G, the definition of its purpose, which implied the specification of its performance P. Likewise, these elements of the contents of the system Σ_I affect other characteristics of Σ_I, for instance, its structure S and the used material M; but right at the beginning of the design of the plane these factors are irrelevant to the definitions of G and P. Yet later, in the course of the development of the final output system Σ_O its structure S, the material M and many other parameters must be specified. There will also be a series of intermediate models produced for tests, like scale models used for experimenting in a wind-tunnel or prototypes needed when the performance of the system is tested out. If finally the initial idea of a fighter plane is realized, systems development results in a prototype used for the production of a series of aeroplanes.

Ex. 4.9 The introduction of a new aircraft series will be escorted by the operation of simulators. A flight-simulator is a separate system, a model similar to the real aircraft-system as far as its performance P is concerned, but its structure S is quite different from that of its *co*-model. The same goes for the material M it is made of and the generic purpose G it serves; and a simulator will not be embedded into airspace the way a real plane is. Simulators react like a black-box-system, generating an output similar to the performance of the original plane when they are fed with corresponding input-data. They will thus signal critical conditions due to imperfect operations without crashing as the real system would. These models serve didactic purposes among others, allowing the training of coping with dangerous situations without endangering anybody. ♦

In the preceding example Ex. 4.9 of a flight-simulator the characteristics of the material the *co*-system is made of are insignificant and the material need not be represented by the model, the simulator. We now want to give an example of modelling where the characteristics of the used *material* are of crucial importance.

Ex. 4.10 In Sect 2.4 we mentioned heart-transplantations. These xeno-transplants (Greek *xenos*: 'strange') – where tissue is transported from one organism to another one – always bear the risk that the transplant will be repelled or afflicted by diseases. So, one prefers to use the patient's own parts whenever this is possible, or artificial substitutes. The former will be the case when knocked out teeth are re-implanted, limbs lost due to an accident are sewn on again or burnt skin is replaced by skin-tissue taken from other parts of the patient's body. Examples of substitutes are artificial hearts and hips or the implantation of artificial teeth. The least problematic procedure is the replacement by the patient's own tissue, for example, skin layers produced by cell cultures taken from the patient or the patient's own blood taken prior to an operation, stored and replacing blood losses occurring during an operation. However, the correspondence of material parameters *M* characterizing the transplant with those of its receiver decides on the success of these operations. ♦

The fifth term in our arbitrary list of features characterizing a system is the way *E* it is embedded into its environment. This embedding will be of relevance to the following example:

Ex. 4.11 The spatial arrangement of the items of a system is of importance, for instance, whenever we document the positions of the subsurface supply channels of a city: the mains for electricity, gas and water, telephone lines, sewers, subways and so on. These maps must describe the positions of those parts within their environment correctly in order to prevent damages caused by building work or road works.

Another example of the importance of spatial embeddings may be found in stereochemistry, the branch of chemistry dealing with three-dimensional arrangements of the components of molecules: chiral (Greek *kheir*: 'hand') compounds are asymmetrical and cannot be superposed on their mirror images, isomeric (Greek *isos* and *meros*: 'equal' and 'part') compounds possess the same number of atoms of each element, but in a different arrangement. Chiral and isomeric compounds are made up of the same elements, but possess different physical properties; they agree in the quality and quantity of these constituents, nevertheless, the differing physical and chemical characteristics of the resulting compounds are entirely due to the way *E* these constituents are embedded in space. Again, we realize that the characteristic elements of the contents of a system are not always unbiased as the embedding *E* of the parts of a system is to some degree related to its structure *S*. ♦

On demand, the list of characteristics of a model may be supplemented in order to meet all the requirements of a special application of modelling. A sufficiently fine-structured list will permit the classification of almost every single special case of modelling.

4.3 Projection

When mapping the *co*-model Σ_I into its image, the model Σ_O, its set C_I of contents will be mapped into C_O, the corresponding ensemble of contents of the model-system Σ_O. This projection p: $C_I \rightarrow C_O$ may be restricted to a subset of C_I; it thus possesses a kernel in C_I which contains those components of the contents C_I of the *co*-model not represented in the model.

The *co*-image of the mapping p is the subset of the contents C_I of Σ_I projected into the contents of the model Σ_O. Let ${}^i c_I$ again be a part of the *co*-image of p and ${}^i c_O$ its image in Σ_O with ${}^i c_O = p({}^i c_I)$ in C_O. This restriction of the projection p to an element ${}^i c_I$ of C_I will in general be a 'homomorphism' itself and only transport parts of ${}^i c_I$ into ${}^i c_O$. The mapping p thus possesses a kernel of its own in ${}^i c_I$. If, for the moment, we regard C_I and C_O as sets and leave their internal structures out of consideration, we may represent the described situation by the following diagram of Fig. 4.2.

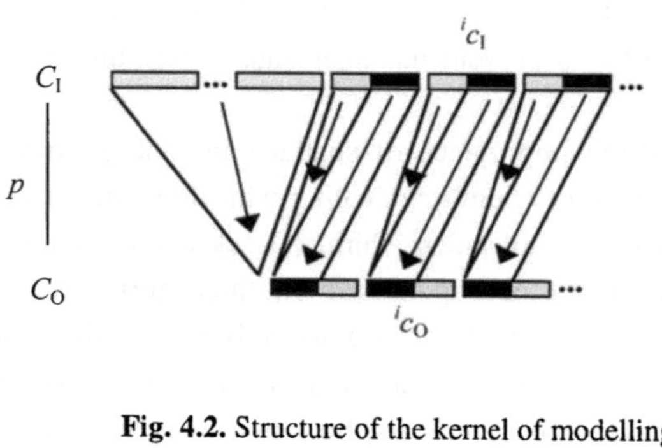

Fig. 4.2. Structure of the kernel of modelling

In Fig. 4.2 we disregard the order of the elements of C_I and put those elements which are completely part of the kernel of p to the left, they are shaded grey. The elements of the *co*-image of p are the rectangles ${}^i c_I$; a part of each of them is shaded grey as it belongs to the kernel of p, too, the rest is blackened, it belongs to the real *co*-image of p. The black parts of the ${}^i c_O$ in C_O constitute the image of the mapping p, whereas the segments of C_O shaded grey represent the *co*-kernel of the projection p.

We realize that the kernel of modelling is structured itself. It includes contents c_I of the *co*-model not at all represented in the model. In addition, the kernel of p also contains those elements of the *co*-model bearing information lost by the modelling process, although they are part of entities transferred to the model-system Σ_O.

Determining the kernel of p or its complement in C_I, the *co*-image of p, is a means to characterize models in great detail. It is impossible to describe each single combination of parts and subparts of C_I constituting a type of model of its own, because there are too many of them. If, for example, we start off with N elements of C_I, we have to deal with 2^N-1 combinations to form the *co*-image of p. And there will be even more if we consider the respective kernel of p within each member $^i c_I$ of its *co*-image and its extent depending on the item $^i c_I$. Nevertheless, some combinations of the $^i c_I$ forming the *co*-images of particular models are significant. They deserve a detailed description, and we want to characterize some of them in the following examples. These scenarios may lead to the conclusion that the characteristic features of a *co*-model are independent of each other, but this is not true as we have already seen in the preceding section; we emphasize the fact that a system is an integral whole of all its parts represented by the sum of the components of its contents. So, our examples serve the sole purpose to describe those characteristics of a system defining its regime.

Ex. 4.12 If the structure S of the *co*-model is the only or the major part of C_I represented by the model, we will design a *simulacrum* of the *co*-model differing significantly in all the other respects mentioned in Sect. 4.1.

We may, for instance, regard a skeleton made of plastic, constructed to represent the various parts of the human bone *structure*. So the generic purpose G the model serves is didactic and quite different from that of a real skeleton. The performance P of the *co*-image will not be mirrored by the model; the material parameters M may be represented by the model to some extent if the dimensions and their proportions correspond with the average proportions of a human being or if the material is the same as that of the *co*-image; this may be the case if human bones were used to construct the skeleton. The embedding E of a real bone structure will be disregarded, because the surrounding tissue is missing, and the process of ageing of a skeleton made of a synthetic material differs considerably from that of a skeleton consisting of real human bones; and the bone structure of a man still alive is subject to a third way of ageing. This kind of a model is a means to *demonstrate the structure* of its *co*-image, and that is the sole purpose *the model* has to serve. ♦

Very often, it will be difficult to distinguish the generic purpose a system is supposed to serve from its performance. To a certain degree the purpose depends on the performance, and *vice versa*, the performance of a system may be designed to serve a certain pre-defined purpose.

Ex. 4.13 If the main concern of modelling is to preserve the generic purpose G of the system to be modelled, we may call the model a *substitute*. So, we may decide to use a tennis-racket instead of a kitchen-sieve if no sieve is at hand. Of course, the purpose a tennis-racket is supposed to serve is different from that of a sieve. But, by redefinition we may – using the fact that their performance is rather similar when they are used to filter out solid particles – substitute a tennis-racket for a sieve. The following martial example is more concrete and more convincing:

Assume a soldier fully equipped with all sorts of weapons, but without any provisions stands by the banks of a river rich in fish. As he does not have rod and line at his disposal, he may decide to use a hand grenade to substitute for this missing fishing equipment; after all, throwing a grenade will serve the same purpose G of catching fish, but its performance is quite different from that of a fisherman's tools. The grenade will be more effective than the traditional combination of rod and line, but its structure S is definitely different as well as its performance P and the material M it is made of, not to speak of its *temporal efficiency*; this is – compared with the slow and contemplative way of an angler –, indeed, a new criterion added to the contents C characterizing a system Σ. ♦

It seems to be impossible to study the performance P of the constituents of a system without taking the structure S of the system or the generic purpose G it has to serve into account. But in some cases the performance of a system is one of its deciding characteristics and thus affects the process of modelling.

Ex. 4.14 In order to demonstrate this, we regard a construction kit or a collection of technical spare parts. If, for instance, we have got a collection D of electronic devices, like resistors, capacitors, transistors and so on, each of them is a separate technical system with its own structure S and serving its own generic purpose G. But if our interest lies in the field of developing more complex electronic structures, we only notice their performance P within a greater system they are part of. Thus this generic electronic system – output by the process of modelling – possesses a structure S_O of its own as well as its own generic purpose G_O, its own performance P_O and material parameters M_O, and it itself may be embedded into further generic systems of higher hierarchical orders. Maybe, not each of the other contents of the *co*-model will be completely neglected as part of the kernel of imaging within this example, but they are of minor importance to the process of constructing the generic model; the fact, for instance, that the parts of the construction kit are three-dimensional as well as the apparatus assembled from them is rather insignificant. ♦

The materialization of a *co*-model within its model may be a characteristic, too. We have already had examples of models representing their various objects of mapping in a different material and on a different scale, e.g. figurines reflecting an object or moulds (cf. e.g. Ex. 1.2 and Ex. 1.4). On the other hand, we also reproduce *co*-models by models of – more or less – the same material parameters. As an example we mentioned the preparation of an organ of the human body used for didactic purposes or for transplants; model and *co*-model correspond in proportions and material.

Ex. 4.15 A digital electronic device, say, a digital computer, may serve as a further example of making up a model and its *co*-model in the same material. Thus, a digital computer can be projected onto another electronic device representing it in an analogous way, say, an analogous computer. Within the digital computer system mathematical entities are modelled by a discontinuous encoding using digits, whereas the analogous system represents them by continuously varying physical parameters. Nevertheless, the same type of material – electronic parts – is used for the assembly of both systems. ♦

Finally, the embedding E of a model and its *co*-model seems to play a predominant role in the way we represent the objects of modelling. Thus, a dynamic system can be pictured by a static snap-shot, or a three-dimensional object can be represented by a two-dimensional draft or by a photography as we pointed out in Chap. 3; or if we remember the supply grid of a city mentioned in Ex. 4.11, we realize that it is in fact three-dimensional and that the vertical distances between the various lines are significant. The two-dimensional map will represent this arrangement in a planar sketch reflecting the proper horizontal distances – but on a different scale – whereas the vertical distances will be digitalized and modelled by a string of numbers.

Thus, when preparing for the process of modelling a given system and projecting it into a model, we must, first of all, decide which of these characterizing features of the *co*-image of the mapping process should be preserved and mirrored by the model and which shall be dispensed of and be part of the kernel of the mapping.

5 The Art of Modelling

When we speak about 'the art of modelling'[1], there is the idea at the backs of our minds that modelling is a creative process (cf. Sect. 2.4). As creativity is a talent and cannot be learned or taught, it seems to be difficult to treat it in a textbook. The artist learns by experimenting, creating works of art, rejecting them if they are not to his satisfaction or trying to improve them so that they come up to his expectations. We for our part use different methods of working; they are more technical than those of an artist, and thus we replace creative experimenting and practice by the description of some select experiments already carried out. These descriptions are precise case studies of models designed to serve various purposes in different fields of research; they are compiled in the appendix to this book.

Of course, an artist will not rely on his talent alone, but will study and, maybe, adopt certain techniques having a critical look at theories. He will only master these techniques to perfection after a long period of practising them. We shall call the creation of works of art under the influence of conservative aesthetics works of *normal* art, while the development and application of new techniques or of new aesthetic ideas may lead to new époques in arts which is tantamount to some kind of *revolution* in artistic perception.

To compare modelling in sciences with the work of an artist may seem far-fetched. But, indeed, we may look upon the field of the 'Theory of Sciences' under similar aspects and detect periods of *normal* sciences and times of *scientific revolutions*. This theory was developed by Kuhn (1962), who analysed the structure of these scientific revolutions which represent the periods of heterostasis in a field of research leading from one homeostatic level of *normal* science to the next (cf. Subsect. 2.2.6).

---

[1] This idiom follows Knuth (1969) who wrote about 'The Art of Computer Programming'.

Modelling, in general, is a part of those periods governed by normal sciences. In these cases models are tools applying the information and methods of working of 'normal' sciences as a means to solve problems occurring in the scientist's everyday life. They are constructed in order to understand, qualify and quantify observed phenomena on the basis of 'normal' sciences, that means by scientific theories accepted as being valid (i.e. *well-founded*). These models have to meet certain quality criteria (e.g. cf Sect. 5.2), standards developed by experience and from ideas handed down from one generation of scientists to the next one. We will mainly be concerned with this sort of modelling. Yet, models are also developed in periods of scientific revolutions; then they mediate between the revolutionary new ideas and the traditional way of thinking. In that case they serve heuristic purposes in order to make new conceptions more comprehensible to the scientific public hanging on to a conservative way of thinking. Those models are out of the reach of these quality criteria.

The process of modelling can be subdivided into two major constituents: there is the phase of the creation of a model and there is the phase of its assessment with respect to the qualitative aspects to be regarded. Our description of the creative phase is reduced to technical terms; it will be a summary and evaluation of the aspects already introduced in the previous chapters. We will present this technical résumé in the following Sect. 5.1.

The discussion of the quality criteria a model – developed in periods governed by normal sciences – has to meet leads to an entirely new field of reflections. On the one hand, we have to define these criteria: some of them have already been touched on in the previous chapters, but a complete list of them is still to come and will be compiled in Sect. 5.2. On the other hand, it will turn out to be really difficult to *decide* in the individual case whether a given model is a 'good' model or not. This leads us to the central problem of decision-making, and as we are not only interested in decisions between the alternatives 'entirely good' and 'absolutely poor', we shall have to develop means of *quantifying* the quality of models in the following chapters.

Generally, the different phases of systems analysis which we have distinguished in Chap. 2 are not clearly separated, and we will come across some of them more than only once in the course of a project. The discussion of the various items in this chapter is based on the following example.

Ex. 5.1 Given a receptacle R_1 (Fig. 5.1) made of some porous material (e.g. of unglazed pottery), installed in some distance above another receptacle R_2 made of a material impervious to water (e.g. steel). The whole set-up is covered to protect it from falling precipitation (e.g. by a sheet of glass). A certain amount A_1 of water will be poured into R_1, and the system will be watched until R_1 is empty again. At this moment (at observation time t_{obs}) R_2 will be inspected: it may, at this instant, still be empty or contain a certain amount $A_2>0$ of water. ♦

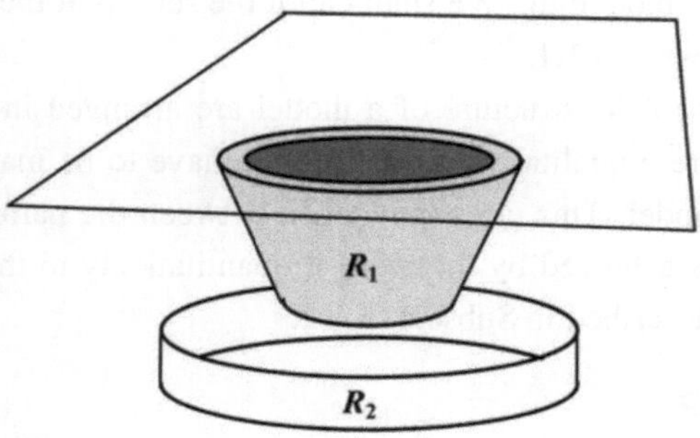

Fig. 5.1. Basic model of Ex. 5.1

We shall now try to explain some aspects of modelling using this scenario. The described case seems to be subject to hydrological and meteorological considerations, and there are many hydrological and meteorological processes affecting this experiment. In order to illustrate the creation and the evaluation of models, we just mention some of them, so we need not enter into a long-winded discussion on the fields of the involved natural sciences. Consequently, our model will neither be *complete* nor will it be *perfect*, as the completeness of a model implies that it covers any process influencing the observed phenomenon and explains it appropriately; and for the perfectness[1] of a model we require that any aspect it covers represents a corresponding aspect valid for its *co*-model, the input-system Σ_I.

---

[1] This criterion of the perfectness of a model corresponds to the *correctness* of models developed in the field of deductive logic; these models of logical theories are either 'true' or 'false' (cf. Chap. 6).

5.1 Creating a Model

Creating a model includes performing analyses of the system to be modelled as well as of the system to be constructed. The modeller will thus define the model-system by its subsystems and the structure combining them, and he will organize the details of his model so that its counter-image – the *co*-model of the process – is adequately mirrored according to the purpose the model has to serve. This is the qualitative aspect of modelling. We shall call it the set-up of the model, and it will be described in Subsect. 5.1.1.

Once the details of the structure of a model are arranged in its set-up the parameters of this purely qualitative representation have to be matched to the counter-image of the model. This correspondence between the parameters of a model and its *co*-model is achieved by *adapting* it quantitatively to the conditions of its counter-image as described in Subsect. 5.1.2.

5.1.1 The Set-up

The set-up of a model covers the qualitative aspects of modelling. They result in a qualitative representation of a system to be modelled; this qualitative model is sometimes called the *conceptual* model, which must be realized and quantitatively adapted to its *co*-model in a second phase of modelling which we describe in Subsect. 5.1.2.

The following list of steps can only give short mentions of some general characteristics of the set-up of models. As this part of the modelling process depends on the particular field of research, it will be carried out according to the field of application. Our list will thus be incomplete regarding the aspects of special models.

Furthermore, the single steps constituting this phase of organizing a model are not entirely independent of each other; there may well be interconnections between some of the stages of model-development described below, and there may even be cyclic repetitions of one or of a whole sequence of these stages.

Definitions

First, the system Σ_I to be modelled must be specified and the purposes the model Σ_O representing Σ_I is supposed to serve have to be defined. As Σ_I will in general be an open system, its boundaries have to be determined. In that case the interactions between the parts of Σ_I and its environment are significant; they determine the interfaces between the system to be modelled and its neighbours. As an interface is an entity shared by the neighbouring systems, it must be clear up to what degree of detailing the interactions between two systems are to be taken into account. This aspect is essential as the actual object of research is Σ_I and not its environment; so we generally only know one side of the interface, looked at from the system to be modelled. Yet, it may sometimes be difficult to find out about these interactions without studying the environment of the system more thoroughly; this may require additional financial resources and a prolongation of the deadline. In the rare cases when system Σ_I is considered a closed system it must be defined accordingly.

Ex. 5.2 We take up the experiment mentioned at the beginning of this chapter (Ex. 5.1) and describe the input-system Σ_I and the model to be developed in full sentences; it could read like this:

> 'There are a pot R_1 of unglazed pottery and a receptacle R_2 impervious to water installed beneath it in a certain distance. An amount A_1 of water will be poured into R_1, and we want to *model* the process of water disappearing from R_1 (and possibly being collected in R_2)'. ♦

In a second step of developing the set-up of the model we define the generic purpose the system Σ_O has to serve. Maybe, it is meant to represent the internal structure of Σ_I only, explaining it, or it is needed to predict future states of Σ_I. In the first case we shall only study the static structure of Σ_I, in the latter case we also have to take the dynamic aspects of Σ_I into account. If the demand is that the model is to be predictive, this requirement will be met by constructing a black-box model, but it may well be that we also want to analyse and mirror the internal structure of Σ_I. In that case the input-system Σ_I must be specified in greater detail, and the accuracy of predictions to be produced by the model-system Σ_O must be kept in mind; these aspects affect the 'quality' of the system Σ_O to be generated, and they will influence the costs of systems design and the demand for resources assigned to the respective project.

Ex. 5.3 In Ex. 5.2 the generic purpose of the model-system Σ_O may, for instance, be to represent any path of transport of water leaving R_1 and entering R_2 as well as any interaction between the system and its components on the one and its environment on the other hand. The objective of modelling in this case is to detect any process effecting the water balance of the system and to evaluate it as meticulously as possible.

But we may also model in order to predict. Then the objective is either the prediction of the point t_{obs} of time when R_1 will be empty, or we try to predict the volume of the amount A_2 of water collected in the receptacle R_2 below R_1 at the moment t_{obs} receptacle R_1 runs empty. Of course, a prediction will not be possible unless we know the processes effecting the water balance, but most probably we will not be able to take each of these processes into account, because some of them will hardly be quantifiable and we must therefore neglect them; some of them may even not have been discovered yet.

This shows that we must reckon with two different conditions governing the process of modelling: the problem is either to predict the values of one or of both of the parameters t_{obs} and A_2 mentioned prior to any experiment and solely depending on (assumed values of) A_1 – in that case there is no observation to be evaluated –, or the modeller has some data at his disposal which have been gained by the observation of a couple of experiments carried out with varying amounts A_1 of water poured into R_1 and under various environmental conditions ♦

Of course, the available resources are of main interest to those who model. Naturally, there is a connection between the standards a model is expected to come up to and the means spent on the development of this system. A financial plan for the project of modelling to be carried out must be agreed upon as well as on the number of staff assigned to this enterprise. Furthermore, schedules must be laid down for the completion of the project and of each of its parts. It may be reasonable to agree upon penalty clauses covering the cases of either changing demands of the client or of the modeller not keeping his side of the agreements.

Documentation

Any agreement about the definition of a project must be documented as well as any succeeding changes in the schedules. With growing insight into the fine-structure of the input-model Σ_I and into the demands on the model-system Σ_O to be generated either the client may wish to expand the range of his orders or the modeller may see the need for additional resources.

The documentation of the progress of the project is also very important. So, each single stage of development has to be recorded, especially the completion of the analysis of subsystems of the generic input-system Σ_I as well as the completion of subsystems of the model-system Σ_O. Finally, when the development of the model is brought to an end and it is either accepted or rejected by the client this decision must be taken down, as well. A rejection must be well justified: an imperfect model may be the result of an inadequate definition of the system to be developed or of changing demands; these reasons are only checkable if there is a neat documentation of the progress of the project. The acceptance of a model by the client may be provisional, because only after a phase of analysing its performance the client can ascertain whether all the quality criteria (cf. Sect. 5.2) are met. Certain qualitative characteristics can be decided on at once, as soon as the project of developing the model is completed, whereas, for instance, the predictive capabilities will only be made out after a longer series of tests. Some aspects of the quality of a model will even never be apparent; for example, whether the developed system can or cannot cope with extraordinary combinations of circumstances which seldom arise and have not yet occurred. Naturally, this phase of analysing the model must be documented thoroughly, any later complaints will be dealt with on the basis of these records.

Systems Analysis

The initial step of modelling is the analysis of the input-system Σ_I. It will be best to start off with trying to detect facilitating (idealizing) aspects within Σ_I if it is not to be represented entirely by a black-box model.

Ex. 5.4 When studying the scenario of Ex. 5.1, we come to the conclusion that there are certain symmetry-properties. The loss of water R_1 can be calculated from the planar model of Fig. 5.2 by integration about its axis of rotation.

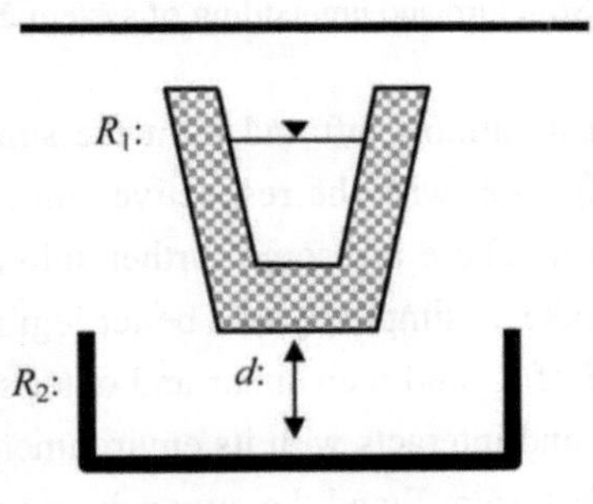

Fig. 5.2. Planar model of Ex. 5.1

In Fig. 5.2 the receptacles R_1 and R_2 are represented by their planar sections, the sheet of glass preventing precipitation from falling into R_1 is the horizontal line on top of the sketch. The symbols are the same as in Fig. 5.1, except symbol d which has been added; it stands for the vertical distance between the bottoms of R_1 and R_2; the horizontal line within R_1 and the black triangle represent the water-level corresponding to the amount A_1 of water poured into R_1. The assumption of rotational symmetry will facilitate systems analysis and modelling considerably and make the system more accessible to the various theories at hand.

We now analyse the system Σ_I to be modelled. By successively opening the black-boxes of subordinate levels we will detect a structure somehow similar to that of Fig. 5.3 showing major parts of Σ_I and their various interconnections with the generic environment of system Σ_I.

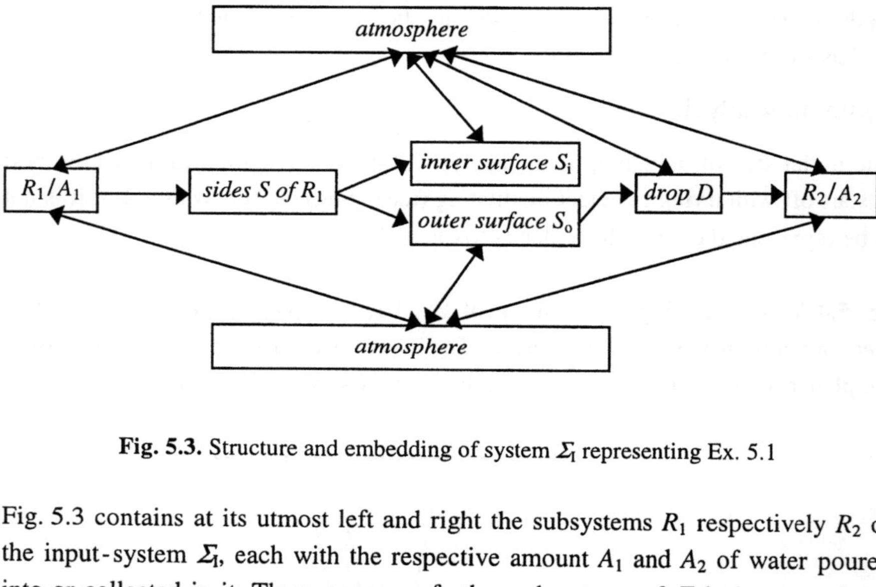

Fig. 5.3. Structure and embedding of system Σ_I representing Ex. 5.1

Fig. 5.3 contains at its utmost left and right the subsystems R_1 respectively R_2 of the input-system Σ_I, each with the respective amount A_1 and A_2 of water poured into or collected in it. There are some further subsystems of Σ_I in between them. We have – for the sake of simplicity and better legibility – restricted our representation to the sides S of R_1 and their inner and outer surfaces S_i and S_o respectively. The system is open and interacts with its environment, the atmosphere. There is an exchange of water between Σ_I and the atmosphere caused by evaporation and condensation. These interactions are represented by the arrows pointing into both of the corresponding directions.

The various parts of the system are connected by simple arrows indicating physical processes transferring water from one subsystem to the next. Thus water will seep into the pores of R_1, pass through its sides S and issue from their outer surface S_o. A certain amount of water will also rise in S under the influence of capillary forces and form a capillary fringe wetting the inner and the outer surfaces of S above the water level. Finally, drops D of water will form on the outer surface S_o of the sides S of R_1, these drops will come off, fall and be collected in receptacle R_2.

We have described the obvious paths of water transportation. We either learned about them when we observed an amount A_1 leave R_1, or we already knew them from experience prior to this experiment. In the latter case we may have heard about them or observed them in experiments carried out before. There may be other phenomena causing water to disappear from R_1 or governing this process, and consequently our analysis may be incomplete, but for the moment we assume that no such phenomenon has yet been observed and studied and that Fig. 5.3 represents the state of the current knowledge intersubjectively[1] available.

Feasibility Study and Selection

When such a complete description has been gained of the input-system Σ_I as well as of its internal structure as of the internal interactions between its parts and those between its parts and its environment, then the items relevant to modelling must be selected, or – put the other way round – those belonging to the kernel of the process must be sorted out, those we want to leave out of consideration.

Within our example Ex. 5.1 described by Fig. 5.3 we may decide to leave the interactions between the drops falling from the outer surface S_o of the sides S of R_1 and the atmosphere out of consideration, because they can hardly be quantified and will be negligible as long as the vertical distance d between the two receptacles R_1 and R_2 is short: the mass of a drop D falling from S_o into R_2 will not be affected considerably, neither by evaporation nor by condensation on a passage of the length $d = 10$ cm (but this will certainly be different if d is some 1,000 m). The same considerations are true for the phenomenon of condensation of atmospheric humidity on the surfaces of A_1 and A_2 and on S_i and S_o. We thus continue with a simplified[2] version of the diagram of Fig. 5.3 which is basic to our systems design:

---

[1] *Intersubjective* means that this knowledge is available to anybody to whom it may concern, who is interested in the matter, has to do with it and is an expert on the relevant subject under consideration.

[2] There are only limited directions of the arrows in Fig. 5.4 when compared with Fig. 5.3 representing the directions of the various reactions.

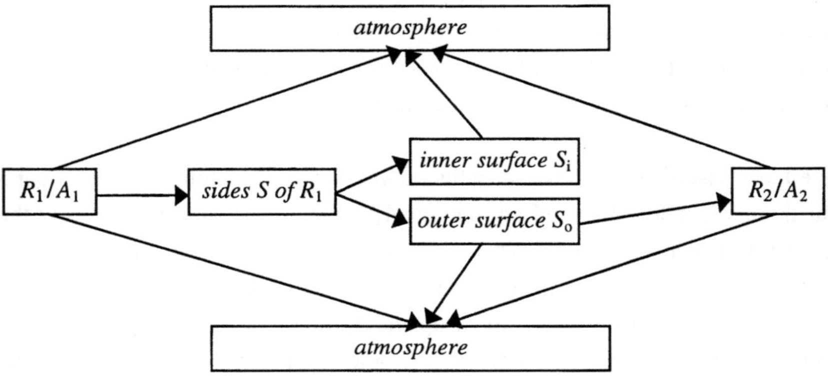

Fig. 5.4. Simplified representation of Fig. 5.3

The problem is now redefined into the more concrete form of modelling the transfer of water from R_1 into the atmosphere and into R_2 regarding the phenomena of seepage, capillarity and evaporation concerning all parts of the system Σ_I with the exception of those effects mentioned above like, for instance, the evaporation from the surface of the drops falling into R_2.

As soon as the system is defined accordingly and this definition is laid down in an appropriate documentation, those features of the scenario to be modelled – the input-system Σ_I – relevant to the project will be determined. This means that the contents of the system Σ_I are selected for modelling. Those items belonging to the contents of Σ_I, but not included into the process of modelling are part of the kernel of this mapping. As we have pointed out in Sect. 4.3 (cf. Fig. 4.2), even those parts of the contents of Σ_I modelled by Σ_O contain features neglected by the modelling process which are thus part of the kernel of this mapping, too. The kernel of modelling may be defined deliberately under utilitarian aspects, or it may be due to a lack of knowledge, which may include phenomena already existing, but not yet detected or phenomena regarded as being insignificant today, but possibly gaining greater relevance in the future.

In connection with the above-mentioned considerations we decide to neglect the three-dimensional embedding E of the set-up of the system Σ_I within our example, the phenomenon of condensation on the parts of Σ_I and to omit the interactions between the falling drops D and the surrounding atmosphere. These disregarded phenomena are parts of the kernel of modelling, and the modeller has intentionally not taken them into account like others, e.g. tidal effects, because they are negligible. There may even be phenomena not yet observed, but influencing the model Σ_I; these may, for instance, be certain chemical or physical interactions between the water penetrating the sides S and the material they are made of. These phenomena are *ipso facto* part of the kernel of modelling.

Furthermore, we have described possible limitations of the project of modelling. These are, e.g. the limited means to predict certain parameters prior to any experiment or after a number of observations only; in our example we should, for instance, also consider the effect of clogging. The minerals contained in the water will precipitate and clog the pores of the sides S. Micro-organisms will develop in the water poured into A_1 and lead to the same effect, and each time the experiment is repeated they are washed further down into the pores of S; they will dry out when R_1 runs empty and finally prevent the transport of water through S if the experiment is repeated with the same hardware.

In conclusion we summarize that the modeller selects parts of the system Σ_I to be modelled as being relevant and decides to mirror them by his model; these parts form the *co*-image of modelling. But there are also parts of Σ_I not represented by the model, they belong to the kernel and they are neglected either unintentionally, because the modeller is not aware of them, or deliberately, because he thinks that they are of minor relevance.

Projection

The contents of the input-system Σ_I thus selected for modelling have to be projected onto the corresponding output-system Σ_O. This mapping is to be carried out rather as an *isomorphism* not leading to too much of a further kernel, entities of the supposed *co*-image of modelling neglected by the mapping. The idea of an isomorphism representing the process of modelling is thus restricted to the model and to that part of its counter-image *significant* to modelling.

The imaging process underlying modelling leads to a model representing its counter-image by its own means. These means depend on the specific field of research; they may, for instance, be computer programmes, analogous set-ups or entirely dialectical representations of reality.

In the analysis of the experiment described in Ex. 5.1 we first confined ourselves to the phenomenon of seepage through the sides S of R_1 and that of evaporation from the various surfaces of the system Σ_I. Yet, when studying these processes more thoroughly, we may decide to take further parts of the input-system Σ_I into account or interactions between this system and its environment; we may also prefer to carry out further idealizing simplifications of Σ_{I}, thus again enlarging the kernel of modelling.

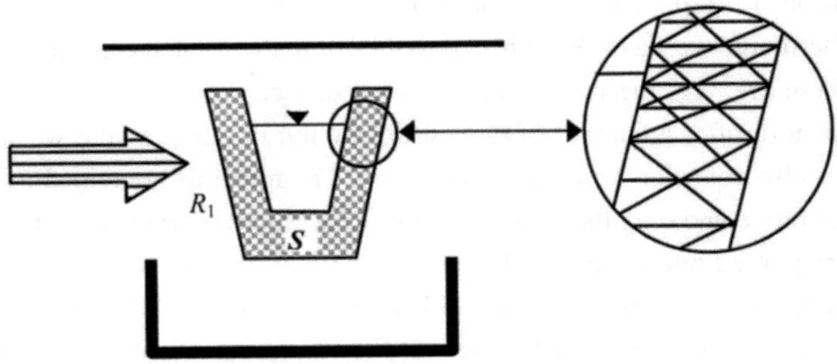

Fig. 5.5. Further assumptions idealizing Ex. 5.1

In order to demonstrate this step of modelling, we attach importance to the movement of the air surrounding the set-up of the *co*-model. Those regions of the sides S of the receptacle R_1 exposed to the wind will dry out sooner than the rest. This increase in evaporation at the outer surface S_{o} of those regions will cool down this part and thus delay evaporation at the corresponding inner surface S_{i} of the side. The shaded arrow on the left-hand side of Fig. 5.5 reflects this phenomenon. We represent it by a constant average of the speed and of the direction of the wind measured at the location. We also try to simplify calculations by *discretization*. In order to discretize, we cut the two-dimensional representation of the sides S into cells of constant conditions governing seepage and capillarity, and we will get a triangulation of the area or a reduction to a combination of polygonal cells; the right-hand side of Fig. 5.5 shows details of a triangulation of the area.

We have thus detected and described the structure of the input-system of the example and of the interactions between its various parts. This analysis has been carried out by opening the black-box of Σ_{I} and any further black-boxes of subordinate levels. With this analysis we may start to construct models serving different purposes, for instance, the purpose of explaining qualitatively why water disappears from the receptacle R_1, or we may want to determine functions describing – quantitatively – how volume A_2, finally collected in R_2, depends on A_1 and the way it is determined by material or meteorological parameters. The latter determination may be performed *ex post* and is then based on the results of experiments; or it may be *ex ante*, prior to any experiment, and will thus be used for predictions.

If it is possible to carry out several experiments and thus observe different amounts A_1 of water disappear from R_1, modelling will be easy, because the internal structure of the process can be disregarded. In that case the whole system Σ_1 is just one black-box which we will not open. In a first attempt we merely determine the relation between varying amounts A_1 of water poured into R_1 as the input into Σ_1 and the corresponding amounts A_2 or observation times t_{obs} as the respective outputs. This relation between, for instance, the values of A_1 and A_2 will either be a closed-form representation by a function $A_2=A_2(A_1)$ or it will be a best-fitting curve. By inter- and extrapolation we can *ceteris paribus* (i.e. for a constant set of values of the further defining parameters $p_1, p_2, \ldots$) derive predictions for any other input-value of A_1, different from the observed scenarios. These representations can be refined to reflect the dependence of A_2 on the set of further defining parameters $p_1, p_2, \ldots$ kept constant at the first attempt, this will lead to a functional representation $A_2=A_2(A_1, p_1, p_2, \ldots)$; the approximation by a best-fitting 'curve'[1] of higher dimensions may be cumbersome. We shall study the details of these methods of adapting a model quantitatively in Subsect. 5.1.2.

In general, it will be complicated to find a closed-form analytical representation of these functions $A_2=A_2(A_1, p_1, p_2, \ldots)$, and readings from graphical representations by best-fitting high-dimensional spaces not accessible to visual interpretations seem to be impossible. So, if we want to take these additional varying parameters $p_1, p_2, \ldots$ into account, we either have to introduce correcting terms into the direct relation $A_2=A_2(A_1)$ between A_2 and A_1 or we must develop generations of functions $A_2=A_2(A_1)$ or best-fitting curves, each one valid for a fixed set of values of these parameters. So we may decide not to use the black-box approach to modelling, though we have the results of observations we can evaluate; in that case we will rather prefer to build the model in order to mirror the internal structure of the system to be modelled, the way we have to do when modelling is carried out *prior* to any experiment.

When no observations are at hand, when we have not yet carried out experiments leading to relations between data input into the system under consideration and its respective output or if we are not able to do so, the strategy of systems development will be quite different. In that case we have to make use of all the knowledge of the fine-structure of the system available to us. Within our example this will be the decomposition of the generic system into its relevant parts accord-

[1] In spaces of general dimensions the analogue of a curve embedded in the plane is sometimes called *hyperspace*.

ing to Fig. 5.4. We also have to determine the values of the parameters reigning the system. In the case of a black-box system, mentioned above, our observations depend on *external* parameters only. Now we have to determine values of *internal* parameters of the system to be modelled, influencing the process which transforms its input into its output. In Fig. 5.4 we describe the interactions between the amount A_1 of water poured into R_1 and the atmosphere as well as those between A_1 and the sides of R_1 as relevant, and we also refer to the transport of water from these sides into R_2 and into the atmosphere. These processes are governed by physical, chemical and meteorological parameters like the evaporation of water with certain chemical characteristics under the influence of certain meteorological circumstances and the seepage through a porous medium. In order to set our model to work, we have to determine the values of all those parameters. Therefore we now have to enter into the internal structure of the system Σ_I under consideration, whereas we have regarded it as an elementary unit in the case of a black-box study.

Of course, some sort of experimenting must be done in order to determine the values of those parameters. But these experiments will be quite different from the generic experiment of pouring an amount A_1 of water into receptacle R_1 and watch it disappear when we constructed the black-box model. Now we must determine material and environmental parameters. We will thus analyse a sample of the material the receptacle R_1 is made of in order to determine its permeability which will – due to clogging – depend on the age of R_1 with regard to the number of experiments carried out. Furthermore, we will, for example, measure the rate of evaporation at the given location under various climatic conditions.

With the knowledge of these values of material and meteorological parameters we can apply the physical laws of percolation and evaporation governing the scenario of Ex. 5.1 to the input A_1 of the system. In order to make use of the available mathematical calculus, it will generally be necessary to transform a model into a computer programme. It may well be that there are already software systems applicable to the special problems the model-system deals with or at least covering a part of these problems. In that case the parameters of the system have to be adjusted. Sometimes various packages must be combined, and if they are not compatible, suitable interfaces must be developed or routines dealing with those aspects of the model-system not covered by standardized software tools. In case no closed software-system dealing with the special mathematical problems of the model-system Σ_O is available, a special 'made-to-measure' system of computer programmes must be developed. Within this package the various parts of Σ_O are represented by routines of their own, interconnected according to the relations governing Σ_0. In the case of the development of extremely complex systems Σ_0 the task of programming will be split up and distributed to a number of separate teams. Under these conditions the interfaces between the various separately developed packages must be well-defined, and it must be ensured that no item of Σ_0 gets 'lost' when the complete system is split up.

From the qualitative point of view we have thus subdivided the set-up of a model into a global and into a detailed design the way we described systems development in Subsect. 2.4.3 and into the realization of these designs by, e.g. computer programmes. We shall call this phase of modelling – comprising global and detailed design and their realization leading to a model in the narrow sense of the word – the *construct*[1] of the model.

5.1.2 The Adaptation

The construct of the model qualitatively representing its *co*-model will now be adapted to the modelled system in order to match its parameters quantitatively. This adaptation is a process lasting for the lifetime of the model. The model must be quantitatively adapted to its *co*-model before it is put into operation and will constantly be tuned while in operation to meet changing demands and the fluctuation of parameters of its counter-image. Adaptation prior to any application of a model is called *calibration* (cf. Subsect. 2.4.3); a calibrated model is capable of producing sets of input-output relations corresponding to those observed in its counter-image. In the course of time the contents of a modelled system, like its structure or performance, may change and the model must be adapted accordingly. We call this the *tuning* of a working model.

Calibration

We have seen in Subsect. 2.4.3 that the etymological roots of the word *calibration* are synonymous to the generic term modelling. Today the word is used in the narrow sense mentioned in Subsect. 2.4.3 and covers merely the quantitative aspects of adapting a model. In this way we calibrate a spring balance within the range of *validity* of Hooke's law where the elongation of a spring and the weight of the body to be measured are proportional. In order to calibrate this balance, standard

---

[1] Some branches of 'empirical' sciences have developed an ample vocabulary naming the components of a model and their quality. We have adopted many of these terms, but have slightly modified them in order to cover a more general usage of the term 'modelling'.

weights are fixed to the spring and the resulting elongation is marked on a scale. By linear interpolation the scale thus gained is refined to meet the demands for the precision of reading. This example of the spring balance is rather straightforward; we prefer to discuss the black-box model of Subsect. 5.1.1 representing the scenario of Ex. 5.1; it relies upon data gained from observations of the input-system Σ_I and measurements establishing a relation between the input into the system Σ_I and its output.

The number of experiments carried out in order to determine this relationship depends on the available resources, e.g. time and money. For our example we assume that there have been five experiments and that the calibration of the black-box model depends on the results of these five observations only. The measured data are represented by the dots of Fig. 5.6 (We did not experiment at all; this is just another of those *Gedankenexperimente* already mentioned). We can now approximate the unknown relation between the amount A_1 input into the modelled system Σ_I and the resulting observation time t_{obs} or amount A_2 of water by a curve 'best fitting' the five measurements of the assumed experiments. 'Best fitting' may, for example, mean that the sum of the vertical distances between the dots and the best-fitting curve must be a minimum. So by idealization and approximation we gain a best-fitting curve. This curve allows – by interpolation and extrapolation – predictions of A_2 depending on any value of A_1 used for experimenting.

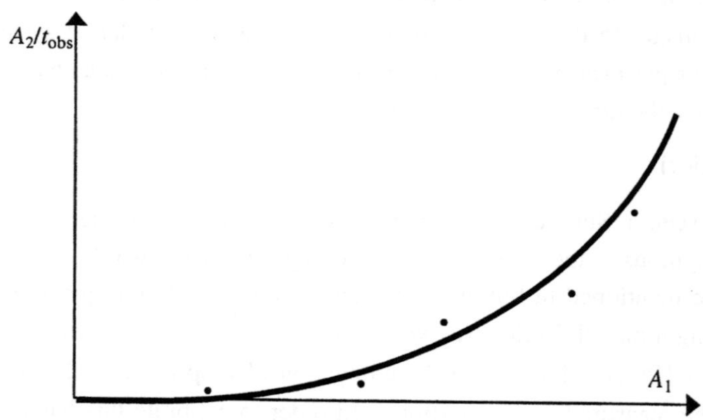

Fig. 5.6. Calibration by a best fitting-curve

Adjusting the curve best fitting the five points of measurement has got something of a game of chance as this procedure depends very much on the particular position of the measured points and on the intuition of the person drawing the curve. Different set-ups of points will influence the construction of the curve, and different people evaluating them will certainly come to different results; the five points of Fig. 5.6 may, for example, also be approximated by a straight line or by a parabolic curve.

Predictions of the variables t_{obs} and A_2 depending on A_1 may be gained in different ways. Under the aspects of applicability the easiest one will be to find a closed-form representation of the curve of Fig. 5.6 reflecting the functional connections between these variables. This means, we have to find mathematical functions $t_{obs}=t_{obs}(A_1)$ and $A_2=A_2(A_1)$ with graphs more or less the same as that of Fig. 5.6. But mostly it will be rather tedious to construct such a functional relation between the dependent variables t_{obs} or A_2 and the independent variable A_1. So we may decide to use the graph of Fig. 5.6 for predictions by *readings*. In that case we locate the position of a special value $*A_1$ of A_1 on the abscissa of the coordinate system, we then determine the corresponding point P on the best-fitting curve located vertically above the value of $*A_1$. Finally, we will gain the respective values of $*t_{obs}$ or $*A_2$ depending on A_1 by looking up the value of the ordinate of P within the coordinate system (cf. Fig. 5.7).

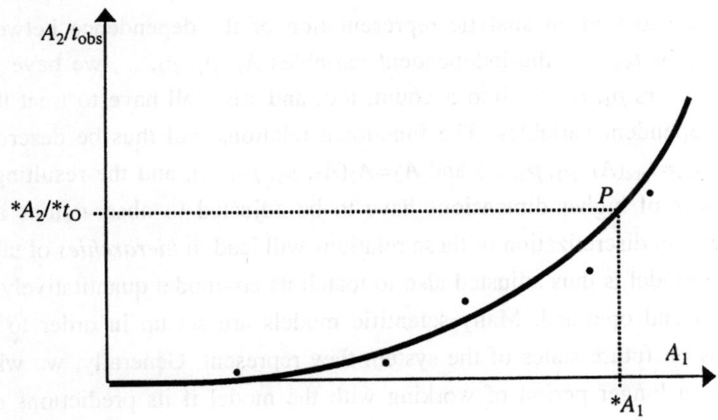

Fig. 5.7. Predictions gained by readings from a best-fitting curve

Depending on the precision of determining the values A_1 on the abscissa and of t_{obs} or A_2 on the ordinate of the coordinate system as well as of the point P on the best-fitting curve, this way of reading will lead to a certain loss of information contained in the continuous distribution of the variables A_1, t_{obs} and A_2; this is due to the *discretization* of the values of the variables (cf. Sect. 3.3).

Therefore we may decide to make use of discretization right from the beginning and subdivide the abscissa into consecutive, mutually excluding intervals covering the range of values A_1 under consideration. The length of these intervals corresponds to the exactness of reading. Let us assume it is $\pm 2.5\,\text{cm}^3$ for A_1, then the corresponding entries in the table are something like ... $100\,\text{cm}^3 \leq A_1 < 105\,\text{cm}^3$, $105\,\text{cm}^3 \leq A_1 < 110\,\text{cm}^3$, and so on. The exactness of reading for A_2 will lie within the same intervals of length $5\,\text{cm}^3$, whereas the exactness of reading for t_{obs} will depend on the physical conditions and range between some minutes and some hours, depending on the speed of the processes leading to the extraction of A_1 out of receptacle R_1. This way of discretizing the observed values leads to a *table*.

Up to now, we have only considered the relationship between the variables A_2 or t_{obs} and A_1. But the function transforming the value of the independent variable A_1 into the corresponding values of t_{obs} or A_2 is determined by further parameters p_1, p_2, ... and their respective values. These parameters may, for instance, describe meteorological phenomena influencing the process of A_1 leaving receptacle R_1 or the process of ageing due to the clogging of the pores of R_1. Consequently, the results of modelling as mentioned above will only *ceteris paribus* be acceptable; and if we try to find an analytic representation of the dependence between the variables A_2 or t_{obs} and the independent variables A_1, p_1, p_2, ..., we have to take these parameters p_1, p_2, ... into account, too, and we shall have to treat them as further independent variables. The functional relations will thus be described by functions $t_{obs}=t_{obs}(A_1, p_1, p_2, ...)$ and $A_2=A_2(A_1, p_1, p_2, ...)$, and the resulting best-fitting spaces of higher dimensions have to be adjusted to observations accordingly; likewise, discretization of these relations will lead to *hierarchies* of tables.

When a model is thus adjusted also to match its *co*-model quantitatively, it can be released and operated. Many scientific models are set up in order to permit predictions on future states of the system they represent. Generally, we will only know after a longer period of working with the model if its predictions are dependable; but sometimes calibration is also used to determine its predictive capability prior to its use in practice. This is done by a *split-sample* test: in our example we derived the graph of Fig. 5.6 from only five available observations. Let us assume now that there had been ten experiments prior to calibration, but that we used only the results of the five mentioned. We now add the additional points representing the data measured in the five experiments not yet considered into Fig. 5.6 and represent them by circles; we thus gain the sketch of Fig. 5.8. We will now conclude that our predicting-best fitting curve is rather good at 'predicting' those additional data and that it is in concordance with the additional observations, or we may think this curve is rather good, but it may be improved by deforming it just a little, or we may even decide for a new curve regarding the evidence based on all the ten experiments.

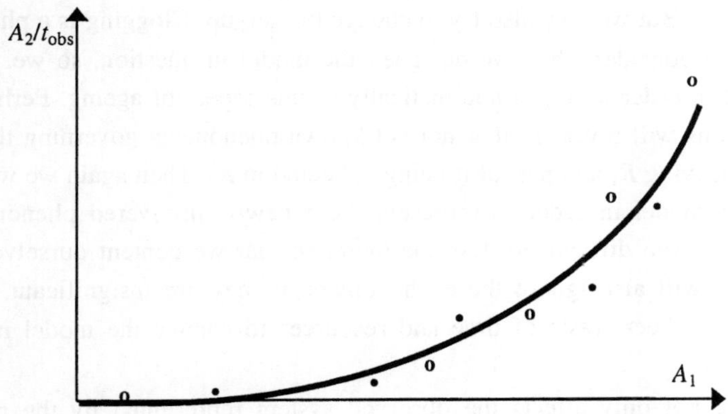

Fig. 5.8. Split-sample test

Sometimes not each single of a series of models is calibrated, but only a prototype, for instance, in the case of mass production of spring balances: a prototype balance is calibrated in a *proxy-test*, and the scale thus derived is transferred to its copy; only spot checks will be made to ensure constant quality. We may, likewise, decide to use one sample of receptacle R_1 for the calibration of a model and another one – of equal size and material – for its operation in order to avoid the phenomenon of ageing prior to the release of the model.

Tuning

When a model is released and run regularly, it will be necessary to readjust it from time to time, because differences between the results of modelling and the observations of the original system will develop in the course of the operation of the system. These deviations may be due to ageing or they may indicate phenomena not considered or apparent when the model was set up. So the spring balance will wear out in the course of time and its scale must be readjusted to its changing elastic modulus.

Likewise, we may detect that the seepage rate determined for the material of R_1 deteriorates in the course of time, because the pores clog up. If we did not take clogging into account when the model was generated, we must tune it accordingly. This may lead to various alterations in the model-system. The easiest will probably be to re-calibrate it in order to make it match changed conditions. Re-calibration will leave the set-up of the model unchanged, it will only take the recent developments of parameter values into account like the lower permeability of the material. But we may also try to change the set-up. Clogging is a phenomenon we did not consider when we designed the model in question, so we will try to modify it in order to adjust automatically to this aspect of ageing. Perhaps future observations will reveal further not yet known phenomena governing the process of water leaving R_1 and part of it being collected in R_2. Then again we will have to adapt our model in order to represent these newly discovered phenomena, but, maybe, it is too difficult to describe them, so that we content ourselves without them. We will also ignore these phenomena if they are insignificant, so that it would be a sheer waste of time and resources to change the model in order to cover them.

Ageing not only affects the observed system represented by the model, the model itself is subject to ageing, too: the underlying theories may be improved as well as the available resources. New theories and methods of calculation and advanced computer systems with a better performance will be developed and will permit the explanation and representation of more complex phenomena than before. Using these advanced theories and methods we will be able to develop models representing their counter-images more adequately. But, in general, the more sophisticated a model, the greater are the demands made on the high standards of the equipment used to operate it. Meteorological models, for instance, require fast computer systems with large storage capacities; as long as these resources are not available even extremely dependable models are worthless, because they cannot be run.

5.2 Quality Criteria

When a model has been created and parts of it have been transformed into a system of computer programmes, if it is run under the control of a computer, it has to meet a couple of criteria certifying its quality. Whether we accept a model as being 'good' or reject it as a 'poor' model depends on the way it comes up to the standards laid down by these *quality criteria*. We have already mentioned some of them in preceding sections – like objectivity and reliability –, but as their significance depends on the varying fields of application we want to describe them more thoroughly in this chapter.

As we have seen in Sect. 5.1, a model may be treated as a black-box model if it represents an input-system Σ_I, permanently observed and with a well-defined relationship between its input and its output. In that case the internal structure of Σ_I will be disregarded, and the quality of the model is determined by the 'best-fitting' representation of this input-output relation. The resulting model-system Σ_O will be accepted as long as its output matches data observed in Σ_I sufficiently well. Some black-box models will 'learn', they will adjust automatically to minor changes and trends in the observed data by adapting to the ever growing number of observations. We have described this process of adjustment in Subsect. 5.1.2 when we adapted the best-fitting curve of Ex. 5.1 constructed on the basis of five observations to an increased set of ten data measured in Σ_I. In that example the five additional observations were known prior to the calibration of the model, but were considered later in order to test the predictive capabilities of the model. We may as well assume that the original set of five observations was all we had for calibration and that the additional ones only occurred during the future operation of the model and we re-calibrated it in order cover them, too.

The capability of a model-system Σ_O of learning and automatically adapting to varying data measured in reality is limited to minor changes and rather 'slow' and steady trends shifting these observations. Abrupt, discontinuous and violent variations in parameters in reality indicate a 'revolution' within the observed phenomena and the processes causing them. In that case we may regard this new input Σ_I into modelling as another black-box system and adjust Σ_O to the changed conditions. But, maybe, we decide to study the system more closely, subdivide it into its relevant components and determine the relations between them, because we think a black-box will no longer represent these observed phenomena adequately. In other words, we will initiate an analysis of the input-system Σ_I and dispense with the black-box approach.

When we enter into the details of Σ_I and analyse the system – either under the condition of a 'revolution' as mentioned above or right at the beginning of the development – the process of determining the dependability of a model is somewhat more difficult. Of course, there are observations and measurements in that case, too; we thus carry out experiments in order to determine material parameters as well as meteorological and others governing the process of water disappearing from receptacle R_1 in our example Ex. 5.1. But often the qualities of these models must be known as soon as their development is completed and prior to any comparison between modelled data and data collected in reality[1]. After some time of modelling there will be observations we can use in order to evaluate the qualities of our model, but in some cases they will be small in number and their evidence of the qualities of a model will be of little significance under statistical aspects.

That is the catch in evaluating models: we lay down certain quality criteria they have to satisfy, and in many cases we must decide prior to any comparison between the model and reality whether they will meet our demands or not; hence these evaluations merely give information about the *prima facie* dependability of the system Σ_O. Even in these cases we may have further evidence at our disposal supporting or refuting the assumption that we can depend on the model in question, but it will still be difficult to decide once and for all whether this model comes up to all the quality criteria as it has to be evaluated *prior* to its application. In this situation we have to cope with the problem of deciding under uncertainty. Some of the evaluations are performed in the confidence that the results will be dependable, while modellers sometimes are merely 'looking for evidence' that the model in question is not definitely a 'poor' model.

The criteria can be classified into groups under logical and formal aspects, but we shall disregard these relations for the moment and discuss them in the order of their appearance in the process of modelling. Parts of the aspects these criteria deal with may overlap when they are applied to certain examples of modelling; besides, the relevance of some of them depends on the field of research using the models. So, for instance, those modelling in social sciences are more concerned (cf. Subsect. 2.4.1) with the objectivity of models than modellers in the fields of technology and natural sciences are – or should be –, because the latter often work within

---

[1] This will be the case when a model is used to support decision making under uncertainty, e.g. when risky innovations are introduced, for instance in space flight or nuclear technology, or when political leaders have to decide on the eventuality of a war.

rather rigid methodologies not allowing much of a scope for biased planning and interpretations of experiments.

5.2.1 Formal Correctness

Aspects of this criterion are frequently called the *verification* of a model. Those who prefer to talk about the verification of a model when we test its formal correctness distinguish between *verification* and *validation*. In our view the validity of a model is the generic quality criterion (cf. Chap. 9) which we will describe in Subsect. 5.2.4 and discuss thoroughly in Chap. 10. There have been semantic discussions about the implications of the terms *verification* and *validation* with respect to the dependability of a model. For this reason and in order to avoid any ambiguity, we have decided to subsume this aspect under the term of the *formal correctness* of a model, instead.

This criterion demands a correct transformation of the input-system Σ_I to be modelled into the model Σ_O. For instance, the used calculi must be correct representations of processes observed in Σ_I, used computer programmes must be debugged and work the way they are supposed to do.

If, for instance, we model the system Σ_I of Ex. 2.5, transforming discontinuous saw tooth signals input into Σ_I into a harmonic trigonometric output, we may do so by a black-box approach or by analysing the transforming process P_I of Σ_I. The model Σ_O will be formally correct if its output-signal O_O equals the corresponding summand of the Fourier series in its frequency as well as in its amplitude. If Σ_O is a black-box model, this will be the case if the frequency and the amplitude of O_O correspond with the parameters p and c of Ex. 2.5 according to equation (2.6). But if we have analysed the process P_I of Σ_I and have, e.g. found out that it is a Wien-Bridge, the corresponding model Σ_O supposed to represent the performance of this circuit will only be formally correct – from this point of view – if it represents this performance appropriately. In case a computer programme is used in the model Σ_O, its being debugged means that it runs, and its – what we called – 'doing what it is supposed to do' implies that its performance reflects the analyst's schedules, that, for instance, its input-output relations are as planned. This is only one facet of the formal correctness of a model. We shall discuss this criterion in full detail in Chap. 9, but we want to describe some of its aspects already in the following Subsection.

5.2.2 Objectivity

An important quality criterion requires the *objectivity* of the modelling process. This demand seems to be so self-evident, that it seems to be unnecessary to mention it explicitly, but there are known cases where modellers offended against the rule of objectivity unwittingly. Quite naturally, we leave those cases out of consideration when these rules are flouted intentionally.

We have already mentioned (in Chap. 2) that the criterion of objectivity demands the independence of an experiment and its output from any external influence, like the observer, the place or the used equipment. The process of modelling must be reproducible: this means that whenever the experiment of modelling is repeated by a different team and at a different location it must lead to the same results. Consequently, a model must be accepted by the influential scientific circle; this acceptance is based on the – intersubjective – agreement that the model mirrors its counter-image Σ_I uninfluenced by subjectivity. Of course, this agreement does not imply that the model is 'true', it only means that it does not reflect an entirely subjective attitude, and sometimes it merely implies that the model is not *entirely* based on these aspects.

The objectivity of a model is of varying importance in the field of modelling. If we accept artistic creation as an act of modelling, we will not expect it to be objective in the sense mentioned above. On the contrary, a work of art is meant to be individualistic and to stand for a *subjective* representation of its counter-image, otherwise it will often be despised as a bread-winning work of art. But if we think of models in natural sciences or engineering, they should be objective *ipso facto*; even if modelling is not always reproducible, this does not matter as long as the scientific community accepts the model as an unbiased[1] representation of its counter-image. This may, for instance, happen when a singular find is used for reproduction, say, by a palaeontologist or a forensic scientist. In those cases no repetition of the reproducing experiment is feasible, at least not at the moment. Here objectivity is achieved as soon as a model is intersubjectively accepted, when experts agree that the palaeontological reproduction is appropriate, or when a court accepts a forensic reproduction as a piece of circumstantial evidence in a trial. But this does not really imply that the model (reproduction) is really *objectively true*.

---

[1] In this case the unbiased representation may be regarded as also being adequate and dependable; therefore regarding these models as objective will imply their acceptance as being *valid* (cf. Subsect. 5.2.4 and Chap. 10) – at least for the time being.

The only examples of models being literally *true* are mathematical and logical models; they are true *by definition*.

When we regard modelling in social sciences, the situation may be different again. Frequently, these models are based on mass data gained by survey leaving room for not being objectively carried out: the questionnaires depend on the persons making them out, and the analysis of the the survey results may be due to subjective interpretations. So objectivity is a criterion positioned somewhere between those diametrically opposing fields of modelling in arts and mathematical models. It is relevant to the evaluation of statistical data models developed in social sciences, medicine and other empirical sciences, and it covers various aspects of the process of building a model and evaluating it. In those cases objectivity must take priority as a characteristic of planning an experiment and of carrying it out as well as of the analysis and the interpretation of the results. This means that all these phases of modelling have to be reproducible or intersubjectively accepted by a group of experts.

There are several reasons why modellers offend against the demand for objectivity. They may do so, intentionally forging the results of experiments; this is a clear offence, but these 'modellers' are hardly ever prosecuted. Rare as these cases may be, there have been some – and definitely too many – reports about manipulated data bases used in empirical sciences. Very similar to these forgeries are experts' reports on request. In general these reports will not entirely be based on forged data; the engaged experts will just make use of different possibilities of evaluating facts and will interpret them in favour of their financial backers. As an example we mention the various contradictory reports on the hazards caused by passive smoking.

Models will not be objective when scientific results gained from them are published before the corresponding experiments have been reproduced. The announcement of a successfully performed cold fusion by Fleischmann and Pons (Ex. 1.19) may serve as an example. Sometimes this kind of offence against the principle of objectivity might be understandable, when we think of the stress put on scientists constantly to publish new results in their fields of research. So, if there seems to be something really sensational – like a successfully performed cold fusion –, and if there is the risk that competing groups working on the same subject may be successful, too, some scientists may be tempted to publish statements they will not be able to maintain in the future.

Quite unintentionally a modeller may leave the field of objectivity whenever his actions affect the system he is studying; this will cause the effects we described in

Subsect. 2.3.2. A commonplace example is the ruling out of outliers when evaluating data gained by experimenting. That means that reality is shaped according to the expectations of the observer, he will cause a Procrustes effect. Of course, the same happens when reports are written on request; but in that case the manipulation of data is performed deliberately. There will often be the temptation to ignore data gained by experimenting when they seem to be insignificant; but there is always the risk that we fail to notice an important phenomenon, because just these of all data were significant to it and we were blinded by our expectations.

The observation of a system may also cause the Pygmalion effect leading to self-fulfilling prophecies or to the Hawthorne effect (cf. Subsect. 2.3.2) where the object of research will look as unsuspicious as possible. In both cases the observer unintentionally affects the system he regards by interacting with it. In the case of a Halo effect, which also results from these interactions, the system gains influence over its observer. Sometimes it is difficult to find out whether a model is still based on objectivity or whether one of those effects – or some other event distorting the observer's recognition – has influenced an experiment. To find out, the results should be reproduced by other means and by different teams – if possible. Hawthorne effects caused by observation can be avoided by disguising the purposes of the research.

When we strive for 'objectivity', we should realize that this is an expectation that can hardly – if ever – be fulfilled. Though speaking of 'objectivity' in reality we have to content ourselves with intersubjectivity – a consensus among the initiated that the way they deal with the model is *not* subjective. Besides there are all those factors conditioning our decisions, like the period of time we live in, the social class or the *Weltanschauung*[1], to mention some. So the Ptolemean universe was regarded as 'objective' at his time, just like we think the same of our scientific work. But with these considerations we leave the field of objectivity and turn to the domain of the validity of a model. We will discuss this criterion in Sect. 5.2.4 and give priority to another aspect in the following section.

5.2.3 Reliability

The criterion of *reliability* covers the precision of measurements of a system. Gauges, for instance, are tested by their manufacturers who will determine their

---

[1] This is another loan-word taken from the German; its meaning lies somewhere between 'philosophy of life', 'world view' and 'ideology', but none of these English words reflects the meaning of *Weltanschauung* correctly.

accuracy in measuring, and this parameter determining the quality of the instrument is known to its user. In the broader sense the term describes the precision of the representation of the results yielded by a model; it thus qualifies the mapping as defined by the modeller. In fact, the reliability of a model does not imply that it represents its counter-image 'adequately', but only that it precisely represents those contents of the *co*-model mirrored by the model even if those contents are irrelevant to the intended purpose of the model. Naturally, reliability requires objectivity: if a model is regarded as not objective, it can *ipso facto* not be reliable.

The determination of the reliability of a system leads to the theory of errors. Repeated measurements will allow to determine the precision and to quantify it by the means of correlation coefficients and variances. A discussion of these statistical methods is out of the reach of this book as these procedures are well-documented in the relevant literature. Nevertheless, even if a model is accepted as reliable, there is still the problem to decide whether it is an 'adequate' image of its *co*-model. As objectivity and reliability are merely partial views of the dependability of a model we conclude that there must still be a quality criterion of overriding importance: the validity of a model.

Both criteria – reliability and validity – must be distinguished and treated separately. If, for instance, we determine the height a body is dropped from by measuring the span of time between the instant the body is released and the moment it hits the ground, an electronic stop-watch – perhaps in combination with photoelectric relays starting and stopping the watch automatically – will be a very *reliable* instrument to measure this quantity. Experiments repeated in a large number will lead to the conclusion that the stop-watch in combination with photoelectric switches is a system measuring time with high precision: there will be no noticeable error in the series of these experiments. So reliability is merely a criterion concerning the quantity the system *really* measures, but not necessarily those features it is supposed to characterize. As long as we do not know the law of free fall, we may think the stop-watch is not only a reliable instrument to measure time, but that this quantity also 'adequately' represents the height we want to determine. According to this information the stop-watch 'pretends' to measure height and does so with great precision. As soon as we know about the law of free fall, we will no longer accept this representation of length by measured time as adequate; we may still use the watch if no other instrument is at hand, but we will convert the time measured during a fall into the corresponding height.

Likewise, a system modelling the scenario described by Ex. 5.1 may be reliable, although we would rather classify it as being 'poor' and 'inadequate' if, for instance, this model leads to the result that A_2, the amount of water finally collected in receptacle R_2 is *always* $A_2=0$ regardless of the amount A_1 of water poured into R_1 and of the further parameters like, e.g the porosity of the material receptacle R_1 is made of. Naturally this model is of high precision in measurements and thus extremely reliable, because repeated experiments will always result in $A_2=0\pm0\,\%$ which is an extremely insignificant error. Nevertheless, we will regard this model as a 'poor' representation of reality, not adequately representing its *co*-model.

The predominant aspect of classifying the quality of a model is the 'adequacy' of its representation of the 'reality' it is supposed to represent. This is the criterion of validity which is a means to qualify and quantify the degree of dependability or 'trustworthiness' we grant a model.

5.2.4 The Criterion of Validity

The *validity* of a model is the predominant means to describe – and measure – its mapping quality, and if a model is regarded as valid, it is accepted as also being reliable and objective. There have been some discussions on the impact of this term, possibly leading to false conclusions about the degree up to which a model can be depended on. As these arguments are mainly based on etymological considerations, as we have already pointed out in Subsect. 5.2.1, we will start with an analysis of the relevant root of the word and study the implications we may derive from a statement like 'this model is valid' and their limits.

The term *valid* is derived from the Latin adjective *validus*, which means 'firm' or 'well-founded' or 'based on firm foundation walls' in English. The word seems to be somehow related to civil engineering and used as a metaphor in connection with modelling in general. So we want to study its original Latin meaning more closely. It is absolutely clear what we mean when we say 'a building is sturdy and based on firm foundations': this implies that it has got a sturdy construction and well-laid foundations, that it will not collapse by itself and that – for the time being – the building will serve the purpose it has been planned for. But this does *not* imply that a 'firm' and 'well-founded'[1] building is built 'to live in it for ever'. A tempest may destroy it, it may be shaken up by an earthquake, it will age and eventually be outmoded and get uninhabitable. If there are any damages, we may decide either to repair them, or we may prefer to demolish the building and erect a

---

[1] Nowadays the metaphor 'well-founded' derived from civil engineering is mainly used in order to describe the quality of lines of arguments; we want to use it in the original meaning of the word, i.e. synonymously to 'erected on well-laid, sturdy foundation walls'.

new one on the site. We may even want to pull down a building still in perfect condition, but considered to be old-fashioned. The master builders of the ancient world certainly thought their buildings were made for ever: some Egyptian pyramids still exist, but the Roman Coliseum is only a ruin, and the Colossus of Rhodes became the victim of an earthquake and no ruins are left. Modern buildings have a rather limited life-expectancy, sometimes they are already demolished some ten years after they were erected.

This introduction may seem to be inappropriate to a discussion on modelling. But the *validity* of a model is the most important criterion used to evaluate its quality, and, maybe, false conclusions will be drawn by the lay public when we call a model *valid*. This is entirely due to semantic considerations; consequently those misunderstandings can easily be avoided when we know the exact meanings of the terms used. The validity of a model – if quantified by *degrees of validity* – is the *degree* up to which we may rely on it to serve its purpose, to use it safely and 'to live in it'. The validity of a model as well as that of a building is the degree up to which it has been adapted to its environment in order to make it stable. If we once more use the metaphor of the foundations of a building, we will conclude that the validity – foundation – of a model depends on the quality of the site it is erected on: if the ground is soft and spongy, the foundation must be extremely sturdy. If the construction is shaken by extraordinary events – tantamount to scientific revolutions when we talk about models – it may either be reconstructed taking these phenomena into account for the future, or an entirely new building may be erected.

So *validity* actually describes the way a model (building) is adapted to reality and 'well-founded' in order to cope with it. Validating thus is the process of determining the validity of the design. Validity may be a *partial* quality only, a construction may be partially valid, i.e. not showing optimum conditions with regard to the way it is adapted to reality, but not totally invalid either; the validity of a model must thus also be measured by *degrees*, and its quality may be improved if it is neither perfect nor absolutely invalid. We even have to live with models known to possess only a limited *degree* of validity, just like people live in houses only partially *valid*, for instance, when they live in regions threatened by earthquakes: whenever the houses are destroyed, they are rebuilt again on the same site. There are also buildings being in constant danger of being flooded – not destroyed – and thus made uninhabitable for a while: the houses located in the fluvial plains where land is submerged several times a year, causing a permanent alternation of periods of living in peace followed by periods when the buildings have to be drained, decorated and white-washed again. Of course, people could give up their lodgings and move up hill, but this would cause them to leave the area which is their home and where they work.

Similarly, we may prefer to stick to a model even if it is only partially valid and try to improve it in order to make it 'better-founded'. Consequently, the conclusions we draw from a validation will not be: to be valid means the model is 'good' and, likewise, an invalid model is not necessarily a 'poor' model. We shall rather trust to a model in *degrees* of its validity determined according to our stage of *knowledge*. We thus interrupt the discussion on the term 'validity' (we will return to it in Chap. 10) in order to give a perspective on the way we *infer* from the available evidence and come to conclusions concerning the validity of a model.

6 Inferences

As we have seen, validating a model mirroring reality means to compare it with its *co*-model, the counter-image which contains the relevant details reflected by the model in question. The validity of a model thus is a means to *evaluate* the quality of the correspondence between reality and its image represented by the model. As a measure validity is determined in *degrees*, and so the quality of a model depends on its *degree of validity*. This means that the conclusion that a model is accepted as being 'good' or that it is rejected as being 'poor' depends on this degree. Conclusions are the results of inference based on the available evidence and they are subject to specific laws. There are very rigid and formalistic rules of *deducing* conclusions from a set of assumptions; this is the way the mathematician proves his propositions and the philosopher founds his statements.

The conclusions inferred from the similarity between a model and its counter-image are not so proven as the theorems valid in mathematics and as the philosopher's propositions. If there is a very close correspondence between reality and its modelled image, this will be a great moment for the modeller, and if there is no correspondence at all, he will suffer a defeat. But, most often, the validation will result in evidence somewhere in between perfect correspondence between reality and its image and none at all. In these cases we have to *induce* a degree of trustworthiness of the model; it will depend on the persuasive power of the evidence supporting the assumption that the created model is a 'good' one. This inference will be well-founded whenever there are many observations in favour of the model, but it should be taken 'with a large pinch of salt' if it has been confirmed only once or twice.

An inference is the transgression from a set of assumptions we work on or the available evidence to a conclusion which we trust to be reliable. This transgression is – on the one hand – based on certain rules of inference and – on the other hand –

on the experience of the person inferring and on the field of research this person works in. We may describe the inferential process by the diagram of Fig. 6.1.

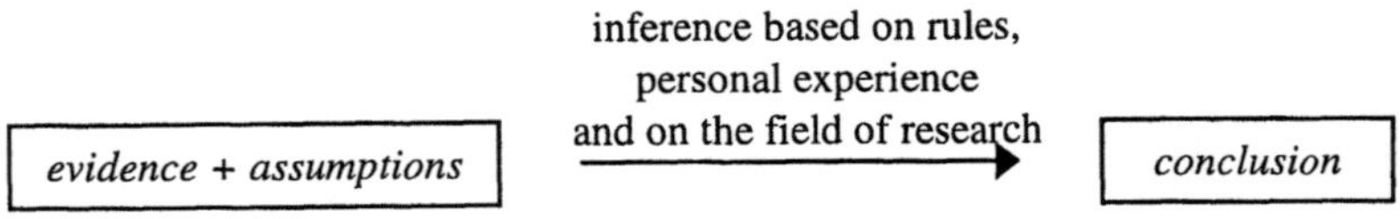

Fig. 6.1. The imaging process of inferring

This diagram looks very much like Fig. 1.7 and Fig. 3.1 representing the way models are gained from an imaging process. Consequently we may interpret it by stating that inferring is also a process of modelling. Here, modelling is the projection of all the available evidence and assumptions into a conclusion we draw using certain rules of inference, our personal experience and our intuition.

Inference in the field of evaluating models is a process related to the inferring *persons* and will therefore inevitably be personal and subjective. Given a certain body of evidence and having agreed on certain assumptions, different people will often come to different conclusions. The reason why the results of inferring processes will differ with the inferring persons is that these persons depend on their individual backgrounds, experiences and psychological dispositions. There may be inferences accepted by a large number of people, so, for instance, many will agree that it is only reasonable to draw certain conclusions from a given body of evidence that is assumed to be true. Consequently all those accepting an inference as feasible must have some personal experience in common as it is illustrated in Fig. 6.1, and they must be endowed with the same way of reasoning basic to the inferential process.

On the other hand, the acceptance of a conclusion may be restricted to a rather small number of initiated. There will, for example, only be a few who will accept the *Yin*- and *Yang* caterogization of numbers and chemical elements we have described in Ex. 1.19. Though the four-colour conjecture has been converted to a theorem; some experts still think that the proof of this theorem is somehow unsatisfactory as it depends on verification using computerized algorithms: a fact violating the mathematician's aesthetics with regard to a mathematical proof (Appel and Haken 1977, Appel et al. 1977); and only a limited number of qualified mathematicians is able to accept the recently given proof of Fermat's theorem as convincing (e.g. Wiles 1995).

Finally, there may be conclusions we have to draw – entirely based on our personal evaluation of a given situation – in order to come to *decisions*. This evaluation may be very intimate indeed and include, for instance, the readiness of the respective person to take risks as well as his personal experience with similar situations and his being influenced by hearsay. There may be situations where, say, a group of ten persons have to decide among ten different ways of dealing with a given scenario and each of the members of this group may reach his own decision not comparable with those of his fellows.

We call those inferences *necessarily* accepted by *everyone* possessing the relevant information about the inferring process *objective*. If a conclusion is accepted by a group of persons having the same degree of evidence and the same experience – i.e. insight into the problems concerning the given evidence – the derived inference is called *intersubjective* with respect to this group. But an inference may also be based on entirely personal considerations, based on the insight of an individual alone; let us call it *subjective*.

The basic ideas of inferring under these considerations are compiled in the following sections. We shall enrich each paragraph by some examples. We admit that there may be overlaps between these sections, because the distinctions between the mentioned criteria are rather fluid and far from being disjunctive.

6.1 Deductive Inference

Deductive (Latin *de*: 'from' and *ducere*: 'to lead') inference is reasoning according to purely logical (Latin *logos*: 'reasoning') arguments. Logical reasoning allows us to proceed from a set of propositions we assume to be true to a set of conclusions. This progression is due to the laws of the logical calculus, and it ensures that the conclusions drawn from true premises are true themselves. In general, deductive inference is directed from the general to the particular case. We shall demonstrate this feature with some examples. The prototype of this way of inferring is the mathematical proof, and, to start with, we shall explain it with the help of examples taken from elementary mathematics.

Ex. 6.1 Pythagoras proved the theorem that the square of the length c of the hypotenuse c of a plane, right-angled triangle (like that of Fig. 6.2) is equal to the sum of the squares of the lengths a and b of the two other sides a and b of the triangle. The hypotenuse c is the side opposite the right angle γ.

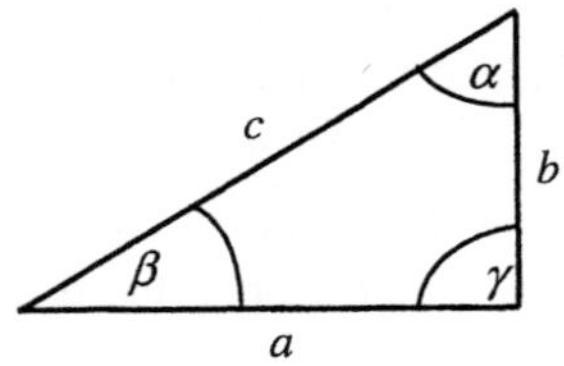

Fig. 6.2. Pythagoras' theorem

Pythagoras' theorem thus states that $c^2=a^2+b^2$ in any plane, right-angled triangle with the respective sides *c*, *a* and *b*. As this is a mathematical theorem, it has been proved and we may trust to it as being *valid* for every plane, right-angled triangle. If, for instance, there is a given triangle of this type with, say, c=5m and a=4m, we are able to compute b=3m using Pythagoras' theorem, and we can be sure that measuring the length of *b* in this special triangle will lead to exactly b=3m. We can trust to this value b=3m without checking our calculation by further measurement, because our conclusion is the result of a computation according to the proven theorem of Pythagoras. ♦

We now reformulate Pythagoras' theorem and present it in a more formalistic way. At first glance this may seem rather farfetched, but the introduced notation will have its merits in our further discussion.

Let *S* be the set of all planar and right-angled triangles with the hypotenuses *c* and the sides *a* and *b*. If *x* is one of these triangles, we call it a member of the set *S* and we denote this fact by $x \in S$, which is a formal representation of the statement '*x* is an element of the set *S* of all planar and right-angled triangles with the hypotenuses *c* and the sides *a* and *b*'. Pythagoras' theorem will thus be reformulated by

$$\bigwedge_{x \in S} P(x) \tag{6.1}$$

This is a very condensed and symbolic representation of the proposition 'for any element *x* of *S* – i.e. for any planar and right-angled triangle *x* – Pythagoras' theorem *P*(*x*) – with respect to *x* – is valid'. We may substitute the theorem *P*(*x*) with respect to *x* in (6.1) by $c^2=a^2+b^2$ – holds for triangle *x* – and reformulate (6.1) by

$$\bigwedge_{x \in S} c^2 = a^2 + b^2 . \tag{6.2}$$

The symbol $\wedge$ is the (*generalizing*) *universal quantifier* and it is the introduction to a generalizing proposition (general law) like '*for any* planar, right-angled triangle $c^2=a^2+b^2$ holds'. The set S – which the universal statement (6.1) has been formulated for – is the *scope* of the quantifier. Deductive reasoning permits to infer any particular case from this general law. Thus, if x^* is any particular planar and right-angled triangle – designated by the asterisk * – we may conclude that $c^{*2}=a^{*2}+b^{*2}$ for the particular values a*, b* and c* corresponding to x^*. Deduction allows to infer a true statement concerning a particular case from a general law. We will reflect this fact by the following diagram with respect to example Ex. 6.1.

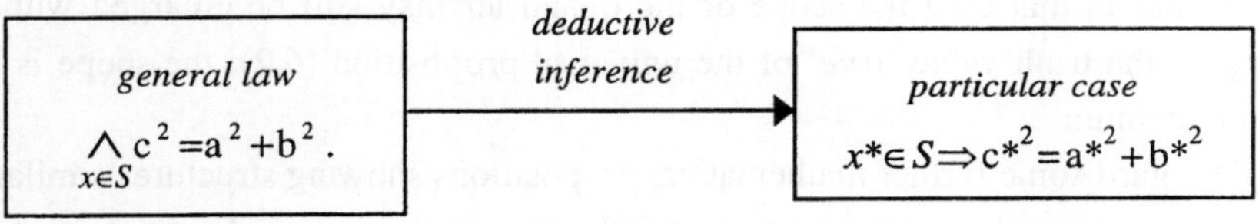

Fig. 6.3. Deductive reasoning using Pythagoras' theorem

In general, the principle of deductive inference can be represented by the diagram of Fig 6.4. According to this representation we may *deduce* from the general law – $P(x)$ is true for every member x of a set S – that $P(x^*)$ is true for any particular element x^* of S:

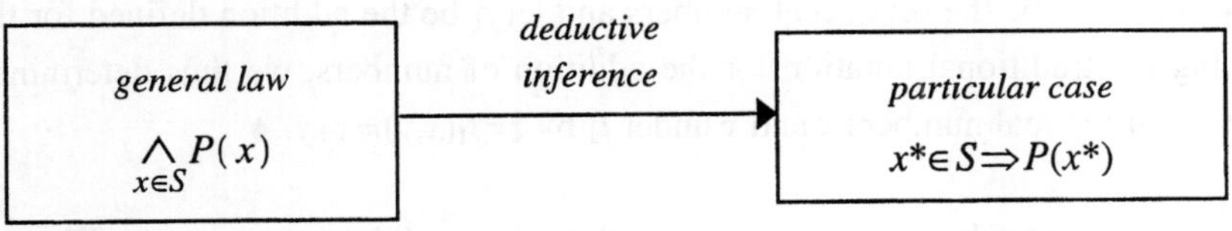

Fig. 6.4. Deductive reasoning, general case

Deductive inference thus is the application of a general law – accepted as being valid, because it has either been proved or accepted as well-founded by experience (like e.g. 'we are *all* mortal') – onto particular cases (like '*this* man is mortal'). According to our definition models are produced by imaging processes. Applying this definition to deductive reasoning we may state that the particular case is a *model* of the underlying general law.

The scope of a universal quantifier will be of great importance when we discuss the validity of models in Chap. 10. In order to elucidate its significance, we study Ex. 6.1 and its representation (6.2) more closely. The theorem of Pythagoras states that the proposition (6.2) is true if the scope *S* of the quantifier is the set of all planar, right-angled triangles. Hence, (6.2) will be false if we define *S* to be the set of all planar triangles, because we realize at once that $c^2 \neq a^2 + b^2$ for any planar, equilateral triangle, in fact $c^2 \neq a^2 + b^2$ holds for any planar and non-right-angled triangle. On the other hand, we may define *S* to be the set of all planar, right-angled and *isosceles* triangles. The generalized statement of (6.2) will be true, of course, but in that case the scope of the quantifier may still be enlarged without changing the truth value '*true*' of the universal proposition (6.2); the scope is not at its maximum.

We regard some further mathematical propositions showing structures similar to (6.1) and (6.2) and the way they exert influence on the scope of a true universal proposition. In order to do so, we introduce a slightly more complicated mathematical notation. Again let *S* be the set of relevant mathematical entities. A mapping f_1 of the ordered pairs (x,y) of elements $x \in S$ and $y \in S$ into *S* itself is a composition of the elements of *S*, and we call it a *binary operator* on *S*. 'The pairs being ordered' means that in general (x,y) is different from (y,x). Mapping these pairs into *S* is tantamount to f_1 assigning a uniquely defined element $z \in S$ to any ordered pair (x,y) of elements *x* and *y* of *S*, and the mapping may be represented by the equation $z = f_1(x,y)$.

Ex. 6.2 Let *S* be the set of real numbers and let f_1 be the addition defined for them. Using the traditional notation for the addition of numbers, we thus determine the image of the real numbers *x* and *y* under f_1 by $z = f_1(x,y) = x + y$. ♦

Let f_2 be another binary operator defined on the set *S* in the same way. These two operators may be connected by the following *distributive* law

$$f_2(z, f_1(x,y)) = f_1(f_2(z,x), f_2(z,y)) \tag{6.3}$$

for any triple *x*, *y* and *z* of elements of *S*.

Ex. 6.3 Let S be the set of real numbers and f_1 the addition + of real numbers like in Ex. 6.2. Let furthermore f_2 be the multiplication · of these numbers. The respective distributive law may be written – using the operators + and · – by the equation

$$z\cdot(x+y)=(z\cdot x)+(z\cdot y)=z\cdot x+z\cdot y. \tag{6.4}$$

This is *the* well-known distributive law connecting the multiplication and the addition of real numbers, and we may formulate it as a general law using the formalism of (6.1):

$$\bigwedge_{x,y,z\in S} z\cdot(x+y)=z\cdot x+z\cdot y, \tag{6.5}$$

where the scope S is the set of all real numbers. There are various ways of defining the real numbers. They may either be introduced hierarchically by consecutively expanding the natural numbers, the integers and the rational numbers; this is the genetic set-up of the real numbers. Or they may be founded entirely on axioms. Whichever method we choose, the distributive law will be *valid*. Within the genetic development it can be proved as a theorem and will thus be valid just like Pythagoras' theorem of Ex. 6.1 is valid as a proven theorem. In the case of an axiomatic introduction of the real numbers the distributive law will be an axiom itself. In that case it is valid *ipso facto* as a defining condition the real numbers have to fulfil. ♦

There are other mathematical structures with binary operators also obeying a distributive law. We want to study one of them:

Ex. 6.4 Let S be a collection of sets (a *set* of sets). Thus $X\in S$ and $Y\in S$ are sets themselves. We define f_1 to be the union $\cup$ of two sets elements of S and f_2 as their intersection $\cap$. For $X\in S$ and $Y\in S$ the respective operators are thus represented by

$$Z_1=f_1(X,Y)=X\cup Y\in S \text{ and } Z_2=f_2(X,Y)=X\cap Y\in S. \tag{6.6}$$

The union $X\cup Y$ of two sets X and Y by definition is the set of all those entities which are elements of X *or* elements of Y. The intersection $X\cap Y$ of X and Y is the set containing all those entities which are elements of X *as well as of* Y. As the conjunction 'or' used in the definition of $X\cup Y$ is not used in the exclusive sense of 'either ... or', the intersection of two sets is by definition a part of their union.

We have illustrated these definitions of unions and intersections of sets in Fig. 6.5.

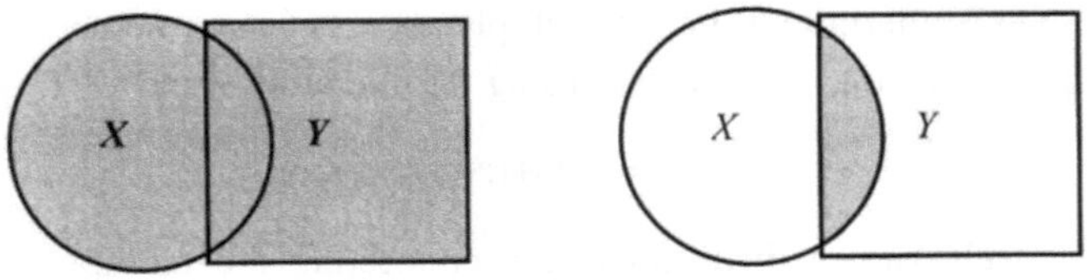

Fig. 6.5. Union and intersection of sets

The union $X \cup Y$ of X and Y on the left of Fig. 6.5 is the region shaded grey and containing all elements of X – represented by the circle – and of Y – represented by the rectangle. If, for instance, X is the set of all the inhabitants of New York and Y is the set of all Catholics living wheresoever, $X \cup Y$ will be the set containing the New Yorkers and in addition all those Catholics not living in NY. We may equally state that $X \cup Y$ is the set of all Catholics living wheresoever in the world plus all the non-Catholic inhabitants of New York.

The intersection $X \cap Y$ of the sets X and Y on the right of Fig. 6.5 is shaded grey, too. And to stick to our example $X \cap Y$ will represent the set of all the Catholic inhabitants of New York (or of all those Catholics living in NY).

The distributive law for unions and intersections corresponding to (6.3) will be

$$Z \cap (X \cup Y) = (Z \cap X) \cup (Z \cap Y). \tag{6.7}$$

This distributive law of equation (6.7) is valid for sets. It is a proven theorem and we may trust to it the way we trust to Pythagoras' theorem. Thus for any three sets X, Y and Z (6.7) holds. If the scope S is defined as a set of sets, we can likewise represent the distributive law – analogously to (6.5) – by

$$\bigwedge_{X,Y,Z \in S} Z \cap (X \cup Y) = (Z \cap X) \cup (Z \cap Y). \tag{6.8}$$

This is to be read 'for all sets X, Y and Z – elements of the scope S of the quantifier used in (6.8) – the proposition of (6.8) is true'.

We can now exchange the two operators $\cap$ and $\cup$ in (6.7) defining f_2 and f_1 of (6.3), and we get another proposition of the – dual – form

$$Z \cup (X \cap Y) = (Z \cup X) \cap (Z \cup Y). \tag{6.9}$$

This statement (6.9) is valid for sets, and it is a *second* distributive law. This means that there is a proof of (6.9). Any triple X, Y and Z of sets taken from the scope S obeys the second distributive law of (6.9), which may symbolically be represented by

$$\bigwedge_{X,Y,Z\in S} Z\cup(X\cap Y)=(Z\cup X)\cap(Z\cup Y). \blacklozenge \qquad (6.10)$$

The interchanging of the operators f_1 and f_2 of (6.3) in Ex. 6.4 has transformed one general law into a second one. We may try to transfer these considerations onto the operators introduced for the real numbers (Ex. 6.3):

Ex. 6.5 If we interchange + and · in (6.4), we gain a corresponding 'distributive law' for the real numbers x, y and z:

$$z+(x\cdot y)=(z+x)\cdot(z+y), \qquad (6.11)$$

which can be written, using the universal quantifier $\bigwedge$ with the scope S of all real numbers:

$$\bigwedge_{x,y,z\in S} z+(x\cdot y)=(z+x)\cdot(z+y). \qquad (6.12)$$

In order to represent a general law, the proposition of (6.12) must be true for each element of the scope of the quantifier $\bigwedge$. We thus study the 'validity' of this equation and start with some simple tests by choosing special values for the variables x, y and z. We immediately realize that for the special value $z=0$ the 'law' represented by equation (6.12) holds for any pair x, y of real numbers. Indeed, $z=0$ leads to equal terms $x\cdot y$ on both sides of the equation (6.12). We conclude that any triple x, y, $z=0$ of real numbers will verify equation (6.11), each of these triples is a *verifying case* for the 'distributive law' described by (6.12) and is thus part of the scope of the universal quantifier of (6.12).

We check (6.12) again using another triple of real numbers. For an easy computation we decide for integers, say $x=2$, $y=3$ and $z=-4$, and we calculate the left-hand side of (6.11) to $-4+(2\cdot3)=2$, the right-hand side yields the same result of the calculation $(-4+2)\cdot(-4+3)=(-2)\cdot(-1)=2$. We conclude that the triple $x=2$, $y=3$ and $z=-4$ of real numbers constitutes another *case verifying* the 'distributive law' of (6.12).

If, on the other hand, we set $x=2$, $y=3$ and $z=4$, the left side of (6.11) will result in $4+(2\cdot 3)=10$, whereas the computation on the right-hand side of this equation leads to $(4+2)\cdot(4+3)=6\cdot 7=42$, a different result. Consequently, the triple $x=2$, $y=3$ and $z=4$ *falsifies* proposition (6.12), it is a *case falsifying* the general law (6.12).

As soon as we detect a single case falsifying a *mathematical* proposition, we are certain that this proposition cannot be proved and will not be a 'law'. We conclude that there is no 'distributive law' like (6.12) derived from the general law (6.4) – valid for the set S of real numbers – by interchanging the operators + and $\cdot$.

But there have been cases – triples of real numbers – verifying equation (6.11). We know for sure that (6.12) holds for any triple x, y and $z=0$ of real numbers, and (6.12) is also verified by $x=2$, $y=3$ and $z=-4$. Perhaps these are the only cases verifying proposition (6.12), but, maybe, we will detect some more. So, we try to determine the maximum scope – within the real numbers – of the quantifier used in (6.12). By elementary arithmetic we transform the proposition used in (6.11) to

$$z\cdot(x+y+z-1)=0, \tag{6.13}$$

and we conclude that this equation is solved by all triples of real numbers with $z=0$, the others x and y chosen arbitrarily or by those x, y and z complying the condition $x+y+z=1$. Our final conclusion will thus be that proposition (6.12) is not generally valid within the set of all real numbers, and the set S of *all* cases verifying it is restricted to $S=\{x,y,z/z=0 \text{ or } x+y+z=1\}$, the set of triples x, y and z defined by $z=0$ or $x+y+z=1$. We may thus state:

$$S=\{x,y,z/z=0 \text{ or } x+y+z=1\} \Rightarrow \bigwedge_{x\in S} z+(x\cdot y)=(z+x)\cdot(z+y) \tag{6.14}$$

as a mathematical theorem. We have proved it and we have determined S as the scope of its quantifier within the real numbers; we call it the *range of validity* of (6.14). This range of validity of a proposition is the set of *all* its verifying cases. Thus, finding cases falsifying a proposition does not necessarily give reason for its overall invalidity; these cases just lie *outside* the range S of validity and we may still be able to determine the more restricted boundaries of S.

We summarize the results we have gained from the examples Ex. 6.2 - Ex. 6.5. The initial proposition is the distributive law $f_2(z,f_1(x,y))=f_1(f_2(z,x),f_2(z,y))$ stated in (6.3). Our analysis leads to the conclusions:

- (6.3) is valid for *any* triple of sets if $f_1 = \cup$ is their union and $f_2 = \cap$ is their intersection or if, *vice versa*, $f_1 = \cap$ and $f_2 = \cup$;
- (6.3) is valid for *any* triple of real numbers if $f_1 = +$ is their addition and $f_2 = \cdot$ is their multiplication;
- (6.3) is valid for *any* triple out of the set $S=\{x,y,z / z=0 \text{ or } x+y+z=1\}$ of real numbers if $f_1 = \cdot$ is their multiplication and $f_2 = +$ is their addition;
- (6.3) is *not* valid if S is the set of *all* real numbers and $f_1 = \cdot$ is their multiplication and $f_2 = +$ is their addition. ♦

These few examples demonstrate the ideas of veri- and falsification of a proposition: the distributive law of (6.3). There may well be other structures (6.3) is valid for.

As soon as a proof of a mathematical proposition has been logically deduced this proof is everlasting. The proposition becomes a theorem, and it is true – at least – for its range S of validity. There may still be further cases outside of S verifying this special theorem, but we can be sure that there are no cases *within S falsifying* the theorem. The everlasting *validity* of a theorem is a characteristic of mathematics.

6.2 Inductive Inference

Outside those theories using deductive inference – like mathematics and certain branches of philosophy – there is no method that can be applied in order to infer particular cases from general laws. A mathematical theory based on axioms and theorems deduced from these axioms is true because of logical reasons, and consequently it is futile to look for falsifying cases for the theory as there are none. Yet, outside mathematics, the direction of inference will frequently be reversed thus giving room to falsifications.

Admittedly, there are further non-falsifiable, non-mathematical theories, but they are of no relevance to modelling. For the sake of completeness we shall describe them roughly, but the general non-mathematical (non-deductive) theory is far more complex than those examples. The cases we are interested in are *empirical* (Greek *empeiria*, *empeiros*: 'experience', 'skilled') and they can be verified as well as falsified. The direction of inference in this case is reverted: that means we proceed by induction (Latin *ducere* and *in*: 'to lead' and 'into') from the particular case to general laws.

Empirical theories are based on the available evidence and on the experience gained by *experimenting*. They are general laws *induced* from 'observation' of relevant particular cases. By 'observation' we mean 'perception by the senses'; hence, empirical laws are induced from experience based on the material world perceptible by the senses. We shall give a few and very general examples in order to explain this idea of induction.

Ex. 6.6 Europeans watching nature had the impression that all those birds called *swan* looked white. Whenever a European child was told, 'this is a swan', it saw a white bird. So from experience and by *inducing* from all those particular white swans the Europeans concluded 'all swans are white'. ♦

In this case the corresponding diagram representing inductive inference looks like that of Fig. 6.6:

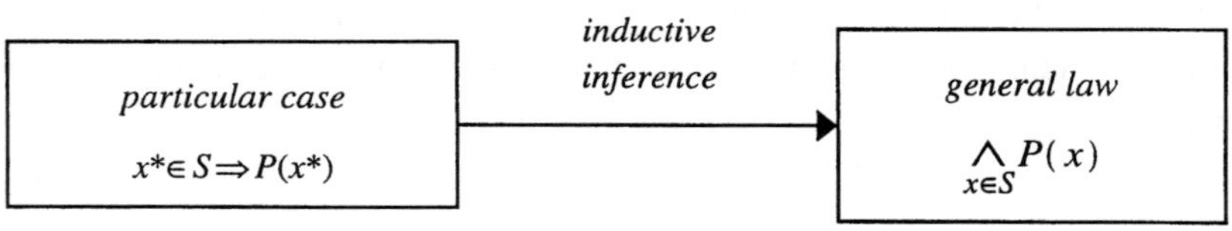

Fig. 6.6. Inductive reasoning

This diagram is *dual*[1] to that of Fig. 6.4 insofar as the direction of inference is reversed. The left side of Fig. 6.6 in our example – at the moment – contains just one special case x^*, but the induction will be carried out whenever a special case x^* (e.g. the observation of a white swan) occurs; so in fact, there may be a multitude of these special cases x^* giving rise to induce the general law of Fig. 6.6.

According to our linguistic rules we would in this case call the general law a *model* of any particular observed case[2]. If, for instance, it is the first time a child sees a big white bird and is told that this is a 'swan', it may say '*this* swan is

---

[1] 'Dual' in the sense mentioned in Subsect. 2.4.2.

[2] As the particular case x^* is the counter-image of this mapping relation, we may again use the duality relation between a model and its *co*-model and conclude that, likewise, any particular case x^* of Fig. 6.6 is a model of the general law on the right-hand side of this diagram.

white'. If the next swans the child observes are white, too, it will probably think that being white is characteristic of a swan. And at the end of its life, having seen hundreds of swans, all of them white without any exception, this now grown-up child will come to the conclusion '*all* swans are white'. When the first white swan was seen by our observer, the colour was just one quality among others. But the more swans he observed, the more significant this particular quality became to him. Having seen nothing but white swans he developed the idea that 'being white' must be characteristic of this species, and he resumed his experience based on a great number of experiments (observations of the bird) and the evidence available to him in the general law 'all swans are white'. In fact, each single white swan observed in the past was a *verifying case* for the *general* law 'every swan is white'. The more verifying cases we observe, the more we trust to this law. But we can never be *sure*, because experience and evidence do not provide rigorous proof of the validity of a general law, which will thus not be as irrefutable as a mathematical theorem.

So, one day the inevitable happened and for the first time somebody saw a black swan, the Australian *Cygnus atratus*. This was a *falsifying case* for the so long accepted general law that every swan was white. And a falsified 'law' is no longer a law. As soon as an empirical theory is falsified it no longer represents a *general* law and seems to be worthless. But, on the other hand, the theory worked pretty well in the past, and after all it was supported by all the observations of swans outside Australia – observed in Europe to be correct. So we may conclude that 'all swans are white' is definitely wrong, but we may correct this false proposition by 'all European swans are white'. And this statement has been supported by each single observation of swans so long. There are only verifying cases for this new empirical theory. According to our experience and the available evidence we think we can trust to the law 'all European swans are white'. And we will trust to this law until somebody observes – say – a blue swan somewhere in Europe. In that case we will have to rework our law so that it covers all verifying cases of white European swans yet detected, but excludes this blue specimen.

The permissibility of inductive inference has been disputed controversially: some argue that induction is not a proper means to gain general laws, others think inducing regularities from single observations is a legal and quite natural way to profit from experience. The problem seems to be the evaluation of the stringency of the conclusions drawn by induction. Assumed, the particular event $P(x^*)$ is the fact that a person wakes up in the morning x^*, in other words he is alive. If there is a consecutive sequence of 20,000 of these events – this is equivalent to the person being some 55 years old – induction might lead to the conclusion that this person wakes up every morning. This is definitely false for time elapsed before his birth, and if we believe that he is not immortal, we are convinced that this statement is also false for the future, because he will not wake up each single morning yet to come. So we have to determine the scope *S* of the induced 'general law' more carefully. If we want this law to describe the fact that the person lives another day and thus lives to awake in the morning of the 20,001st day after his birth, we may define *S* to contain the events of 20,001 waking ups, or – as we know for sure that 20,000 of these events already occurred in the past – just the one event of the person living to see one more day: let us call it S_{1d}. Likewise, S_{7d}, S_{1y}, S_{10y} or S_{40y} are the scopes of induction leading to a survival of the following 7 days, 1, 10 and 40 years respectively. Induction is always permissible, but the credibility of the induced general law depends on its scope *S*: the further *S* reaches into the future, the less we trust to the law. So, there is a *credibility* of an induced 'general law' – it is not absolute as it is supposed to be in the deductive case of Sect. 6.1 – and in addition this credibility is affected by further evidence. The law will gain credit if the person under consideration has not just survived a heart transplantation and is not nursed in an intensive care unit or if he does not live in a country at war. But even if all those circumstances are in favour of the person and if we limit the duration of the scope *S* of the law to, say, a second, the law may become extremely credible, but it may still be false, because the person may die just at that very moment.

This scope of any general law gained by induction is the domain of entities the law is *true* for, but we do not know its boundaries. In our example of a person still alive the scope of the proposition 'he lives at that instant' is made up of all those moments the person is alive, but nobody knows their number. So, in the case of inductive inference we would prefer to talk about the *range of validity* of a proposition, the ensemble of instances the law is valid (well-founded) for, whereas the range of validity of general laws deductive inference is based upon is well-defined and may be determined by proofs. The situation now has become somewhat more complicated. There is still a range *S* of validity, but its boundary is unknown. As an example we take up the statement 'all swans are white'; it was based upon observations before the Australian swan was detected, but even then nobody really knew for sure whether all further observations would be in concordance with this proposition. So there is another domain *V* of all verifying cases detected so far and its boundary is known to us from experience. As *V* is a subset of *S*, the described situation may be represented by Fig. 6.7 where, in general, the boundary *Bd V* of *V* lies completely within *S* and its boundary *Bd S*.

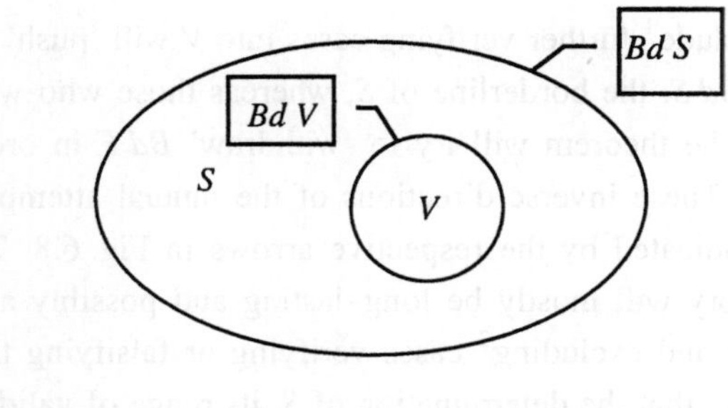

Fig. 6.7. Range S of validity and the domain V of verifying cases

The problem we have to deal with is to determine *Bd S* and thus the range of validity of a given empirical law. We will try to do this by expanding V the set of known verifying cases to its extreme so that it coincides with S. This will be achieved by pushing the boundary *Bd V* of V towards *Bd S*, the boundary of S, until they coincide. Expanding the boundary of V means to experiment and to detect further verifying cases in favour of the law in question.

As S is the set of all cases verifying the empirical theorem under consideration, no matter whether they have already been detected or not, we may likewise try to determine all the cases falsifying the theorem, thus determining *Bd S* starting from the complement of S. If F is the set of known falsifying cases with its boundary *Bd F*, we may illustrate the situation thus arising by Fig. 6.8.

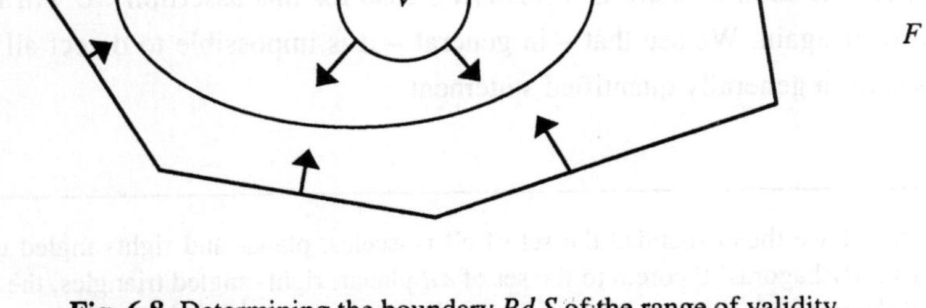

Fig. 6.8. Determining the boundary *Bd S* of the range of validity

Those who try to include[1] further verifying cases into V will 'push' the borderline $Bd\,V$ of V towards $Bd\,S$, the borderline of S, whereas those who want to exclude falsifying cases for the theorem will try to 'withdraw' $Bd\,F$ in order to make it coincide with $Bd\,S$. These inverse directions of the mutual attempts at verifying and falsifying are indicated by the respective arrows in Fig. 6.8. Thus verifying and falsifying a theory will mostly be long-lasting and possibly a never-ending process of including and excluding[2] cases verifying or falsifying the conclusion. This is due to the fact that the determination of S, its range of validity, may be an infinite procedure, infinite in space or time, and we may regard any inductively gained law as being valid as long as it is not falsified. To demonstrate this situation we return to Ex. 6.6 and derive the following generalizing quantification:

Ex. 6.7 'All swans are white'. It will be impossible to verify this proposition *completely*, i.e. to determine every verifying case in favour of this proposition. This is due to the fact that there is a large number of swans that are still alive or lived or will live wheresoever, and not all of them can be inspected, certainly not those who lived in the past, not every single one living today at any part of the globe or will live there in the future. And nobody knows if there are not swans somewhere else in the universe. Therefore we narrow the statement to be validated down and put it like this: 'all terrestrial swans living now are white', and even then we will not be able to inspect all the swans living on the globe at this instant, though there is only a finite and limited number of them.

We already know that this proposition is false because of the black Australian swan. We may expect the case to be settled as soon as we find a falsifying case for the statement under consideration, but we prefer to interpret a detected falsifying case as a mere expansion of the range F of falsifiability of the statement. We will thus adapt our proposition to this situation and reformulate it to 'all European swans living now are white'. We are still not able to verify this proposition completely, and as soon as there is a falsifying case for this assertion we will have to think it over again. We see that – in general – it is impossible to detect all verifying cases for a generally quantified statement.

---

[1] In Sect. 6.1 we thus expanded the set of all isosceles, planar and right-angled triangles verifying Pythagoras' theorem to the set of *all* planar, right-angled triangles, the scope S of that theorem.

[2] This process is somehow related to the 'Principle of Inclusion and Exclusion' used in combinatorial mathematics (e.g. Liu 1968).

On the other hand, we may avoid any uncertainty in this case, when talking about white swans, by using a simple trick. This is just a matter of definition: we may decide to call a white swan a *swan*; hence, our statement 'all *swans* are white' will be definitely true. Likewise, a black swan will be called a *sban*, a *sblan* if it is a blue one, and a green swan will be a *sgan*; we must only use different words to describe different types. ♦

Even if a proposition is spatially extremely restricted, it may be impossible to verify it completely because of its temporal extension:

Ex. 6.8 'All swans living in the *ZOO* of a certain city are white'. We may look up the records of the past and inspect the swans living there at present. Say, there is no non-white swan now and never was one in the past. Nevertheless, the proposition will never be completely verified, because we do not know what may happen in the future. In that case we are content with the *completely* verified proposition 'All swans that lived in the *ZOO* of this city or live there now are white.'. ♦

There are quantified statements we trust to, though we can neither verify them completely nor falsify them:

Ex. 6.9 So, for example, we accept the existential experience made by man that all creatures are mortal as a fundamental, irrefutable law. In the past scientists were convinced that – due to biological considerations of ageing – a human organism – under optimal conditions – possessed a maximum lifespan of some 135 years, which could under no – yet known – conditions be prolonged. Now, the not retarded growth of life-expectancy, observed during the past, gives rise to the assumption that its maximum may be much more than only those 135 years; but in spite of all it is finite.

We also know from *Genesis* 5 that each of the *Patriarchs* lived for some hundreds of years with a maximum of 969 years, the final age of *Methuselah*. Maybe, this part of the *Genesis* is just a *legend*, the *Patriarchs* either did not live at all or they died at an age of less then 200 years. It may also be that the original text of the *Genesis* was misinterpreted: perhaps they did not count in the decimal system in those days and 969 was the equivalent of some decimal 96. Maybe, the ages mentioned in *Genesis* are just *symbolic*, the way we tell our children 'haven't I told you a thousand times ...' when we really told them only twice. So, Methuselah's 969 years, for instance, may stand for 'really old, indeed' and therefore wise and experienced in life. On the other hand, time may have been measured differently in those ancient days: possibly a 'year' was different from our definition; perhaps it only covered a span of some 50 days, or the 'year' was measured the same way we do now, but perhaps the rotation of the earth was ten times faster. Perhaps *Methuselah* and the other *Patriarchs* were a species of their own, not compatible with modern men who will never survive 200 years; it may even be that the genus *Homo* has degenerated considerably ever since *Methuselah* lived, his average age being reduced from some 500 to some 80 years today.

But all of them died, and we accept the statement 'we are all mortal' as irrefutable, though it will never be verified completely, nor falsified. ♦

There are further examples of laws induced from evidence and experience gained by experimenting. Each of these empirical laws possesses a limited range of validity, but – for the time being – we *trust* to them and regard them as sufficiently credible:

Ex. 6.10 We accept empirical physical laws as credible. We trust to the law of the conservation of mass, even though we know that matter is annihilated when it gets into contact with antimatter. Nevertheless, when experimenting we think a conflict between matter and antimatter to be improbable, and we assume that the mass of a rock of 1,000 kg dropped from a height of 1 m will still be 1,000 kg when it hits the ground (if it shatters, the mass of the collected fragments will still equal 1,000 kg).
Furthermore, we trust to the transportation laws of Sect. 1.3 and their special formulations as Ohm's, Fick's and Darcy's laws. We must admit that these laws are not completely verifiable in the sense mentioned above, and we are sure that the *strict* mathematical equation (1.1) will never hold, but it is a good hypothesis to work with.

We also talk about *natural constants*, and we introduce them into physical calculations as being constant, and this is also a good working hypothesis. But there seems to be evidence that some of the natural constants vary in the course of time or depend on the local *co*-ordinates of the observer within the universe. So, Webb et al. (2001) gained evidence that Sommerfeld's Fine Structure Constant has obviously been subject to an evolution (cf. Sect. 3.3, too), thus being far from being *constant*. As the value of this parameter is entirely determined by other physical constants some of them cannot be regarded as being constant, i.e. having a *constant* value independent of the temporal – or spatial – situation of the observer. But as far as our daily work depending on physical constants is concerned, we may think that it does not really matter whether these 'constants' are *absolutely* constant or not, we may content ourselves with some average values regarded as being representative and sufficiently invariant for any application of physical theories. ♦

Obviously, we are willing to accept empirical theories as 'true', though we are well aware of the fact that we will never be able to verify them *completely*. We agree that 'we are all mortal', that Ohm's law is 'true', and we are convinced that 'all European swans are white' – we even, by definition, accept the statement 'all *swans* are white' as being 'really true'. We accept empirical laws as 'true', even when there is evidence of their falsity. Thus, we think physical calculations using physical 'constants' can be relied on, even if these constants vary in time and space. Even though possibly falsified we still accept the assumption that the natural constants are sufficiently 'constant'. If we find out that the charge of the electron is time- and space-dependent, we will still consider it as a 'constant' unless the field of research is just this relation between charge, time and space. Within the temporal and spatial sphere we live, in the variations of the 'constant' will be regarded as not perceivable and thus negligible. And whenever one of these laws is falsified, we will think that it is just a phenomenon we must try to explain. If, one day, we detect a person being some 5,000 years old, living somewhere on the globe, scientists have to give an explanation. This case seems to falsify the theory 'we are all mortal'. But the person may die within the following 1,000 years, and we may still agree that everyone living under conditions different from those this man lived in is mortal and will die in the course of some 135 years. If an electric conductor is discovered disobeying Ohm's law under normal conditions, this will also be a revolution. We shall try to explain it, and we will gather insight into this new physical phenomenon. Nevertheless, Ohm's law was a good tool in the past and, in general, we will not question those calculations carried out so far and being based on this law.

So induction, though not permissible for logical reasons, is a tool in order to evaluate experience and the evidence available to us and to profit from them. Everybody creates his own model of the world he lives in and he will constantly check this image of reality by looking for cases verifying or falsifying it. Cases verifying an empirical theory will confirm it, whereas falsifying cases will cause a restriction of the range of validity of this personal model, they will cast doubt on it. But anyhow, it seems to be quite natural that we mirror reality and strive to adapt the image to reality by experimenting and inducing laws supported by reality.

We now want to introduce another logical quantifier bearing some relevance to the field of verifying and falsifying empirical theories. It is in a way *dual* to the universal generalizing quantifier, it is the (*particularizing*) *existential quantifier*, and its symbolic representation is

$$\bigvee_{x \in S} P(x). \tag{6.15}$$

This string of symbols is an abridged version of the proposition 'among the elements of the set S – the *scope* of the particularizing quantifier – there exists at least one for which the proposition P is true'. Thus $\bigvee$ stands for 'there is' or 'there are'. We shall give an example of the usage of this quantifier to make our point clear and for that we return to our favourite proposition about swans.

Ex. 6.11 Let S be the set of all swans living in the *ZOO* of London. Let furthermore $P(x)$ be the proposition 'x is blue'. With these interpretations of S and P (6.15) reads 'there is a blue swan (at least one) living in the *ZOO* of London'. As there is no limit to the observation time we may state even more precisely, 'there is a blue swan (at least one) in the *ZOO* of London; it lived there in the past, lives there now or will live there in the future'.

We may easily check the first two parts of this proposition. We inspect the swans of the *ZOO* living there, and we look up the files recording the past. Of course there never was and never will be a genuinely blue swan in the *ZOO* if we disregard those swans dyed blue for show-effects. But anyhow, we cannot be positively sure that there will not be blue swans kept in this *ZOO* in the future. So this proposition may easily be verified as soon as the first blue swan moves into the *ZOO* of London. But we shall never be able to falsify it, because there may always be a future point of observation verifying it. ♦

Within this example the only possibility of falsifying the statement 'there is a blue swan (at least one) living in the *ZOO* of London, or was in the past or will be some time in the future' is to limit the future. Say, Dec. 31st, 3050 the *ZOO* of London will be closed. If we have not discovered a blue swan there up to that day, we may conclude that (6.15) is false. But, generally speaking, propositions using a particularizing quantifier ranging over an infinite scope S cannot be falsified even if they are false[1]; they can only be verified. This situation is *dual* to that of the gen-

[1] We owe this – evident – statement to Popper (1964).

eralizing quantifier which we may falsify, but probably never verify completely, like the proposition 'we are all mortal'. We follow Popper (1964) and call propositions using the existential quantifier *metaphysical statements*. Some well-known examples of metaphysical and because of that not falsifiable propositions are 'there is a God', 'there is a life after death', 'there is intelligent life somewhere in the universe'.

Though certain propositions may never be verified – like 'there is intelligent life somewhere in the universe' – and not be falsified either, we, nevertheless, may trust to them. We also trust to the statement 'some time in the future there will be no human beings on the earth any more'. This statement cannot be falsified, and it will only be verified the moment when the last human being dies; nevertheless, we already believe in it now.

The non-falsifiability of these *empirical* statements should not be muddled up with the propositions of Sect. 6.1. Those, if proved by logical reasoning, are also non-falsifiable, but, as we already stated in Sect. 6.1, a proof of those theorems is final and *everlasting*, and corresponding propositions are *true* because of logical reasons, whereas the mentioned empirical statements are not falsifiable by definition. Metaphysical statements may even be false, but not falsifiable, like 'there are blue swans'. Let us assume there are *really* no blue swans, nowhere in the universe, there never were and will never be. This proposition will never be falsified, and every swan we see and whatever colour it is will support this proposition, as it does not *falsify* it.

Consequently, when modelling we are mainly interested in those propositions which are empirical on the one hand, and not metaphysical on the other. We are neither interested in those statements true because of logical reasons – they are called *tautologies* – nor are we interested in metaphysical statements as neither of these propositions is falsifiable. What we are really interested in are those statements which can be falsified. But as we have pointed out in this section, falsifying an empirical non-metaphysical theory is just – like verifying it – a means to limit the range of its validity. Consequently, by successively discovering verifying and falsifying cases of the empirical theory in question, we determine an area A containing the *real* boundary $Bd\,S$ of the set S of validity of the theory. This area causes some uncertainty about the way the real boundary $Bd\,S$ runs (cf. Fig. 6.9). We thus have to agree upon a certain *range of validity* of a scientific empirical theory, upon degrees of its dependability. This agreement is stochastic; this means that it is only yielding values of probability for the exact boundary line of S. On the other hand, if scientific theories are concerned, then they must – though limited only by stochastic considerations – be accepted by the relevant scientific community. The insight into the cases verifying or falsifying a theory under consideration is shared by various persons, we therefore call their acceptance *intersubjective* (cf. Subsect. 5.1.1); its quality depends on the available evidence.

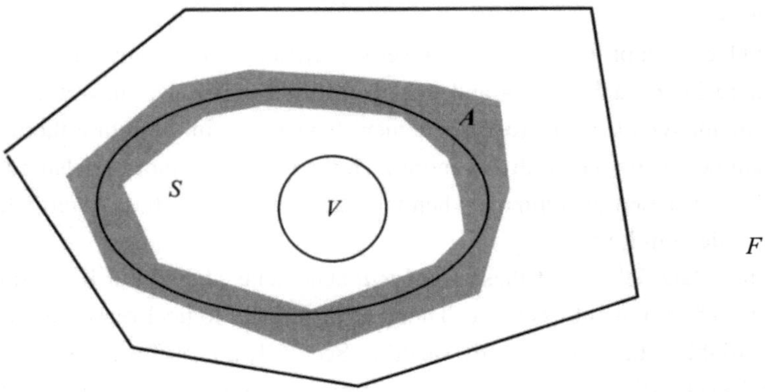

Fig. 6.9 Undeterminedness of the boundary of the range *S* of validity

'General laws' derived by inductive inference may be the basis on which deductions are based themselves. If, for instance, we are convinced that all European swans are white, and if we are pretty sure that every human being is mortal, we may *deduce* from the observation 'a man saw a European swan' that the swan was white and that its observer is mortal. We may even by definition make a more watertight proposition if we modify it to 'a man saw a *swan*' whereby a *swan* is by definition white. Now, we may still doubt whether the observer of the swan is mortal, but we know for sure that the observed bird was white by definition. In that case the proposition 'a *swan* is white' is a tautology (Greek *tauto* and *logica*: 'the same' and 'word'); we may, likewise, define human beings to be mortal and immortality to be superhuman and thus deduce for sure that the (hu-) man who saw a *swan* is mortal and that he saw a white bird.

The number of the members of the community evaluating a scientific theory is of importance. The more agree on its acceptability, the better the chances of a theory to be generally accepted. Nevertheless, sometimes we rely on our personal experience when we judge a situation. Having started this chapter with the everlasting proofs of deductive-logical theorems and after having studied the empirical intersubjective propositions, we will now turn to purely subjective and personal inference.

But the conclusions derived from induction may be misleading. If, for instance, you get muddled when using a motorway and enter it using an exit, you will be a ghost-driver driving in the wrong direction. Let us assume no other car is approaching you for a while and you will certainly not be overtaken. So after two or three kilometres you may induce you are on a safe journey whereas it may be your last one.

6.3 Personal Inference

In the first section of this chapter we have discussed the irrefutable and reliable deduction of conclusions from general laws; in the second section we have shown how these 'general laws' can be induced from experience and – if necessary – with the help of suitable definitions. But there are also inferences not resulting in generally accepted laws at all.

Sometimes decisions have to be made without any or only little available evidence or experience to be founded on. Consequently, these decisions – made under uncertainty – are entirely subjective as they are based on the personal evaluation of the prevailing situation. One might think that everybody would avoid making decisions if there is no factual evidence to back them up. But there may be circumstances which demand an immediate decision without giving you time to look for credible evidence. We want to illustrate these ideas with an example:

Ex. 6.12 Imagine the following situation: you are imprisoned in the chamber *C* of the ground plan of Fig. 6.10, and there are only two possible ways to leave it: the doors *A* and *B*, and there is no visible difference between them. As soon as you open one of these doors, the other one will be locked for ever. You have been told that only one of the doors will lead into freedom, but not which one. There is absolutely no clue, but you must make up your mind and open one of the doors if you do not want to die from thirst or hunger. This elemental fear of death is stronger than any reflection and will finally compel you to act at all costs and thus refute the commonly held opinion that a decision without any logical evidence is not feasible.

In order to stimulate your decisiveness, we make the situation more threatening. We are indebted to Poe (1952) for this idea. So we assume that the wall opposite the doors is movable, and it is steadily moving towards you thus making the area of *C* smaller. This additional element is represented by the arrow in Fig. 6.10 indicating the direction the wall will move.

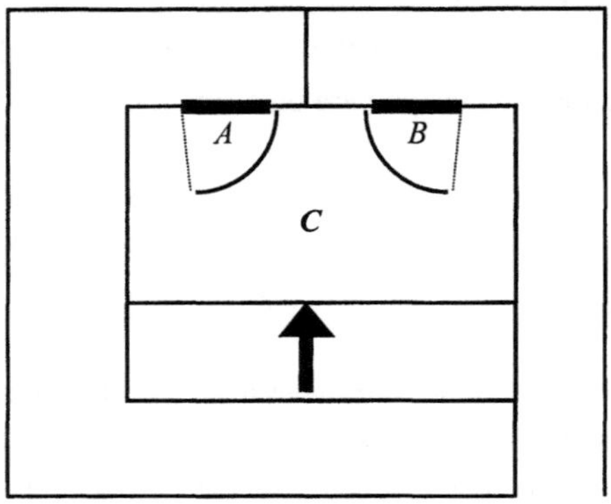

Fig. 6.10. Groundplan of the scenario of Ex. 6.12

As your situation is getting more desperate any minute, you must come to a quick decision. This step may be influenced by very personal parameters, such as you always prefer the right-hand choice to the left-hand or something similar. Or you may decide to support your choice of the door to be opened by the counting-out rhyme '*eeny meeny miney mo*'. Then, of course, there will still be the problem where to start with counting – the door you will finally open is determined by this decision. You may come to the conclusion that there is an equal chance of survival, no matter which door you open. You may even try to support your decision by transferring the responsibility onto an additional experiment. You may thus – as there are equal chances of deciding for the door leading into freedom – toss a coin for it. In this case you depend on the outcome of an experiment carried out prior to the decision you have to make. This is a *relaying* experiment (Richter 1956). We will return to the idea of relaying experiments in Subsect. 7.2.3 when we discuss intersubjective degrees of rational belief and personal probabilities. ♦

As we have seen in the preceding example, we may support the process of coming to a decision by auxiliary considerations like relaying experiments. This may be the case when there is no evidence in favour of one single decision. The process of deciding will get more complicated as soon as there is some evidence, because it may well be that we cannot tell whether this further information – so helpful at first glance – is really in favour of the one or the other choice. In order to demonstrate this scenario, we make our example Ex. 6.12 even more complicated:

Ex. 6.13 The plot is the same as in Ex. 6.12, but this time the two doors *A* and *B* are marked. One of them bears the label 'Emergency Exit', the other one 'No Exit'. At first glance it seems to be clear how to decide. But you may be suspicious and think that those who brought you into this desperate situation certainly have no reasons to be helpful. So, maybe, they labelled those doors with the ulterior motive to confuse you and drive you to the wrong decision: maybe, the 'No Exit'-door is the one you should choose.

But perhaps – you suspect – those who are after your blood and put you into that room ***C*** already anticipated your doubts and so put the correct label 'Emergency Exit' onto the door *B* leading into freedom. In the end you feel lost like in a maze; and the situation is well-mirrored in the sentence 'If he thinks that I think that he thinks that ...', well-known to gamblers. So, after having lost all your decisiveness because of endlessly tossing and turning the pros and cons, you shift the responsibility onto that ominous coin or you do like Alexander the Great with the famous Gordian knot – but risk your life. ♦

As we have seen, evidence available to us when we have to decide may be of little help. Or we are not able to interpret it correctly in order to benefit from it. A slight modification of Ex. 6.13 will illustrate this:

Ex. 6.14 The two doors *A* and *B* now carry different marks. Say, one is coloured red and the other one is blue, or *A* is marked with the left one of the pictograms of Fig. 1.9 (the arrow pointing to the left), whereas *B* is marked with the pictogram on the right-hand side of this figure (representing an escalator). In that case it will probably be even more difficult to come to a decision. We may be tempted to interpret the colour 'blue' as more promising than 'red' or an 'escalator' as being preferable to an 'arrow pointing to the left'. ♦

These are only few examples, but they show clearly that when we have to come to a decision relying on our personal insight there may not be much of an evidence in favour of a particular alternative and it may sometimes just be a *game of chance*. This somehow unsatisfactory and frustrating situation does not mean that a lack of evidence will release us from the necessity of making decisions, but we must be aware of the fact that our decision may not be optimum. On the other hand, if there is no convincing evidence at hand, *nobody should be blamed* if a decision agreed upon under those vague circumstances finally turns out to be the wrong one.

Ergo, general laws cannot be derived from personal inferences as they are rather induced from the intersubjective state of insight into a field of research. But as personal inference may be repeated several times or may be carried out by different persons, it is inevitable that this subjective inference is diverse and finally gets intersubjective. In that case it may serve as a source of induction of general laws from which we may again deduce statements concerning special cases.

Just like inductive inference, introduced in Sect. 6.2, personal inference depends on *degrees of belief*, that e.g. a personal decision is justified or that the induction of a general law is well-founded on evidence. These degrees of belief are measured in terms of probabilities giving room to ambiguities, because they only represent our conviction that 'we are right' or, to put it more precisely, rather that 'we are not completely wrong'. We must admit that the only way of reasoning leading to a fully trustworthy result is logical deduction as it is used in mathematics or some branches of philosophy, for example.

7 Probabilities

The dependability of a model is determined by its validity, the way it is well-founded. The quality of the foundations of a model becomes obvious in certain *degrees* depending on the evidence available to the modeller. There may be observations in favour of a model, thus supporting and confirming it, and there may be observations not in concordance with a model. The latter will certainly reduce the belief in the dependability of a model and cast doubt on it; in the worst case its futility will be proved and it will be refuted. This information will help us to induce a *belief* in the dependability of a model, which can be measured in degrees. This does not mean that we trust to a model blindly, but that our evaluation is founded on evidence and induced from experience gained from experiments. It is thus a *degree of rational belief* and may be quantified in terms of probability.

Probabilities govern inductive and personal reasoning. Although this is not a textbook on probability theory, we have to explain some of its basic ideas. Furthermore, we want to show that some approaches to the concepts of probability seem to be more appropriate than others when we have to deal with model theory.

7.1 Deductive Stochastic Inference

Statements based on probability considerations are stochastic (Greek *stochazesthai*: 'to guess', 'to suspect'). Some stochastic propositions may be gained from deductive inference. According to our definition in Sect. 6.1 deduction leads from a general law to a proposition covering individual cases. This inference may cover general laws of probability theory and individual *stochastic* cases as well. We want to illustrate these facts with the following example.

Ex. 7.1 Let U be an urn containing ten balls; four of them are white, the rest of six are coloured black. We have been informed about this ratio between white and black balls and we start to experiment by taking balls from the urn *with replacement*. This means we take a ball at random (without looking into the urn), note its colour and put it back into the urn before we take another one.

Let W be the event of taking a white ball and B be the event of taking a black ball. If we repeat this experiment many times, then we expect W to occur in about four out of ten of these experiments and B in about six out of ten. This is of no consequences to the first takings: the first ball taken out is either white or black, and the second taking results either in a total of two balls of the same colour or in one black and one white ball. But as we proceed to a hundred and more experiments we expect *about* 40 percent of all taken balls to be white. With the increasing number N of takings the absolute frequency H of getting a white ball should come closer and closer to the value $0.4 \cdot N$. If, for example, $N = 100{,}000{,}000$, we then expect H to be 'rather close' to the value $H^* = 40{,}000{,}000 = 0.4 \cdot N$, which means that we do not expect H to differ from H^* for more than some, say, 100,000; but we certainly cannot demand that the value of H is exactly $H = 40{,}000{,}000$.

As we have stated above, expecting 40% of the balls to be white in a series of many experiments is without any consequences to the first takings out; nevertheless, we can extrapolate the formula we used to calculate H^* for the value $N = 1$ being representative of a single taking. This calculation leads to the value $0.4 \cdot 1 = 0.4$ which, of course, does not represent the number H of white balls taken in one experiment; in that case the value of H will either be $H = 1$ or $H = 0$. We call this value extrapolated from the general considerations to cover the case of $N = 1$ (a single experiment) the *probability* p of the (occurrence of) the event W, and we write $p(W) = 0.4$. ♦

In the previous example we have *deduced* the probability of the occurrence of a particular event (W) in a single experiment. This deduction was based on general assumptions, on our knowledge of the contents of the urn U and the evidence available to us, that is the taking of a ball being performed at random and with replacement. As we knew that 40% of the balls in U were white, we deduced from this fact that some 40% of the balls taken in a long series of N experiments carried out with replacement would be white, too. With a large number N of takings, we conclude, the number of occurrences of event W will come close to the value $p \cdot N$ with the *probability* $p = p(W)$ of the occurrence of W in one taking. This probability p is by definition a real number $0 \leq p \leq 1$ – with values between 0 and 1.

When we expect a large number N of experiments to result in a number H of observations of the event W coming close to $H^*=0.4\cdot N$, we are willing to accept deviations between the actual absolute number H of these occurrences of W and the hypothetical value H^*. In Ex. 7.1 we proposed an acceptable maximum difference of some 100,000 if $N=100{,}000{,}000$. In fact, we have to accept any difference between H and H^*, but if this difference – let us call it ΔH – is, say, $\Delta H=1{,}000{,}000$, it will be surprising, and even more surprising if it is $\Delta H=50{,}000{,}000$. We will think that there is not much of a chance for $H=0$ or $H=100{,}000{,}000$, but we must admit that both results are not really *impossible*; it will just be strange if we never see a white ball among 100,000,000 balls taken out of U or if they are all white. Though neither $H=0$ nor $H=100{,}000{,}000$ for getting a white ball is in concordance with our expectations, they are not inconsistent with our deductions $p(W)=0.4$ and $p(B)=0.6$. We now want to modify Ex. 7.1 slightly:

Ex. 7.2 We still assume the same ratio between white and black balls in the urn U as in Ex. 7.1, but this time *another* person takes 100,000,000 balls with replacement and we do not know the result of each single experiment. We are just told that for an event X representing either $X=W$ or $X=B$ the experimenter has observed $H=0$ or $H=100{,}000{,}000$, and we are asked to guess which event is represented by the unknown X; this means we are told that all the 100,000,000 taken balls are of the same colour and we have to guess which colour that is. We feel impressed by this remarkable outcome of the experiment resulting in $H=0$ or $H=100{,}000{,}000$. We may guess that $H=0$ is more like $X=W$, because there are less white than black balls in the urn, whereas $H=100{,}000{,}000$ looks more like $X=B$, but it will be very difficult to come to a decision between $X=W$ and its alternative $X=B$. ♦

In the case of Ex. 7.1 we *deduced* the value of a stochastic parameter p from the known characteristics of the set-up of the experiment: the knowledge of the contents of the urn and that the takings were performed with replacement. Likewise, we deduced a predominance in Ex. 7.2 being in favour of either the event $X=W$ or of its alternative $X=B$ depending on the observations resulting from the experiments.

On the other hand, there may be the problem to find out about the contents of an urn U if we have only been told that it contains balls, but do not know their colour or the ratio between the number of balls coloured differently. In that case we have to perform the experiments of taking balls from U and to study the results. When interpreting the outcomes of these experiments, we may *inductively* infer stochastic propositions concerning the contents of the urn. There are two different ways of stochastic inference. One of them is founded on frequency considerations, the other one allows stochastic inferences from single observations or from a series of a few experiments only. We shall explain both methods in the following sections. We shall start with the better known frequency aspect of probability, though we think that we can often base the evaluation of the quality of a model only on a small amount of circumstantial evidence.

7.2 Inductive Stochastic Inference

Let, for instance, U again be an urn, once more containing a certain number of balls. This time we assume that neither the number of balls in the urn nor their colours is known to the observer. In order to get information about the contents of the urn, we again start taking balls from it. We will thus gain evidence enabling us to come to certain conclusions. These experiments may either be carried out with replacement – the way we experimented in Ex. 7.1 – or without (in that case the balls will be removed from the urn and not replaced into it). If we take the balls without replacing them into the urn, it is merely a question of time until we know for sure what the contents of the urn is like: we just have to repeat the experiment until the urn is empty in order to know for certain how many balls were in it right at the beginning and of what colour they are. If we replace the balls, we will never get any information about their number; but even in that case the experiment of taking out balls will give us some information about the composition of the contents of the urn.

In general, it will be impossible to find out the number of balls in U unless we empty it completely, and our stochastic inference will be restricted to the various colours of the balls and to the ratio between the number of balls having the same colour. Let X_i be an event characterized by the colour of a ball taken from the urn U, like, for instance, $X_1=W$ – a white ball is taken from U – or $X_2=B$ – the ball is black – and so on if U contains balls of other colours as well. There will be different ways of induction leading to general statements concerning the contents of the urn U.

7.2.1 Frequency Considerations

We assume, the experiment of taking balls from the urn and replacing them is carried out N times; here N is a natural number (one of the numbers 1, 2, 3, ...). Let

$H_i=H(X_i)$ be the absolute frequency of the occurrences of the event X_i in this series of experiments, i.e. the number of times the occurrence of event X_i – the taken ball showing the ith colour – is observed in the course of these N experiments. Obviously the value of H_i lies within the range limited by the integer numbers 0 and N, i.e. $0 \le H_i \le N$, and the numbers of all the observed H_i add up to N, the number of performed experiments of taking a ball out of U.

We now define the relative frequency $h_i=h(X_i)$ of the occurrences of event X_i in the series of N experiments as the quotient of H_i divided by N:

$$h_i = \frac{H_i}{N}. \tag{7.1}$$

Obviously h_i lies within the range $0 \le h_i \le 1$, and the sum of all derived relative frequencies h_i is 1. We may now try to infer the composition of the contents of the urn from the various values of $h(X_i)$ for each event X_i observed while experimenting.

Ex. 7.3 Let us assume, we carry out $N=100{,}000{,}000$ of these experiments of taking a ball from that urn U and replacing it afterwards (there is no information about the number of balls in U or about their colours). Say $X_1=W$, $X_2=B$ and so on, this means we may define further events X_3, ... representing other observed colours if necessary.

Let, for instance, the result gained in this series of 100,000,000 experiments be $H(X_1)=39{,}998{,}674$ and $H(X_2)=60{,}001{,}326$, $H(X_i)=0$ for any other value of i. This means that within 100,000,000 experiments 39,998,674 times a white ball was taken and 60,001,326 times a black ball, and no other colour was observed in these experiments. We compute the respective relative frequencies for these events to $h(X_1)=0.39998674$, $h(X_2)=0.60001326$ and $h(X_i)=0$ for $i>2$.

The following conclusions can be drawn from that result: first, we infer that there must be at least two balls in the urn, a white one and a black one. We may furthermore guess that some 40% of the balls in the urn are white whereas the rest of some 60% are black. We may even come to the conclusion that it is rather improbable that there are any other balls in the urn U but white and black ones.

Let b be the number of balls in U, and let us now assume we are informed that $b=2$, so there are two balls in the urn. In that case we infer from the results of our experiments that U contains exactly one white and one black ball and nothing else – and we are *absolutely* sure about this. But if, for example, we learn that $b=3$, we will be 'pretty sure' that there are one white and two black balls in the urn and nothing else. We are 'pretty sure' in this case, but we are not perfectly sure: maybe, there is one black and one white ball and the third one is of a different colour. Yet, it seems rather improbable that this third ball – neither white nor black – should not have occurred in 100,000,000 experiments; but life is strange.

With $b=10$ we will rather vote for four white and six black balls than for any other combination of colours. If we know that there are $b=100$ balls in the urn, we may think that it will be a good guess to assume that 40 of them are white and the rest of 60 are black. But when we learn that $b=100{,}000{,}000{,}000{,}000{,}000$, the situation becomes rather unpredictable. We still assume that the percentage of the white balls in the urn is *some* 40% and that *some* 60% of the balls are black, but this 'some' still leaves room for balls of other colours. There may be a few red, blue, green balls and balls of any other colour among those 100,000,000,000,000,000 balls in the urn, but they just happened not to be detected in the course of 'only' 100,000,000 taking outs.

So by inducing from the outcome of our experimenting we guess that about 40% of the balls in the urn are white and that ca. 60% of them are black and that for any other colour X_i the ratio is $\varepsilon_i \approx 0$ – insignificantly low – and can therefore be ignored. From these induced values we may now *deduce* probabilities for further experiments. We set $p(X_i) \approx h(X_i)$. This leads to $p(X_1)=0.4$, $p(X_2)=0.6$ and $p(X_i)=0$ for $i>2$. $p(X_1)=0.4$ means that we assume the probability of getting a white ball is 0.4 and that within a great number of subsequent experiments about 40% of the taken balls will be white, that the remaining 60% of those balls will be black and that we do not expect any other colour. ♦

We defined the probability $p(X_i)$ of an event X_i to be the relative frequency of its occurrence in a *large* number of experiments. $p(X_1)=0.4$ means that we are convinced that ca. 40% of the balls in the urn are white and that with the number N of further takings out approaching 'infinity' – this will be represented by $N \rightarrow \infty$[1] – the number of taken white balls will come close to $0.4 \cdot N$.

This 'coming close to' must not be equated with the convergence as defined in calculus. If exactly 40% of the balls in the urn are white, the number of white balls detected in N experiments may differ considerably from $0.4 \cdot N$; it is even possible

---

[1] Actually there is nothing like ∞ 'infinity', because we can only repeat the experiment a finite number of times. But we think that a large number N of experiments, say, $N>10{,}000$ comes sufficiently close to an 'infinite' number of repetitions.

that no white ball is taken at all in, say, 1,000,000 experiments, but this is very *improbable*.

That is the weak point in our considerations: we define the term *probability* by *frequency limits*. This means we think it is very *probable* that the difference between the relative frequency $h(X_i)$ of the occurrence of the event X_i and the probability $p(X_i)$ of X_i will approach zero with a growing number N of takings out. The definition of probabilities by frequency limits is *circular* and based on probabilities itself. In order to avoid this kind of circular definition, we will now introduce probabilities axiomatically.

In our example with the urn two different events X_i and X_j, $i \neq j$, are (mutually) *exclusive*. This means that two of these different events X_i and X_j cannot occur at the same time as each ball is of one definite colour. In other words, the event $X_i \cdot X_j$ – abbreviated by X_iX_j – X_i *and* X_j, getting a ball with both colours determining X_i and X_j is *impossible*, no ball in the urn will, for instance, be coloured black and white, and we conclude that $p(X_iX_j)=0$ in our example whenever $i \neq j$.

If n is the number of available colours, this means that there are only n different ways to colour a ball. Furthermore, we assume that the balls in the urn U are all marked *one* colour, this means they may be black or white or any other of the n colours mentioned, but not, e.g. striped. We do not know what colours are represented within U; we are only sure that whenever we take a ball it will show exactly one of those n colours. In other words, we are sure the event $X_1+X_2+ \ldots X_n$ – the ball is of the colour number one *or* it is of the colour number two ... or of the colour number n – will occur whenever we take a ball: $X_1+X_2+ \ldots X_n$ is a *certain* event, and its probability will consequently be the maximum value $p(X_1+X_2+ \ldots X_n)=1$ a probability can have. As any pair X_i and X_j – $i \neq j$ – taken from the generic set $X_1, \ldots, X_n$ is mutually exclusive, the whole set $X_1, \ldots, X_n$ of events is called exclusive itself.

Notations frequently used are $\varnothing$ when $p(\varnothing)=0$ – $\varnothing$ is an *impossible* event –, and Ω when $p(\Omega)=1$ – Ω is a *certain* event, it defines the *sample space*, the *universe* of the experiments[1] carried out.

The probability p of an event in our example obviously fulfils the following *axioms*:

– with every event X a real number $p(X)$ is associated with $0 \leq p(X) \leq 1$,

---

[1] There may be infinite universes with the number n of exclusive events tending to infinity ($n \rightarrow \infty$).

– if X_1, X_2, ... are exclusive then $p(X_1+X_2+ ...)=p(X_1)+p(X_2)+ ...$ [1] and
– $p(X_1+X_2+ ...)=1$.

Defining probabilities by these axioms helps to avoid the circular definition we started with. Using the terminology of modelling, we might say that the theory founded on these axioms is a *model* of our intuitive understanding of probability and chance. If we induce probabilities from frequencies, that approach is tainted with a further disadvantage. In order to determine relative frequencies $h(X)$, we must experiment. These experiments must be repeatable, they must even be repeatable for an infinite number of times, as the probability of an event is defined by *limits* of relative frequencies for the number N of these experiments tending to *infinity*.

Actually, we cannot be expected to perform an experiment endless times; but we must be certain that for any given natural number N – however large it may be – *we are able* to perform the experiment N times. Yet, sometimes we must even associate probabilities direct with events when a support by experimenting is not possible. We try, for instance, to calculate the risk of failures of expensive or dangerous enterprises, the risk of a severe accident in a nuclear power plant, of a breach in a dyke or of an explosion on board of a space shuttle. We will certainly try to induce the probabilities of the occurrence of those catastrophes without causing them in an experiment. And if we gain information from experiments, that is from catastrophes having happened in the past, nobody will demand their repetition – not to speak of endless repetitions. We thus have to develop an approach to probability allowing us to induce probabilities without experimenting at all or from only a limited number of experiments. In order to do so, we have to introduce some further definitions; we want to illustrate the underlying ideas by some minor changes in Ex. 7.1.

Ex. 7.4 The ten balls in urn U are of different brands now. We assume that five of the balls – a white and four black ones – are manufactured by firm Y_1, the other five by Y_2. As this seems to be somewhat confusing we represent the situation by the diagram of the following Fig. 7.1:

---

[1] The '+'-operator on the left side of this equation represents the 'or'-connection between the events X_1, X_2 ,... and so on, whereas '+' on the right stands for the addition of the real numbers $p(X_1)$, $p(X_2)$, ... ; we also assume that the sums on both sides of the equation converge, i.e. that there are, for example, only a finite number of summands in them.

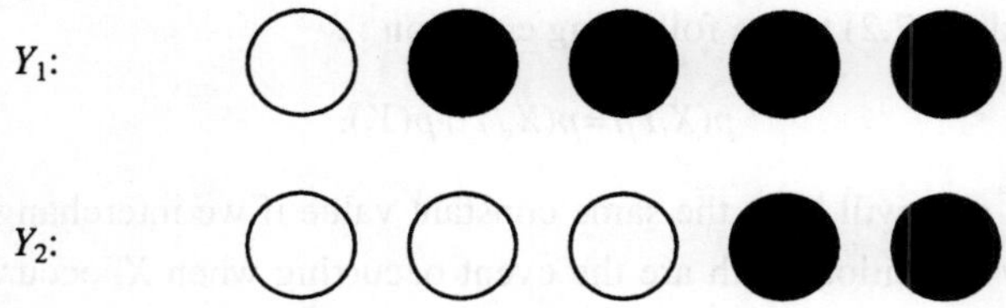

Fig. 7.1. Contents of urn U in Ex. 7.4

The rows of this diagram are assigned to the manufacturers of brands Y_1 and Y_2, the balls are represented by circles of the respective colour. ♦

The events Y_1 and Y_2 are mutually exclusive with $p(Y_1Y_2)=0$, but the X_i and Y_j are not exclusive as $p(X_1Y_1)=0.1$, $p(X_1Y_2)=0.3$, $p(X_2Y_1)=0.4$ and $p(X_2Y_2)=0.2$.

We introduce the brand of the balls as a *conditioning* event. As an example we start to take balls with replacement again, but we only take notice of balls of, say, brand Y_1. There are two ways to fulfil this condition. We may either remove all the balls produced by Y_2 from the urn, U will thus contain only five balls, one of them being white while four are black. Or we leave the urn as it is and ignore any taking out of a Y_2-ball; we replace it into the urn and the result of this experiment does not count, it is neither tallied in the absolute frequencies $H(X_1)$ and $H(X_2)$ nor in the number N of performed experiments.

Under these conditions only one of the five balls of brand Y_1 will yield the occurrence of X_1, whereas event X_2 will occur whenever one of the remaining four balls of brand Y_1 is taken. The respective *conditional* probabilities are $p(X_1/Y_1)=0.2$ and $p(X_2/Y_1)=0.8$; this stands for 'the probability of event X_1 under the condition Y_1 – or under the conditioning event Y_1 – is 0.2, and the probability of X_2 under the condition Y_1 is 0.8'. Likewise, we deduce $p(X_1/Y_2)=0.6$ and $p(X_2/Y_2)=0.4$. These *conditional probabilities* can be computed by the following quotient

$$p(X_i/Y_j)=\frac{p(X_iY_j)}{p(Y_j)} \tag{7.2}$$

provided $p(Y_j)\neq 0$. Similarly, we may compute conditional probabilities $p(Y_j/X_i)$ with values $p(Y_1/X_1)=0.25$, $p(Y_2/X_1)=0.75$, $p(Y_1/X_2)={}^2/_3$ and $p(Y_2/X_2)={}^1/_3$.

We now modify (7.2) to the following equation

$$p(X_iY_j)=p(X_i/Y_j)\cdot p(Y_j). \tag{7.3}$$

In (7.3) the left side will have the same constant value if we interchange X_i and Y_j, as $X_iY_j=Y_jX_i$ by definition, both are the event occurring when X_i occurs as well as Y_j. On the other hand, the right-hand side of (7.3) becomes $p(Y_j/X_i)\cdot p(X_i)$ when the events X_i and Y_j are interchanged. We thus deduce the equation $p(X_i/Y_j)\cdot p(Y_j)=p(Y_j/X_i)\cdot p(X_i)$ which we may divide by, say, $p(Y_j)$ provided $p(Y_j)\neq 0$. This leads to the well-known Bayesian formula which we shall frequently use in the subsequent chapters:

$$p(X_i/Y_j)=\frac{p(Y_j/X_i)p(X_i)}{p(Y_j)}. \tag{7.4}$$

It is convenient to derive a more detailed formulation of Bayes' formula. Let us say that the events $Z_1, Z_2, \ldots$ *exhaust* a given event Y if $Y=Z_1+Z_2+\ldots$. Let thus $Z_1, Z_2, \ldots$ be exclusive and exhaust the universe Ω. We conclude that the events $YZ_1, YZ_2, \ldots$ are exclusive for any event Y, too, and that they exhaust Y; according to the axioms we thus get

$$p(Y)=p(YZ_1)+p(YZ_2)+\ldots. \tag{7.5}$$

Using (7.3) we may substitute each summand on the right-hand side of the equation (7.5) by the corresponding product. We thus conclude

$$p(Y)=p(Y/Z_1)\cdot p(Z_1)+p(YZ_2)\cdot p(Z_2)+\ldots, \tag{7.6}$$

and we deduce

Bayes' Theorem: If $X_1, X_2, \ldots, Y_1, Y_2, \ldots$ and $Z_1, Z_2, \ldots$ are events and if in addition $Z_1, Z_2, \ldots$ are exclusive and exhaust Ω, then for any X_i and every Y_j satisfying $p(Y_j)\neq 0$

$$p(X_i/Y_j)=\frac{p(Y_j/X_i)p(X_i)}{p(Y_j/Z_1)p(Z_1)+p(Y_j/Z_2)p(Z_2)+\ldots}. \tag{7.7}$$

7.2.2 Degrees of Rational Belief

With our aforementioned aim of defining probabilities not depending on experiments needed under frequency considerations we tread new paths. We now start with *conditional* events X/Y, this means that we are only interested in occurrences – or non-occurrences – of event X under the *condition* that event Y occurs or has occurred. Conditional events X/Y have already been introduced in Sect. 7.2.1 where we defined conditional probabilities.

We will now interpret the *conditioning* event Y in a very special way. Subsequently it will always be the *evidence available* to us and we shall reflect this fact by writing $Y=E$ (for evidence). X/E thus denotes the event X regarded under the aspects of the known evidence E, and the basic definition of a probability is

$$p(X/E). \tag{7.8}$$

While $p(X/Y)$ was defined in (7.2) of Sect. 7.2.1 by means of the *absolute* probabilities $p(XY)$ and $p(Y)\neq 0$, now $p(X/E)$ is the basic term we start with in this section. We limit the values of $p(X/E)$ to the interval [0;1] within the real numbers, i.e. $0\leq p(X/E)\leq 1$, and we use $p(X/E)$ to quantify the belief in the occurrence of X under the evidence E available to us. The larger the value of $p(X/E)$ the more we believe in the occurrence of X; if we set $p(X/E)=0$, we think it is (almost[1]) impossible that X occurs, likewise, if $p(X/E)=1$, we regard X as a (almost) certain event.

Of course, the evidence we rely on may also be gained from frequency considerations and by carrying out a large number of experiments; consequently the definition of probabilities based on conditional events is more general and comprises the frequency-limit-considerations as a special case. In Ex. 7.3, for instance, we took balls from an urn for 1,000,000,000 times and we observed 39,998,674 occurrences of the event X_1, a white ball being taken; furthermore, we learned later that, e.g. $b=10$, the number of balls in the urn was ten. In that case the evidence E available to us is that 'there are ten balls in the urn U' and that 'taking a ball out of U for 1,000,000,000 times in 39,998,674 cases led to the event X_1'. We thus gain the rational degree of belief $p(X_1/E)=0.4$; we believe that the probability of taking a white ball is 0.4, and that four of the ten balls in the urn are coloured white, the others are black.

As we already mentioned when speaking about the attempts to induce probabilities of the occurrence of catastrophes (cf. Sect. 7.2.1), we may have to gain $p(X/E)$ without experimenting at all. In these cases experts will calculate the risks by studying the technical details only relying on their expertise[2]. They will cer-

---

[1] The meaning of this additional 'almost' in brackets will be explained in Sect. 7.3 where we distinguish discrete and continuous stochastic variables.

[2] The expertise may to some degree be gained by *modelling*.

tainly do without experiments resulting in those accidents. We now want to illustrate the idea of the rational degree of belief in the occurrence of a conditional event:

Ex. 7.5 In Fig. 7.2 a ball B will be dropped onto a disc D. The ball B is made of steel and its diameter is 1 cm. It is dropped from a height of 120 cm and out of a vertical position above the centre of D. The diameter of D is 20 cm, its thickness is 3 cm, and it is made out of the material M.

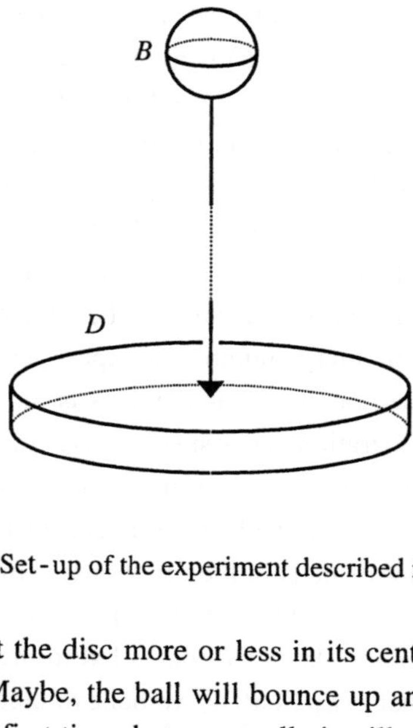

Fig. 7.2. Set-up of the experiment described in Ex. 7.5

We expect the ball to hit the disc more or less in its centre if it is dropped out of the described position. Maybe, the ball will bounce up and down for a while after it has hit the disc for the first time, but eventually it will come to rest. Let X be the event 'B finally lies on the disc', the complement $\overline{X}$ to this event X is described by the proposition 'the final position of B is not on D'.

The evidence E available to us is that 'there is a ball B falling ... '. Let further evidence be that the material M of the disc is steel, the same material the ball B is made of. We are *not* going to really carry out the experiment, we will just go through it as another *Gedankenexperiment* and estimate $p(X/E)$ and $p(\overline{X}/E)$. Given the aforementioned conditions we will expect B to bounce up and down for quite a while after it has been dropped and suppose that there will be hardly a chance for it to come to rest on D. This assessment is represented by the relation

$p(X/E) \ll p(\overline{X}/E)$, $p(X/E)$ is much smaller than $p(\overline{X}/E)$; we are almost certain that X cannot occur under those conditions.

We now change the test conditions. This time the disc is made of cardboard. If it is rather hard, we may still think $p(X/E) < p(\overline{X}/E)$, $p(X/E)$ is smaller than $p(\overline{X}/E)$, but both values do not differ as much as they do when the disc is made of steel. If the material M is rather soft, we even consider a relation $p(X/E) > p(\overline{X}/E)$ as appropriate, and in the case of a medium quality of the cardboard we may accept $p(X/E) \approx p(\overline{X}/E)$, the two probabilities to be more or less the same, i.e. approximately 0.5 each.

Eventually, if M is extremely soft – let us assume D is made of fresh, moist clay – we are almost certain that B will sink into the disc and not get out again. We thus expect $p(X/E) \gg p(\overline{X}/E)$, $p(X/E)$ will be much larger than $p(\overline{X}/E)$. ♦

In the preceding example we derived degrees of belief in the occurrence of X or of its complement $\overline{X}$ by falling back upon our experience; we have profited from our knowledge of the reactions of colliding bodies. If we had never seen two bodies made of steel collide or a solid body sink into a lump of clay, we would never have been able to derive any degrees of *rational* belief in Ex. 7.5. Of course, our expertise must lie in the field we want to evaluate. If it does not, we have to contact experts on that matter, and if we have asked each of them and weighed their expectations, then, finally, we will be capable of developing a rational belief justified by the experts' experience.

Frequency considerations will not take us any further in this case, because the ball will not be dropped at all, the experiment is planned as a *Gedankenexperiment* only and is thus entirely fictitious. Maybe, we decide to perform the experiment, but then only once. In that case relative frequencies are either zero or one, and they are no reasonable values for probabilities; they just characterize the impossible and the certain event respectively, the limits of probability extending to the certain knowledge of the outcome of experiments.

By experts' ratings considering all the estimations of the experts on the particular problem we may come to a value of $p(X/E)$, the probability that event X will occur given the available evidence E. These estimations are means to quantify the degree $p(X/E)$ of rational belief in the occurrence of X given E. So far we have just used it as a means of comparison when we, for instance, state that it is more probable to observe X given evidence E than $\overline{X}$. We may support the quantifying estimation of numbers out of the interval [0;1] by carrying out analogous experiments.

These experiments are stochastic models of the situation in question. The following example will illustrate the idea.

Ex. 7.6 Fig. 7.3 shows a 'wheel of fortune': a needle is balanced and supported in the centre of a round board. When the needle is pushed, it will rotate round this centre for some time, but eventually its point will come to rest in one of the sectors marked Y and $\overline{Y}$. Let Y stand for the *event* of the needle coming to rest in the sector Y and let it represent event X; likewise, $\overline{Y}$ stands for the sector $\overline{Y}$ and for the event of the needle finally pointing to $\overline{Y}$, and it is to represent event $\overline{X}$.

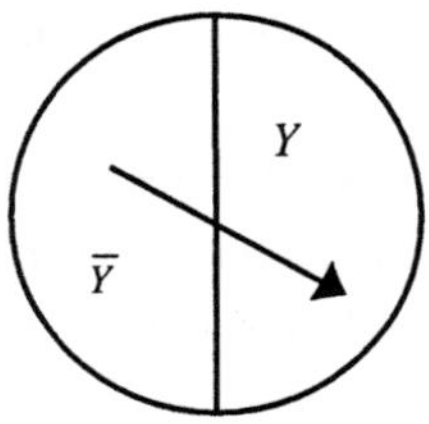

Fig. 7.3. Experiment supporting the quantification of degrees of rational belief

Let us start with the initial arrangement of Fig. 7.3 where the sectors Y and $\overline{Y}$ are of equal size. This evidence E^* of equal sizes of Y and $\overline{Y}$ results in equal probabilities of $p(Y/E^*)$ and $p(\overline{Y}/E^*)$ of 0.5 each; this means that we regard the probabilities of the needle finally resting in Y or in $\overline{Y}$ to be equal and thus estimate them to 0.5 each. We apply this model to Ex. 7.5 with a disc D made of hard cardboard. We want to find out whether the probability of the occurrence of X given E – hard cardboard – is the same as the probability of Y given E^* – equal sizes of Y and $\overline{Y}$ '? Maybe, we think the probability $p(X/E)=0.5$ – a fifty-fifty chance for the ball dropped onto a disc of hard cardboard to come to rest on the disc – seems to be too great, we may expect $p(X/E)<0.5$. In that case we will cut the sector Y in half and mark one of the halves Y the other half $\overline{Y}$; consequently, 25 % of the full circle are assigned to Y and 75 % to its complement $\overline{Y}$. We now turn to our 'wheel of fortune' again in order to find out whether the probability $p(Y/E^*)$ equals $p(X/E)$. If we think it should be smaller, we subdivide Y by further bisections, if it should be larger, we enlarge Y accordingly, i.e. we divide $\overline{Y}$ in half. So by successively changing the areas of Y and $\overline{Y}$ according to our assessments we may eventually reach a satisfying ratio when we say $p(Y/E^*)$ of Y under the condition E^* is the same as $p(X/E)$ of X under E. This final situation is reflected in Fig. 7.4.

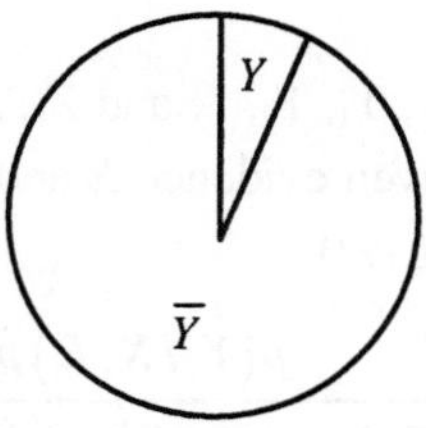

Fig. 7.4. Final position of the 'wheel of fortune' quantifying degrees of rational belief

The ratio between the area of sector Y and that of the full circle may serve as an estimate of what we assume the probability $p(X/E)$ to be like. This estimate can be made as precisely as necessary by adjusting the boundary between Y and $\overline{Y}$ iteratively until we get the impression that it represents our assessment correctly. We stress the fact that this analogous experiment used to quantify probabilities has never been carried out and is a *Gedankenexperiment* just like the original experiment of a ball dropping onto a disc. ♦

As we have pointed out, the evidence E may be induced by frequency considerations, but this need not necessarily be so. The degree $p(X/E)$ of rational belief in event X given evidence E is more general than the approximation $p(X)\approx h(X)$ by frequency limits. The other remarks and definitions of Sect. 7.2.1 are easily translated into the terminology of degrees of rational belief:

Thus, the events $X_1, X_2, \ldots$ are called *exclusive given evidence E* if according to E no two of them can occur together, that is $X_iX_jE=\varnothing$ and $p(X_iX_jE)=0$ for any pair X_i, X_j with $i\neq j$. The corresponding axioms are

- with every event X a real number $p(X/E)$ is associated, and $0\leq p(X/E)\leq 1$,
- if $X_1, X_2, \ldots$ are – exclusive given evidence E – then $p((X_1+X_2+\ldots)/E)=p(X_1/E)+p(X_2/\mathrm{E})+\ldots$,
- $p(X/(E=X))=1$ and
- $p(XY/E)=p(X/YE)\cdot p(Y/E)$.

The fourth axiom must be included into the list. It is analogous to proposition (7.3), and it is an abbreviated notation of $p((XY)/E)=p(X/(YE))\cdot p(Y/E)$.

Bayes' theorem (7.7) can now be reformulated for *degrees of rational belief.* In these terms it will be:

Bayes' Theorem: If $X_1, X_2, \ldots, Y_1, Y_2, \ldots$ and $Z_1, Z_2, \ldots$ are events, and if in addition $Z_1, Z_2, \ldots$ are exclusive given evidence E and if they exhaust Ω, then for any X_i and every Y_j satisfying $p(Y_j/E)\neq 0$

$$p(X_i/Y_jE)=\frac{p(Y_j/X_iE)p(X_i/E)}{p(Y_j/Z_1E)p(Z_1/E)+p(Y_j/Z_2E)p(Z_2/E)+\ldots}. \qquad (7.9)$$

This theorem is a means to infer degrees of rational belief from an increasing base of evidence. It describes the process of 'learning by experiment' and the Bayesian inference is the corresponding way of inferring; we are going to study it in the following section.

7.2.3 Bayesian Inference

In this section we aim at a reinterpretation of Bayes' formula (7.4) of Sect. 7.2.1 in a rather particular way:

$$p(X_i/Y_jE)=\frac{p(Y_j/X_iE)p(X_i/E)}{p(Y_j/E)}. \qquad (7.10)$$

Subsequently the event X_i will be the event that a given *hypothesis* H is 'good' and therefore acceptable: we equate $X_i=H$. A hypothesis H is a suggestion made as a base for further research. In general, right at the beginning, there is no evidence E that H is *obviously* 'true' or 'false'.

As an example we regard a planar and right-angled triangle T with the hypotenuse c and the sides a and b. Let evidence E be 'T is planar and right-angled, the length a of a is a=3 cm and the length b of b is b=4 cm'. The 'assumption' c=5 cm is obviously true, because we know from Pythagoras that for this – as well as for any planar, right-angled – triangle the equation $a^2+b^2=c^2$ holds, in our case this is $3^2+4^2=5^2$, which is true indeed. The proposition c=5 cm is true for the triangle under consideration for *logical* reasons; thus the proposition c=5 cm is *not* a hypothesis as defined above, we rather call it a *tautology* (Greek *tautos* and *logos*: 'the same' and word). If we had 'assumed' c=6 cm, this proposition would have been inconsistent with Pythagoras' theorem and thus be false given the available evidence E. This statement c=6 cm is false for *logical* reasons. It is called a *contradiction*, because $3^2+4^2 \neq 6^2$ contradicts the *theorem* $a^2+b^2=c^2$ for any planar and right-angled triangle. Thus $c=6$ is *not* a hypothesis either.

After having excluded tautologies and contradictions from being hypotheses, we may be quite 'generous' with the definition. Any probabilistic proposition *possibly* being true or false will be regarded as a hypothesis. *H* may even be a prejudice like, for instance, 'women are more skilled in driving a car than men are' or *vice versa.* We may even subsume more complex systems of propositions under the generic term *hypothesis* and thus call a complex *theory* a hypothesis, too. Some authors distinguish between the terms 'theory' and 'hypothesis', but we interpret a 'theory' just as a more *complex hierarchical* system of 'simple' hypotheses, and as we have already pointed out in Subsect. 2.2.2, we admit that the meaning of the term 'complex' is rather relative.

We will even regard metaphysical statements as being hypotheses. 'God exists', 'there will be an eternal life after we have gone the way of all flesh', 'there are guardian angels' are hypotheses. They definitely do not bear scientific examination, but we think they may render good examples of the development of 'personal probabilities'.

We have defined *H* to be the event that the hypothesis is a 'good' one and will be accepted. The letter *H* thus stands for the hypothesis *H* itself as well as for the proposition '*H* can be accepted' and for the evidence that it is really 'good'. As we have excluded true or false propositions because of *logical* reasons, this means that we are more or less *convinced* that a hypothesis *H* is acceptable, that it is 'good' or – on the contrary – that it is 'poor' and consequently not acceptable. At best, we only believe that we 'can trust' to a given hypothesis *H* or that we cannot. The *quality* of a given hypothesis is determined by *testing* it, scrutinizing it under the aspects of finding cases *verifying* or *falsifying* it.

There are hypotheses easily to be falsified. As an example we assume that fair-skinned people resist extreme sunshine better than dark-skinned. The evidence *E* for this working hypothesis may be based on our knowledge of physics: that light surfaces reflect light and heat much better than dark ones do. We certainly do not need many experiments in order to falsify this hypothesis. Probably further evidence available to us *prior* to any tests will confirm that this hypothesis cannot be maintained, because we know from experience that most southerners are dark-skinned whereas northerners are fair-skinned and easily get terrible sunburns.

But often it will be difficult to determine the degree up to which we can accept a theory, and it may well be that it can never be verified or falsified *completely.* The examples illustrating these hypotheses range from mathematical conjectures (e.g. Riemann's conjecture[1]), over propositions like 'there is intelligent life somewhere in the universe' to 'there are guardian angels'.

As soon as we have put forward a *scientific* hypothesis, we shall try to test it in order to determine how far it can be trusted. In general, *tests* are experiments we perform, but the evaluation of metaphysical theories may be based entirely on the observation of events occurring by chance. A car-driver may, for instance, have survived three severe accidents. On the basis of the evidence available to *him* – his personal experience – he may be convinced that the hypothesis 'there are guardian angels' is 'good' and 'dependable'[2]. Even though, we think, he will not test this hypothesis *deliberately*, this means he will not provoke any further accidents or test his hypothesis by experimenting like jumping from the top of a skyscraper without using a parachute. There may be further evidence supporting this theory of the existence of guardian angels. Maybe, one day somebody shoots at him and misses him, the building he works in collapses and he is the only one to survive among some 800 people and so on. But if he really tries to *test* this hypothesis of guardian angels deliberately by jumping from the top of a high building without the aid of a parachute, he might be lucky and survive, because accidentally a sufficient number of sun-blinds are opened on the lower floors of the building. If this rather spectacular event should happen, the theory of the existence of guardian angels would be confirmed once more. But it will be more realistic that this 'gambler' hits the pavement, and this will put an end to his theory. If the experimenter survives, we may still be sceptical and argue 'this is not really a *proof* of the existence of guardian angels, it is just another case of *good luck*'. In case the experiment fails, those believing in the existence of guardian angels may still conclude that 'he has worn out the patience of his personal guardian and was dropped, because he tempted providence'.

Testing a hypothesis H means to test it deliberately by experimenting. At best we may gain a verifying result, but we must also take the risk of a falsifying outcome into account. In general, the belief in the dependability of a hypothesis is based on probabilities accepted by a community of people working with it, but they may also be entirely personal. The procedure of inferring these probabilities and quantifying them is described by Bayes' theorem (7.9).

---

[1] This important hypothesis about the distribution of the complex zeros of the ζ-function has neither been proved nor falsified yet; nevertheless, the assumption that it is a theorem has let to a number of fruitful considerations.

[2] Cf. Sect. 2.4 and our remark on the dependability of models.

Before we demonstrate the details of Bayesian inference we would like to explain some linguistic and formal rulings. We have substituted the event X_i of (7.10) by the proposition 'hypothesis H is acceptable', and we also represent this proposition by the letter H. Consequently, the complementary event to X_i will be the proposition 'hypothesis H is *not* acceptable', the negation of H, which will be symbolized by $\neg H$. Further explanations will be based on the example of one of the prejudices mentioned above, which we want to discuss now:

Ex. 7.7 Let hypothesis H be: 'Women are more skilled in driving a car than men are' and $\neg H$ is: 'Women are *not* more skilled in driving a car than men are'. This latter negation $\neg H$ of hypothesis H is tantamount to $\neg H$: 'Women are equally or less skilled in driving a car – compared with male drivers'. ♦

One day scientists may find a genetic 'proof' of this hypothesis. There may, e.g. be a genetic disposition of the female sex making them superior in driving cars; maybe, they are the 'born' motorists. On the other hand, as long as there is no such 'proof' we have to test our hypothesis H by observation and gaining further evidence by experimenting. Let us assume we watch a man – not on drugs – steer his car erratically along a country-road; obviously he is a poor driver and his style of driving shows his inability: he accelerates at the wrong moments, changes into the wrong gears and almost loses control of his car in a slight bend, endangering an oncoming car driven by a woman, who skilfully avoids the fatal crash. Well, we regard all this as *supporting* our hypothesis H. What we have just observed is a case *verifying* H and we now trust to H more than we did before. The *range of validity* of H has grown – even if only a little – because it comprises one more verifying case. This is in concordance with our terminology used in Sect. 6.2. We attach importance to this interpretation of the terms *verification* and *validity*. They do not imply that hypothesis H is true. We refer to Ex. 6.4 and Ex. 6.5 where we discovered *verifying* and *falsifying* cases for an algebraic relation. Likewise, we observe a case *falsifying* hypothesis H – verifying its complement $\neg H$ – when we interchange the drivers of those two cars, the woman now losing control and the man skilfully avoiding a collision.

Objections have been raised to this usage of the terms verification and validation, and it has been suggested to use them only in those cases where they established the absolute truth of a theory. Likewise, a falsification should prove that the theory under discussion was absolutely false. The reason for these demands is the fear that the lay public might misunderstand terms like verification and validation as too absolute. The Latin roots *verus* of 'verification' and *valis* of 'validation' mean 'true' and 'well-founded', and 'to falsify' is derived from the Latin adjective *falsus*, equivalent to 'false'; thus these terms – when applied to a hypothesis – do suggest an absoluteness of its quality that is not justified. In order to avoid these problems, other terms are frequently used. Our verifying case may as well be called a case *supporting*, *confirming*, *corroborating* or *vindicating H*, a case *in concordance* with hypothesis *H*. A falsifying case invalidating a hypothesis thus is *not in concordance* with *H*, it *weakens* the hypothesis, *casts doubt* on it and is *inconsistent* with it. Sometimes the fear of being too rigid when applying these terms even leads to rather tortuous formulations. If, for instance, there has been evidence in favour of a hypothesis *H* under research – we would call it a *verifying case* – we might say 'this event *does not invalidate H*' or 'it *does not falsify* this hypothesis'. In general, these scruples are restricted to those cases favouring a hypothesis. The reason may be not to inspire too much confidence in a hypothesis thus confirmed and to avoid taking responsibility in case there are future falsifying cases giving evidence of a rather restricted range of validity of the working theory.

These considerations are rather *pragmatic* (cf. Sect. 9.2); they take possible misinterpretations by laymen into account. They try to avoid any positive-sounding terms insinuating the impression that the quality of a theory might be better than it really is, and there is also a very guarded vocabulary avoiding possible misunderstandings yet. Among the members of a scientific community the terms *verification* and *validation* used as defined – and frequently, but not uniformly used in various fields of research – should not lead to any misunderstandings. So, what we propose is to use these terms in the following linguistically justified sense: the *validation* of a working hypothesis *H* is the procedure of testing it by experimenting and thus determining the foundation of *H*. Any case *verifying H* will put trust to it and make it even more justified (more *valid*). Likewise, any *detected falsifying* case will reduce the trust to *H* (cast doubt on it) as an acceptable working hypothesis and weaken its foundation. Thus, by validating a hypothesis its range of validity is determined: this is the set of verifying cases it is founded on.

If *H* is a hypothesis, it must be tested (validated) by experimenting, and a degree of validity may be inferred from the results of these tests. Bayesian inference is based on the evidence gained from a series of successive experiments. Sometimes the available evidence may be overwhelming and the hypothesis may thus be easily accepted or refuted. For instance, the hypothesis that women are better at driving a car than men are can easily be proved right or wrong by using frequency considerations. In order to do so, we will evaluate mass data recorded by insurance companies and the authorities, and from these official statistics we will induce a degree of validity of this hypothesis formulated in Ex. 7.7.

On the other hand, if this is not possible, we shall have to rely on single observations. We will thus perform a series of single tests T_j, with $j=1,2,3, \ldots$. If, for instance, the hypothesis of Ex. 7.7 is entirely based on personal experience, a first test T_1 is the observation of two cars – one driven by a woman, the other one driven by a man – almost colliding. Each test T_j leads to a result (recognition) R_j. In Ex. 7.7 R_1 is 'this woman is a more skilful driver than the man in the other car'. In general, we set $Y_j=R_j$ in (7.10). As the available evidence is changed by experimenting and by gaining deeper insight into the theory under research, we have to introduce a subscript to E, too. The E of (7.10) will thus be substituted by E_{j-1}, and we derive a new representation of Bayes' law:

$$p\left(H/R_j E_{j-1}\right)=\frac{p\left(R_j/HE_{j-1}\right)p\left(H/E_{j-1}\right)}{p\left(R_j/E_{j-1}\right)}, j=1,\ 2,\ 3,\ \ldots\ . \tag{7.11}$$

According to this formula each test T_j and its result R_j will lead to a new degree of rational belief in hypothesis H. As it depends more on the outcome R_j of the test T_j than on this test T_j itself, we represent the increase in evidence and the changing rational belief in the hypothesis by the following diagram:

$$p\left(H/E_{j-1}\right) \xrightarrow{R_j} p\left(H/R_j E_{j-1}\right)=p\left(H/E_j\right)\ j=1,\ 2,\ 3,\ \ldots\ . \tag{7.12}$$

In this representation $p(H/E_{j-1})$ is the degree of belief in hypothesis H *prior* – or *a priori* – to the test T_j and the recognition R_j gained from it. Thus, the degree $p(H/E_{j-1})$ of rational belief prior to R_j is based on the available circumstantial evidence prior to T_j. As soon as T_j has been carried out and has yielded its result R_j, the available evidence will be R_jE_{j-1} – R_j *and* E_{j-1} – resulting in the evidence E_j *posterior* – or *a posteriori* – to the test T_j and its outcome R_j. The jth test T_j and its result R_j thus transform evidence and the degree of rational belief in the hypothesis H from prior into posterior degrees with respect to T_j (and to R_j).

The terms transforming[1] prior to posterior degrees of belief in (7.11) and (7.12) are $p(R_j/HE_{j-1})$ and $p(R_j/E_{j-1})$. The factor $p(R_j/HE_{j-1})$ in (7.11) is not at all a prob-

---

[1] Cf. (2.3) and (3.2), too.

ability with regard to theory H, it is rather a probability of the result R_j of the experiment T_j carried out *under the condition* of H. With respect to the hypothesis H the term $p(R_j/HE_{j-1})$ is called the *likelihood* of the hypothesis H. The likelihood of a hypothesis H is the degree of belief in the occurrence of the result R_j of an experiment under the condition (assumption) that this hypothesis is a 'good' one and that the general prior evidence $E=E_{j-1}$ is available to the experimenter. Finally, there is no relation between the hypothesis H and the denominator $p(R_j/E_{j-1})$ of (7.11) at all. Relative to H $p(R_j/E_{j-1})$ is a mere normalizing constant, and we may reformulate equation(7.11) by the proportion

$$p(H/R_jE_{j-1}) \propto p(R_j/HE_{j-1}) \cdot p(H/E_{j-1}), \tag{7.13}$$

this means that the posterior belief in H is proportional to the product of its likelihood and its prior probability – all with respect to the performed test T_j and its result R_j.

The transformation of prior into posterior probabilities represented by the Bayesian formula reflects the inductive procedure of 'learning by experiment'. This becomes more evident when we link a sequence of the transitions represented by (7.12) to form the following chain:

$$\ldots p(H/E_{j-1}) \overset{R_j}{\rightarrow} p(H/R_jE_{j-1}) = p(H/E_j) \overset{R_{j+1}}{\rightarrow} p(H/R_{j+1}E_j) = p(H/E_{j+1}) \ldots . \tag{7.14}$$

This is one of those *I-P-O*-chains (2.3) we have already studied in Subsect. 2.3.1 (cf. (3.2), too). The jth test T_j and its result R_j transform its input $p(H/E_{j-1})$ prior to the test into the output $p(H/E_j)$ posterior to it which is an input itself into the following test T_{j+1} transforming it to $p(H/E_{j+1})$ and so on. The development of available evidence from E_{j-1} to E_j and so on is a process of learning by growing experience gained by successively experimenting. We shall explain this process of gaining new insights by experimenting with a simple and very technical example. As before this example deals with the removal of balls from different urns, but again it is only a *Gedankenexperiment*, actually we will not take out any balls at all.

Ex. 7.8 Let U_4 and U_6 be two urns. We cannot see their contents, but we know that each one contains ten balls. Furthermore, we know that in U_4 four of those balls are black while the remaining six balls are white. Conversely there are six black and only four white balls in U_6 (cf. Fig. 7.5).

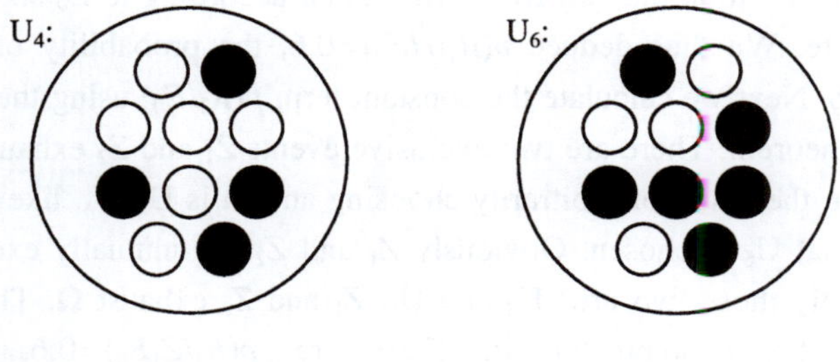

Fig. 7.5. The urns of Ex. 7.8 demonstrating Bayesian inference

We are free to choose one of the urns; we call it urn U. We do not know whether this is U_4 or U_6. But once we have chosen U, we are allowed to carry out a limited number of tests like removing balls from U without replacing them. As soon as we have taken nine balls from U we can be absolutely sure about its contents, because at least five of those balls are of the same colour. With some luck we may even be able to tell after only five removals whether $U=U_4$ or not; this will be the case if those five balls are of the same colour, but this event is seldom. Now we continue with our test and assume that only a very limited number of removals is possible, say, a maximum of two or three, because each taking out – for example – costs an enormous amount of money. But at the same time it may be of vital importance to decide whether $U=U_4$ or $U=U_6$.

As there is no evidence which of the two urns is U_4 and which is U_6, our choice will be completely random. We put forward the hypothesis H: $U=U_4$, its alternative will thus be $\neg H$: $U=U_6$ (we might as well have defined H to be $U=U_6$ and $\neg H$ by $U=U_4$).

The evidence E_0 prior to any experiment is our knowledge of the contents of U_4 and of U_6, the rules set up for testing – removal without replacement –, and the arbitrariness of our choice; so the chances to be chosen should be the same for $U=U_4$ and for $U=U_6$. Given evidence E_0 we thus deduce prior probabilities $p(H/E_0)=p(\neg H/E_0)=0.5$. We call these prior probabilities the *initial* degrees of rational belief in the hypothesis H: $U=U_4$ and in its alternative $\neg H$: $U=U_6$. If we cannot afford to experiment because of the costs, we may as well guess or toss a coin in order to decide for $U=U_4$ or $U=U_6$. But let us assume we can well afford to pay for one or two experiments (i.e. removals of balls from the chosen urn).

We perform the first test and take a ball from urn U resulting in, say, $R_1=W$, the event W of a white ball being taken. The likelihood $p(R_1/HE_0)$ is the probability of taking a white ball given hypothesis H: $U=U_4$ and the available evidence E_0. According to H we took the ball from urn U_4 and according to E_0 six of the balls in U_4 are white. We thus deduce $p(R_1/HE_0)=0.6$, the probability of the observed result is 0.6. Next we calculate the constant term $p(R_1/E_0)$ using the notation (7.9) of Bayes' theorem. There are two exclusive events Z_1 and Z_2 exhausting Ω. $Z_1=H$ is the event, the result of arbitrarily choosing an urn is $U=U_4$, likewise $Z_2=\neg H$ is the event that U_6 is chosen. Obviously Z_1 and Z_2 are mutually exclusive, and as there are only those two urns U_4 and U_6, Z_1 and Z_2 exhaust Ω. Thus the various terms of the denominator in (7.9) are $p(R_1/Z_1E_0)=0.6$, $p(Z_1/E_0)=0.5$, $p(R_1/Z_2E_0)=0.4$ and $p(Z_2/E_0)=0.5$, and we deduce

$$\begin{aligned} p(R_1/E_0) &= p(R_1/Z_1E_0)\cdot p(Z_1/E_0)+p(R_1/Z_2E_0)\cdot p(Z_2/E_0) \\ &= 0.6\cdot 0.5+0.4\cdot 0.5=0.5. \end{aligned} \tag{7.15}$$

We are now in a position to induce the posterior degree of rational belief in our hypothesis H: $U=U_4$ from the first experiment and its result R_1. We do so by using Bayes' formula (7.11):

$$p(H/R_1E_0)=\frac{0.6\cdot 0.5}{0.5}=0.6=p(H/E_1). \tag{7.16}$$

According to this equation (7.16) – after the first experiment has been carried out and has resulted in a white ball being taken from U and based on evidence E_0 – our degree of rational belief in hypothesis H rises from a prior value of 0.5 to a posterior value of 0.6. This experiment supports (confirms etc.) our theory, it is a *verifying case* in favour of H. The evidence available now is $E_1=R_1E_0$, it unites the result R_1 of the first experiment and the initial evidence E_0 mentioned above.

In case we have run out of money and cannot afford to buy any further tests, we have reasons to guess $U=U_4$ if we have to guess at all, because the hypothesis H: $U=U_4$ with $p(H/E_1)=0.6$ is more dependable than its alternative $\neg H$: $U=U_6$ with only $p(\neg H/E_1)=0.4$. But let us assume that there is enough money for one more experiment, one more removal of a ball from U resulting in R_2, say, $R_2=W$, a white ball being taken again.

The probability $p(H/E_1)$ of H prior to the second experiment is $p(H/E_1)=0.6$, the probability we induced from the first test. The likelihood $p(R_2/HE)$ of H given the available evidence is the probability of getting a second white ball given hypothesis H: $U=U_4$ and evidence E_1 that already a white ball has been taken from U without replacement. As there are only nine balls left in U for the second test, five of them white and four being black due to H: $U=U_4$ and E_1, we deduce $p(R_2/HE_1)={}^5/_9$. The constant denominator of (7.9) is accordingly calculated to

$$p(R_2/E_1) = p(R_2/Z_1E_1)\cdot p(Z_1/E_1) + p(R_2/Z_2E_1)\cdot p(Z_2/E_1) \qquad (7.17)$$
$$= {}^5/_9\cdot 0.6 + {}^3/_9\cdot 0.4 = {}^{42}/_{90},$$

with $p(R_2/Z_2E_1) = {}^3/_9$, the probability of drawing a second white ball out of U_6 – under the condition $\neg H$: $U=U_6$ – if a white ball has already been removed from this urn U_6 in the first test, and the probability $p(Z_2/E_1) = p(\neg H/E_1) = 0.4$.

We once more apply Bayes' formula and induce, this time starting from R_2, the posterior degree $p(H/R_2E_1) = {}^5/_7$ of rational belief in hypothesis H. Again, we define $p(H/E_2)$ by equating $p(H/E_2) = p(H/R_2E_1)$, posterior evidence E_2 with respect to the second experiment comprising its result R_2 and prior evidence E_1 with respect to this test T_2.

The second removal again supports our theory H: $U=U_4$; its result $R_2 = W$ is another verifying case for H. Thus the degree of belief in H has risen during the experiments from an initial value of 0.50 over an intermediate value of 0.60 to a final value of 0.71. If we had to choose between H and its alternative $\neg H$ now, we would definitely prefer to vote for H.

On the other hand, if the outcome of the second test had been $R_2 = B$, a black ball falsifying H, this would have been a set-back and would have reduced the belief in H to $p(H/E_2) = 0.5$, the initial value we started with. All the money we spent on experimenting would have been wasted. In that case we would have to toss another coin for a decision between H and $\neg H$, if we could not afford to pay for any further tests. ♦

This Ex. 7.8 shows that 'learning by experiment' is a step-by-step procedure leading to varying degrees of belief in a given hypothesis H. The development of the dependability of H follows Bayes' formula transforming prior degrees into posterior degrees relative to the result of a performed experiment.

Again, we stress the fact that degrees of rational belief depend on the available evidence and that they are prior or posterior with respect to an experiment and its result. With this restricted application of the terms *a priori* and *a posteriori* we avoid objections to the usage of these terms raised by philosophers who apply them with a different meaning insignificant to our considerations (cf. e.g. Subsect. 10.4.1).

There are still further objections to the assumption of *initial* probabilities, degrees of rational belief in a hypothesis prior to *any* experiment carried out in order to test them. In Ex. 7.8, for example, we selected an urn at random and decided for a prior belief in the hypothesis H – this urn is of the type $U=U_4$ – of $p(H/E_0)$ given the initial evidence E_0 of a random choice. There may be the argument that the choice might not have been made quite arbitrarily, that some people may automatically prefer the left object if there are two arranged side by side. In order to answer this argument, we will introduce a *relaying experiment* (Richter 1956) carried out prior to the experiments of Ex. 7.8 (cf. Ex. 6.12). We may, for example, – if we regard both outcomes of the choice of U – $U=U_4$ and $U=U_6$ – to be of equal probability – toss a coin: 'heads' will lead to a decision in favour of, say, the left urn and 'tails' in favour of the right urn thus justifying the assumption of equal initial chances for U_4 and U_6 to be chosen.

These suppositions of prior probability are essential as long as there is only one experiment supporting our decision, or if the number of those experiments is small. Furthermore, the initial evidence E_0 must be determined with great care. We would like to confirm this fact by a well-known example:

Ex 7.9 (*The Goat Problem*) The scenario is a TV-Show: there is a choice of three closed doors. The prize – an expensive car – is hidden behind one of these doors, the other two doors cover blanks, there is a goat behind each of them. There is a *candidate* who must decide for one of these doors which will not be opened for the moment; we call it *C*. The *presenter* of the show knows the door hiding the car. He opens one of the two doors left after the candidate has made his choice. We call this door *M*, and let us assume that the candidate sees a goat. The presenter now offers to switch from door *C* to the *third* door *T*, still closed just like door *C* the candidate has chosen.

There have been long discussions about the advantages of switching. Some people think, there is a better chance of winning when we switch to *T*, others think it does not matter whether we switch or not, the chance of winning will be the same in both cases. This conflict can be settled by diverging assumptions about the initial evidence E_0. Certainly event *C*, the candidate has decided for door *C*, is part of E_0, but the candidate does not know the presenter's strategy. Maybe, he decided at *random* for one of the remaining two doors, or, maybe, he made use of his *knowledge* of where the prize is hidden and opened the wrong door, because he wanted the show to go on. We call the former event *R*, the latter will be *K*. So, the corresponding formulation of Bayes' formula is:

$$p(T/ME_0)=\frac{p(M/TE_0)p(T/E_0)}{p(M/E_0)}. \tag{7.18}$$

In (7.18) *T* is the event 'the prize is hidden behind the third door, neither chosen by the candidate nor opened by the presenter'; *M* is the event 'the presenter decides for a certain door and there is a goat behind it'. Let initial evidence be $E_0=CR$, 'the candidate decides for door C, and the presenter chooses M at random from the remaining two doors and detects a goat'. Thus $p(M/TE_0)={}^1/_2$, $p(T/E_0)={}^1/_3$. For the calculation of $p(M/E_0)$ we use the three events $Z_1=P_C$, $Z_2=P_M$ and $Z_3=T=P_T$, the prize is hidden behind the doors C, M or T respectively; obviously Z_1, Z_2 and Z_3 are exclusive and they exhaust Ω. We thus calculate $p(M/E_0)$ to

$$\begin{aligned} p(M/E_0) &= p(M/P_CE_0)p(P_C/E_0)+p(M/P_ME_0)p(P_M/E_0)+p(M/P_TE_0)p(P_T/E_0) \quad (7.19)\\ &= {}^1/_2\cdot{}^1/_3+0\cdot{}^1/_3+{}^1/_2\cdot{}^1/_3={}^1/_3; \end{aligned}$$

and we induce a posterior degree $p(T/ME_0)={}^1/_2$ of belief in the hypothesis '$T=P_T$, the prize is behind door T'. In this case, when we assume the presenter chooses door M at random, there is no rational reason to switch to the third door T.

On the other hand, let initial evidence be $E_0=CK$, 'the candidate decides for door C and the presenter deliberately decides for that door M, which will not show the prize, but the other blank'. In that case we get $p(M/TE_0)=1$, $p(T/E_0)={}^1/_3$. And with Z_1, Z_2 and Z_3 as in the first case we calculate

$$\begin{aligned} p(M/E_0) &= p(M/P_CE_0)p(P_C/E_0)+p(M/P_ME_0)p(P_M/E_0)+p(M/P_TE_0)p(P_T/E_0) \quad (7.20)\\ &= {}^1/_2\cdot{}^1/_3+0\cdot{}^1/_3+1\cdot{}^1/_3={}^1/_2; \end{aligned}$$

and we induce a posterior degree $p(T/ME_0)={}^2/_3$ of belief in the hypothesis '$T=P_T$, the prize is hidden behind door T'. In this case, when we assume the presenter makes use of his knowledge when he decides for door M, it will consequently be reasonable to switch to the third door T. ♦

In the preceding example the initial evidence E_0 has been of great importance for our evaluation of the result of the single experiment M and for the decision whether to switch to T or not. If the candidate is clearly informed about E_0, the way the presenter selects door M, he decides according to our calculations. If there is no information about the presenter's intentions, if he just announces his decision for M presenting a goat with the words 'I am going to show you something ...', the candidate may think that he wants to keep the show going by making use of his knowledge. As switching does not reduce his chances whichever way the presenter selects the door M, he may conclude it might be advantageous to change to T in any case. So the decision whether to change or not is based on personal assumptions. This becomes even more evident when we introduce a price for switching to the third door, say, 50% of the value of the car; in that case the candidate has to decide whether it is worthwhile to invest that sum.

There is a further important aspect of coming to decisions under stochastic considerations. In order to demonstrate it, we return to (7.15) and (7.16) and study the values of the various variables and parameters on the right-hand side of the equation more closely. In this special case there is a value of 0.5 common to prior probability $p(H/E)$, to $p(Z_1/E)$ and thus also to $p(Z_2/E)$; it may be cancelled in the quotient.

Quite often we may have to do with an equation like (7.16) involving a finite number n of exclusive events Z_i exhausting Ω and equal distributions of probabilities of the various $p(Z_i/E)$, all being constant with the same value $p(Z_i/E)=p=n^{-1}$:

$$p(H/RE)=\frac{p(R/HE)p(H/E)}{p(R/Z_1E)p+p(R/Z_2E)p+\ldots+p(R/Z_nE)p}. \tag{7.21}$$

Here we may put p outside the brackets and combine it with the known (and constant) prior probability $p(H/E)$ of hypothesis H. Thus there will be the proportion

$$p(H/RE)\propto p(R/HE) \tag{7.22}$$

between posterior probability $p(H/RE)$ of hypothesis H with respect to the outcome R of the experiment and its likelihood $p(R/HE)$ with respect to this experiment, too.

This is the basic idea of the *Maximum Likelihood Method* (*MLM*) of deciding for a hypothesis H regarding its likelihood; the larger the likelihood of a hypothesis the larger will be its posterior probability, and we tend to prefer it to every competing hypothesis. The message of *MLM* is to decide for the hypothesis with maximum likelihood respective to the result of an experiment. So why bother about the rather complicated calculations needed for Bayes' theorem if there is the easier way of using *MLM*? The answer is, because *MLM* relies on – more or less – equal probabilities of the alternative Z_i-s according to (7.21). If these probabilities show a great variance, *MLM* may nevertheless be used if the number N of experiments is large, say, $N>100$. In Ex. 7.8 we decided for the hypothesis H: $U=U_4$, the urn showing the greater likelihood for a white ball being removed, because prior probabilities $p(H/E)$ and $p(\neg H/E)$ both equalled 0.5. The following example will demonstrate how an unjustified application of *MLM* leads to irrational decisions:

Ex. 7.10 We modify the relaying experiment of Ex. 7.8 so that the prior probability $p(H/E)$ of the hypothesis H: $U=U_4$ will be $p(H/E)=0.001$, the prior probability of its alternative $\neg H$: $U=U_6$ will thus be $p(\neg H/E)=0.999$. We may, for instance, assume a relaying experiment using a third urn containing 1,000 balls, 1 of them is marked with the numeral '4', the others with '6'. If a presenter takes the ball marked '4' from this urn, he will select U_4 and hand it to us, otherwise U_6 is assigned to us. Of course, we do not know whether $U=U_4$ or $U=U_6$, but we are informed about the set-up of the relaying experiment leading to initial probabilities $p(H/E_0)=0.001$ and $p(\neg H/E_0)=0.999$ for H: $U=U_4$ and $\neg H$: $U=U_6$ respectively.

Let again the result R_1 of the first experiment be $R_1=W$, the removed ball is white. Again we calculate $p(R_1/HE_0)=0.6$, and we know $p(H/E_0)=0.001$. Furthermore, we calculate

$$\begin{aligned} p(R_1/E_0) &= p(R_1/Z_1E_0)\cdot p(Z_1/E_0)+p(R_1/Z_2E_0)\cdot p(Z_2/E_0) \\ &= 0.6\cdot 0.001+0.4\cdot 0.999=0.4002. \end{aligned} \tag{7.23}$$

This leads to a posterior probability $p(H/R_1E_0)$ of hypothesis H calculated to

$$p(H/R_1E_0)=\frac{0.6\cdot 0.001}{0.4002}=0.0015=p(H/E_1). \tag{7.24}$$

So, in this case taking a white ball from the unknown urn U raises the degree of belief in the hypothesis that U is U_4 from prior 0.0010 to a mere posterior 0.0015, though the value of likelihood is clearly in favour of U_4. Taking into account our knowledge of the relaying experiment and the small value of prior probability of the hypothesis $U=U_4$, we would rather accept the verdict of Bayesian inference $p(\neg H/R_1E_0)=0.9985$ and guess that $U=U_6$ instead of judging under the aspect of maximum likelihood. ♦

7.3 Certainty and Prior Probability

As we have set out in preceding chapters, modelling means being creative. Modelling is a mapping generating an image, and evaluating the quality of the mapping and the trustworthiness of this image is a procedure founded on stochastic considerations. Within the theory of probability there are many obvious conclusions accepted to a large extent; but there are also strange aspects, and some of them are still in dispute among the experts. In this section, we want to outline some of them.

Almost

When dealing with discrete random variables, the estimation of probabilities seems to be quite natural and will – in general – not lead to any controversial discussions. If, for instance, we toss an ordinary coin, we grant each of the two sides an equal chance of occurring upside and accordingly estimate equal probabilities of 0.5 to 'heads' as well as to 'tails'. This is the Laplacian approach to these experiments, giving equal probabilities to their results if there is no further evidence in favour of a subset of them. But Laplace's probability assumption has led to some discussion; we shall deal with this theory below. When applying it to 'continuous' probability distributions, it gives rise to some strange observations we want to discuss now.

Ex. 7.11 There is a tape-measure of length $l=100$ cm and one point on the measure has been selected arbitrarily. We now want to determine the probability of this point lying within the interval [9.5;10.5], this means $9.5 \leq P \leq 10.5$, where P represents the position of the point marked on the tape.

As our considerations are entirely based on the assumption that the point has been chosen *arbitrarily*, we guess that the probability p_0 of P lying within the interval [9.5;10.5] will be $p_0={}^1/_{100}$, as the length of this interval is ${}^1/_{100}$ of the total length of the measure. If we now bisect the length of the interval [9.5;10.5] in order to determine the probability p_1 of P being arbitrarily chosen within the interval [9.75;10.25], we may use the same line of arguments and estimate a value $p_1={}^1/_2 \cdot {}^1/_{100}$ of p_1. Continuing the process of bisecting the length of the interval under consideration, we – after the nth bisection – calculate a value of the probability p_n of P being positioned within the interval $[10-(0.5)^{n+1};10+(0.5)^{n+1}]$ of $p_n=({}^1/_2)^n \cdot {}^1/_{100}$. With a growing number of bisections and n approaching infinity, i.e. $n \to \infty$ we derive the limit $p_\infty=0$ for the probability of $P=10\pm0$. Thus, the possibility of choosing $P=10$ is zero, and likewise – following the same line of arguments – we conclude that the possibility for any value of P within the range [0;100] is zero. On the other hand, if a point has been chosen on the tape, it necessarily has a definite position P; as this position results from the experiment of marking a point, it cannot be 'impossible', i.e. not exist at all, though its probability $p(P)=0$ is zero; we call this event *almost impossible* (but not – entirely – impossible).

Likewise, the negation $\neg P$, the complement of the position P, will be evaluated by $p(\neg P)=1$, assigning the maximum value of probability p to this event $\neg P$; nevertheless, it is not entirely 'certain' as the position may well be P; we thus call it the *almost certain* event $\neg P$. ♦

We conclude that – when dealing with continuous stochastic variables – the limiting values of probabilities p like $p=0$ and $p=1$ do not imply that the corresponding events are 'impossible' respectively 'certain'. In this case only statements like 'these events are *almost* impossible or *almost* certain' are justified by the stochastic evidence $p=0$ or $p=1$ (cf. Subsect. 2.3.2.1).

Laplacian Probabilities

We use Laplace's method of constant distributions of probabilities when there is reason to assume that all relevant events are equally possible to occur. These considerations are also pertinent to those cases where no evidence is given in favour of a subset of the relevant events. This is Laplace's *principle of insufficient reason*. There have been controversial discussions whether it is justified to apply this criterion and to use Laplacian probabilities under these conditions.

We have already discussed examples reflecting the scenario of having to come to a decision without any evidence to rely on, for instance, Ex. 6.12 - Ex. 6.14, where we had to decide between two doors, only one of them leading into freedom (in fact there was some evidence – like different colours of the doors – but it was of no help for the decision which door to open). If there is no evidence supporting the process of coming to a decision, we may as well toss for it – or, for instance, throw dice for it if six different decisions are feasible. These relaying experiments – tossing coins or throwing dice – may help us to decide, but the procedure only shifts the problem from one stochastic process to another one, substituting decisiveness by experimenting. These experiments, performed to stimulate a decision, are the relaying experiments mentioned in Sect. 6.3 and Subsect. 7.2.3, and in our example – where we use a coin or a dice – they will just lead to Laplacian probabilities, granting equal probabilities to each alternative.

The argument that probabilities can only be estimated if this estimation is based on evidence is not very convincing, because we must also come to decisions (cf. Ex. 6.12 and Ex. 7.9) under the condition of uncertainty. These decisions are based on personal degrees of belief which may be more or less rational. If we cannot evaluate a scenario and develop a personal belief, we will not be able to see chances and decide at all. So we will invent relaying experiments, which may lead to Laplace's probability distributions if there is no further evidence in favour of a feasible alternative, or we may use analogous experiments like the wheel of fortune of Ex. 7.6 to quantify our estimates.

Our argument that in the case of no evidence in favour of one of the feasible alternatives the Laplacian distribution is the most appropriate is also supported by considerations of information theory. Laplace's principle of insufficient reason leads to a maximum of the average information content of a given problem in the field of decision making under uncertainty (cf. e.g. Mag 1990).

The Initial Probability

The problem of determining the initial probability within a chain of experiments as defined in Subsect. 7.2.3 is linked to the question whether Laplacian probabilities may be used or not. Frequently objections are raised to the assumptions of prior probabilities in any inference following Bayes' theorem if they cannot be founded on frequency considerations. We have represented the process of 'learning by experiment' by a chain (7.14) of Bayesian inferences, and within this chain each prior probability is posterior to the preceding induction – and is thus based on the evidence gained from this transformation – except for the very first one, which we call the *initial* probability of the chain.

So, when dealing with a chain of consecutive applications of Bayes' law the essential prior probability is the initial probability: it is prior to the first performed induction and part of any further calculated probability. If there is no evidence in favour of one of a set of alternative events, we may just toss for it and thus use Laplace's criterion. This is the aspect of deriving personal probabilities calculated under uncertainty.

When evaluating scientific models, for the most part the problem will not be to quantify personal convictions, but rather to find an intersubjectively based degree of rational belief in a hypothesis apt to be used as the initial probability in a chain of Bayesian inferences. This leads to pooling the personal knowledge of the initiated. Because of practical reasons we will not demand to record the personal estimations of each individual expert in the field of research, but we suggest to select a representative from each significant school within it to form a panel. They have to formulate their judgements and estimate the initial probability needed to start a chain of experiments they can learn from. The panel may use various techniques in order to come to a decision, one of them is the *Delphi Method* where each member of the panel puts down his estimation; then the results are collected and made public to the panel. In a second round the members of the panel can use this information and change their votes accordingly. This procedure may be repeated several times, and successively the intersubjective vote will perhaps narrow down to a rather definite assessment with an acceptable degree of variation. This will be the intersubjectively defined initial degree of rational belief prior to the first inference according to Bayes' theorem. Delphi Method may likewise be used to gain experts' ratings of further parameters needed for inductive inference, for instance, likelihoods.

Getting a panel of experts to reach an agreement may be tedious and dragging along. In order to spur on personal decisiveness and to speed up the process of gaining an unanimous estimate, bonus points may be offered as a reward or penalty points may be imposed. This is a matter of agreement; we may thus fine the members of the panel if they do not achieve an acceptable degree of unanimity within, say, three rounds, and – in order to make them work properly – we may also fine them if their verdict will not be in concordance with other observations. Or, in order to be more benevolent, we may give them an award if they soon come to a conclusion and if it agrees with reality.

So there are means to eventually gain the parameter values needed for the process of 'learning by experimenting' and their quantification by degrees of rational belief. Quite naturally, these values are estimates, *bona fides*, and certainly not *true* values. We may trust to them, because they have been determined according to ratings representing the best of available expertise; even though these estimates must be re-checked regularly and adapted to changing foundations of evidence available to the panel in charge of laying them down.

8 Tolerance

As we have seen in the previous chapters, the dependability of a given model is a rather relative measure. At best it depends on intersubjective agreements, at worst it must be based on entirely personal estimations if normative criteria are not at hand. At the same time modelling and the quality of the imaging process leading to the model are influenced by the kernel of the mapping – this is all the information getting lost when the real system under consideration is mapped into the model. Any statement about the dependability of a model is probabilistic and thus affected by uncertainties. To some degree the loss of information due to mapping as well as the stochastic process of the evaluation of a model are founded on the evidence available to the modeller and to the relevant scientific community. We thus have to grant a certain *tolerance* to the performance of a model when we compare it with its counter-image (cf. Subsect. 2.4.3), and we will even try to characterize it by *degrees* itself.

8.1 The Qualitative Aspect

We will start with another *Gedankenexperiment* in order to explain what is meant by the term *tolerance*, and we return to Ex. 5.1 where we described the components of the running empty of a receptacle made of a porous medium.

We have described the input-system Σ_I to be modelled by an *I-P-O*-diagram $I_I \rightarrow P_I \rightarrow O_I$. In our example of Ex. 5.1 the input I_I contains any action and condition given at the beginning of the *process* P_I (of the *process* of water disappearing out of receptacle R_1, part of it being finally collected in R_2) and during its course. Thus, for instance, the amount A_1 of water poured into R_1 is part of I_I, the geometry of R_1 and the characteristics of its porous material belong to I_I as well as the vertical distance d between R_1 and R_2 and all the meteorological parameters regulating evaporation. These parameters are, for instance, the heat (we still avoid using the term 'temperature'[1]) and the atmospheric humidity inside and outside of R_1. The development of these parameters also belongs to I_I as long as water disappears out of R_1 in whatever way, furthermore any other given condition like, for instance, the definition of point t_{obs} of time when R_1 will be empty again.

In our example P_I is the process causing water to disappear from R_1, part of it being collected in R_2. This process is governed by the conditions given in I_I and includes seepage through the porous medium (with falling water level in R_1), the development of a capillary fringe within the sides of R_1, evaporation from the sides inside and outside of R_1, evaporation from the water surfaces in R_1 and R_2, the dripping of water from the (lower boundary of the) bottom of R_1 as well as evaporation of the dripping water. The output O_I of system Σ_I, finally, is the set of all the phenomena we observe at time t_{obs}. By definition of this observation time it will be the very moment when R_1 runs empty. At that instant we will observe the phenomenon that either a certain amount A_2 of water is collected in receptacle R_2 or that R_2 is still empty.

In the terminology of *probability* theory a set of input-*phenomena* – like a certain amount A_1 of water poured into R_1 and the meteorological conditions at the location where the experiment is carried out – is an *event*. We represent this event by the symbol E_{in}. Likewise, an observed output-phenomenon (e.g. a certain amount A_2 of water collected in R_2 at the moment t_{obs} when receptacle R_1 runs empty) will be the *event* E_{out}. These two *events* are not (*stochastically*) independent; they are connected by the *process* P_I mentioned above. The *probability* that E_{out} will occur – given E_{in} – is the *conditional probability* $p(E_{out}/E_{in})$. Here the transforming *process* P_I is no longer mentioned separately, it may be regarded as being part of the slash between E_{out} and E_{in} (or we may regard it as part of the input-event E_{in}).

These events making up the input into and the output from the process governing the system Σ_I to be modelled are described qualitatively: we have mentioned amounts A of water poured into or collected in a receptacle, meteorological conditions and moments when observations are made. In many cases models will represent these qualitatively described phenomena by other phenomena. This will be the case whenever a scale-model is generated like, for instance, in Ex. 1.1 of Sect. 1.1 when, for example, an automobile is modelled in order to be tested in a wind-

---

[1] The term 'heat' describes the *phenomenon*, whereas 'temperature' is a quantitative representation of degrees of heat by measurement (cf. the definition of random variables in this section).

tunnel or when a flight-simulator only reflects the performance of a certain aeroplane. But very often a counter-image is modelled in quantities, i.e. in *degrees* of the observed qualities reflected by a scale. Even in those cases where a model represents its *co*-model only phenomenologically, there may well be intermediate stages of quantitative modelling of the counter-system. So the automobile modelled to be tested in the wind-tunnel mirrors the design of the original car and certain features of its bodywork which will be gained by measurements; likewise, a flight-simulator will output phenomena – alarm signals and a jolting of the device – derived from the evaluation of measured *data.* And even the painter and the sculptor depend on measurements of the proportions when they represent their prototype-models.

So one step of *modelling* is to make the various *events* suitable for calculations. As we have seen, the importance of this quantification is different with different processes of modelling, but it is always of relevance. This quantification of the observed phenomena is achieved by assigning real numbers to the *degrees* of each of them. The assignment is done by measurements (e.g. of the *volume* v_1 of the amount A_1 of water, of the *temperature* T_1 corresponding to the heat inside R_1 and of further real numbers representing the rest of the meteorological phenomena affecting the experiment). Such a function of *events* E_{in} into the (real) numbers is a *random variable* (*r.v.*), and we represent it by the symbol i (resembling the word *input*). In fact i is a *multivariate r.v.* $i=(v_1, T_1, ...)$ defined by its components v_1, T_1 and so on. Analogously the *r.v.* o corresponds to E_{out}. Of course, the degree of exactness up to which the values of a *r.v.* can be determined depends on the exactness of measuring and reading. So the *phenomenon heat* will be represented by the *r.v. temperature*, an *amount* by a *volume* and an *instant* by a *point of time* determined by *measurements.*

When we have finally gained a real-number-representation of all *events* relevant to the input and the output of the *co*-model, we – in a second step of *modelling* – have to develop algorithms, means of calculation representing the physical *phenomena* making up the transforming *process* P_O of the model. We call this set of algorithms the *construct* of the model. The *construct* is *the model* in the narrow sense of the word. It contains the conceptual qualitative aspects of the model as well as the quantitative means used to manipulate the data input into the model-system Σ_O. Process P_O is the simulator, and in our example it consists of various mathematical representations of physical laws and of mathematical algorithms. These mathematical tools – and, if necessary, their representations by computer programmes – are used to manipulate the *input r.v.* i according to the *state of the art*, i.e. according to the given degree of physical knowledge and mathematical capability, thus generating the *output* O_O of the *model* Σ_O.

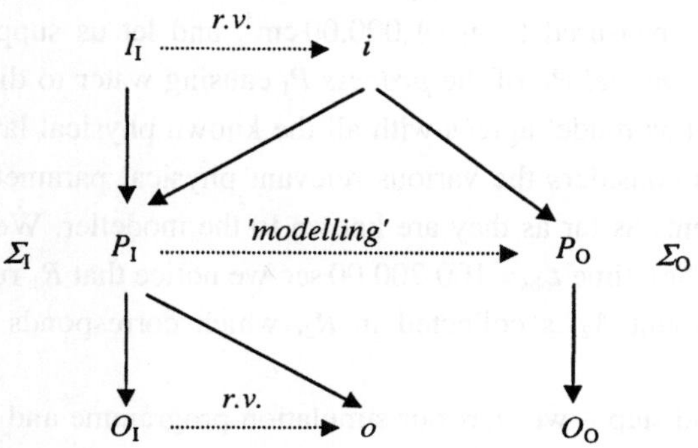

Fig. 8.1. Scheme of modelling

The schematic diagram of Fig. 8.1 reflects these ideas, the original $I_I \rightarrow P_I \rightarrow O_I$-system Σ_I is shown vertically on the left side. The dashed arrows represent the two levels of *model*-set-up: the derivation *r.v.* of random variables and *modelling* in the narrow sense. In general, we use the term *modelling* in a broader sense containing *both* phases: the definition of random variables and the development of the construct P_O. The *model* thus contains the right-hand diagram $i \rightarrow P_O \rightarrow O_O$ as a whole; it is the model-system Σ_O. In practice the transition of *modelling* $P_I \rightarrow P_O$ is sometimes split up into two successive parts: $P_I \rightarrow P_C \rightarrow P_O$. The first transition corresponds to the construction of a (rough) qualitative conception as we have described it in Subsect. 2.4.3. Likewise, the second transition corresponds to the phase of a fine conception where all the details and quantifications are settled. In that case P_C is the *conceptual model*; in our example it contains the physical set-up and the *processes* underlying P_I and their interconnections (cf. Sect. 5.2).

When the development of a model Σ_O is completed, we will try to test its quality. These tests are carried out under the aspects of the quality criteria we compiled in Sect. 5.2; some of them are performed in order to find cases verifying or falsifying the hypothesis H that Σ_O is a dependable 'good' model. Usually the decision whether a case is verifying depends on the observer, and there is the problem to determine the borderline between cases verifying H and those falsifying this hypothesis. The decision whether to accept H or not is affected by utilitarian aspects.

For an illustration of these ideas we return to the above-mentioned example and go through the following *Gedankenexperiment.*

The *r.v.* representing the amount A_1 poured into receptacle R_1 is its volume v_1. Let us assume A_1 is measured to $v_1 = 1{,}000.00\,\text{cm}^3$, and let us suppose we have been able to design a *model* P_O of the *process* P_I causing water to disappear from R_1. We assume that this model agrees with all the known physical laws governing it and that it always considers the various relevant physical parameters and their temporal developments as far as they are known to the modeller. We also assume that after an observation time $t_{obs} = 100{,}200.00$ sec we notice that R_1 runs empty. At that moment an amount A_2 is collected in R_2, which corresponds to a volume $v_2 = 734.52\,\text{cm}^3$.

Now – in a second step – we start our simulation programme and input each of the values measured for v_1 and further relevant parameters – like temperature and other data representing the meteorological conditions – as well as their registered temporal developments. P_O will compute a corresponding output O_O, part of Σ_O. Let us assume O_O contains – among other data – the calculated observation time $^*t_{obs} = 100{,}195.02$ sec from the beginning of the experiment up to the instant R_1 has run empty and the modelled volume $^*v_2 = 734.58\,\text{cm}^3$ of water collected in R_2 at that moment. If the model produces those results, that will be very satisfactory and confirm that it is 'a very good model'. But if we wonder, whether these results really *verify* Σ_I, we will be rather disappointed, as for the observation time t_{obs} there is $100{,}195.02 \neq 100{,}200.00$ and for the volume v_2 of water finally collected in R_2 we must admit $734.58 \neq 734.52$. Comparisons of the computed data with those observed consequently lead to the conclusion that the *model* has been *falsified* and is thus *invalid*, i.e. this test is a falsifying case with respect to the model.

We may now try to refine our model in order to compute output-data O_O so that they equal those measured in reality. But this will prove to be a futile undertaking as we know from Sect. 7.3 that the probability $p(x_1 = x_2)$ of randomly selecting the same value $x = x_1 = x_2$ of a continuous *r.v.* x twice is $p(x_1 = x_2) = 0$. So, a complete correspondence between x_1 and x_2 is *almost* impossible – though not entirely impossible.

On the other hand, who really bothers about $x_1 = x_2$ when measuring and testing an *empiric* theory? The equation $x_1 = x_2$ means that x_1 and x_2 are equal up to the 'last' (infinite) digit following their respective decimal points; it is a mathematical equation like $6 \cdot {}^1/_3 = 2$ and true because of logical reasons. The equation $6 \cdot {}^1/_3 = 2$ means that the multiplication of the real numbers 6 and ${}^1/_3$ on the left side of this equation results exactly in the natural number 2. This equality is with *zero-*

tolerance, the result of the left-hand side of the equation $6\cdot{}^1/_3=2$ and the number on its right are identical and they agree up to the 'last' infinite digit. But when dealing with physical data, *zero-tolerance* cannot be measured, as any measurement provides *finite* numbers of digits only and any computer-controlled calculation does so, too. Thus when we measure a temperature of $T=20.0050037\,°\mathrm{C}$ we are *almost certain* that T is not the real number 20.005003700000..., this is the sequence of the digits 2, 0, 0, 0, 5, 0, 0, 3, 7 followed by an infinite sequence of noughts.

Let $Ver_0(\Sigma_O)$ be the *verification* of a *model* Σ_O under these conditions (with *zero-tolerance* between the simulated data and those actually measured in the *co*-model Σ_I of Σ_O). We conclude that $Ver_0(\Sigma_O)$ is not impossible, but its possibility is zero: *models* cannot – with possibility greater than zero – be *verified* with zero-tolerance.

Consequently, we have to be more *tolerant*, and we introduce the *verification* $Ver_\tau(\Sigma_O)$ of a model Σ_O with respect to a *level* τ *of tolerance*. There is of course still the problem of quantifying τ; the analogue of this problem in the theory of *probabilistic inference* is the quantification of *levels of significance* used in statistics. The great importance of these considerations becomes more obvious when we study the *scenario* described in Ex. 5.1 more closely. Under the mentioned conditions ($*t_{obs}=100{,}195.02$ sec and $*v_2=734.58\,\mathrm{cm}^3$) we stated that we regarded the model as being very dependable and 'good' and that we accepted the described result as a *verification* of our *model*. If the result of the simulation had been some $*t_{obs}=100{,}100.00$ sec and $*v_2=736.00\,\mathrm{cm}^3$, we would still have regarded it as dependable and acceptable. In that case we would have said that our *model* had been *verified* by observation. With computed results, like $*t_{obs}=90{,}000.00$ sec and $*v_2=800.00\,\mathrm{cm}^3$, we might still be 'proud' – as long as there are no other models leading to more accurate results. Results like $*t_{obs}=90{,}000.00$ sec and $*v_2=800.00\,\mathrm{cm}^3$ output by the model can still pass as some sort of *verification* of the *model*, and so on. If the results of modelling are values computed to some $*t_{obs}=20{,}000.00$ sec and $*v_2=20.00\,\mathrm{cm}^3$, we will probably be slightly disappointed, because computed data and measured data differ considerably. Nevertheless, we may still take this for a (weak) *verification* of the *model*, manifesting some sort of a trend, and we will, maybe, accept it as a means of modelling the described scenario if there is no better model at hand.

8.2 The Quantitative Aspect

So there is obviously the problem of quantifying the *tolerance*, and this problem must be analysed in view of the question up to which degree the *model* solves the underlying problems or contributes to their solution and what its *problem-solving capability* is when compared with other *models*. If there is no *model* as yet reflecting the problems to be analysed, we will certainly accept a greater discrepancy between the data o output by process P_I and O_O, the result of the simulation, than will be the case if this model Σ_O is competing with existing *models* already *tried and tested*. And even in that case we will probably tolerate greater deviations between the results of different *models*. This will be the case when a newly designed *model* is more complex and models a great deal more of the physical phenomena compiled in the *real process* P_I than any other *model* does which – though leading to better results – is a less sophisticated approach to the underlying problem. This does not necessarily imply that we always prefer the more complex model to the more 'straightforward' (less complex) one. The model should be adequate and meet our demands, and we may prefer more complex modelling if our main interest lies in studying those phenomena leading to the complexity of the model, even if they contribute to the phenomena observed in reality only marginally.

In a first test of a model-system we may content ourselves with a mere trend represented by O_O like $*t_{obs}>0$ and $*v_2>0$ in our example. Of course, we can neither work with this model nor predict values of t_{obs} or v_2 competently, but the calculated trends indicate that we are on the right way and that we may improve this model so that it will reflect the process P_I governing reality more precisely. And in addition, we must realize that the tolerance we claim is not just a question of differences between data *output* by the model and those derived from measurements in 'reality'. The underlying processes P_I and P_O also differ, and the means we use for modelling are mere abstractions and idealizations; so it is only natural that they lack many characteristic features of those processes P_I observed in 'reality'. But deviations between the processes in Σ_I and Σ_O can hardly be quantified; we consequently have to restrict our considerations with respect to the quantification of tolerances to those aspects covering *data*, either input into or output by these models; the *methods* used for the mapping processes – modelling in the narrow sense – and their tolerance are beyond the reach of quantification. As we may determine the data input into a model as precisely as necessary, the aspect of tolerance is mainly a criterion concerning the output of a given model.

If we have thus lowered our expectations concerning the results gained by modelling, we shall agree on some preset (maximum) *level* $\tau > 0$ *of tolerance* conceded to the *verification* of a model. We may represent this quantifying degree τ by the relation

$$0 \leq |o - O_O| \leq \tau. \tag{8.1}$$

Of course, in (8.1) τ depends on the particular modelled data as well in magnitude as in dimension. We can easily develop a relative measure, for instance, by setting

$$-1 \leq \tau_- \leq \frac{O_O - o}{o} \leq \tau_+ \tag{8.2}$$

whenever the output O_O of the model and the value of the random variable o are non-negative ($O_O \geq 0$, $o \geq 0$). If, for instance, we measure physical parameters on an absolute scale, o and O_O will always be positive or zero. In that case the relative deviation will never be less than -1, and if it is zero, modelled data O_O and those measured in reality will be the same – $O_O = o$ – with the reservation of the above-mentioned tolerance granted to 'identity'. Negative values indicate that modelled data fall below those measured; likewise, positive values indicate that O_O exceeds the output o observed and measured in reality.

Under these circumstances we can *verify* a model *Mod* under various 'conditions' C_1, ..., C_n (in various experiments), each constituting a verifying *case* c_v (with respect to a preset value τ of tolerance). We denote these individual *verifications* of Σ_O by $Ver_\tau(\Sigma_O/C_1)$, ..., $Ver_\tau(\Sigma_O/C_n)$. If in the $(n+1)^{st}$ test of the model, performed under the condition C_{n+1}, we cannot *verify* Σ_O with respect to τ any longer, we may first of all be tempted to raise τ to a level leading to a further *verification* of Σ_O. This is certainly an inappropriate approach to the problem and inconsistent with the fact that we agreed *a priori* – with regard to the test under C_{n+1} – on a *maximum* level of τ, representing the desired precision of the verifiability of Σ_O. Consequently, τ must not be manipulated *posterior* to this test.

Next we may try to 'tune' the model. 'Tuning' means to modify the physical and mathematical calculi compiled in Σ_O in such a way that there are still the verifications $Ver_\tau(\Sigma_O/C_1)$, ..., $Ver_\tau(\Sigma_O/C_n)$ and, in addition, $Ver_\tau(\Sigma_O/C_{n+1})$. In case we succeed, the 'tuned' model will have been *verified* once again with respect to τ by another *verifying case*. This process of 'tuning' is reflected by the process of *adjusting* a model to match *historical* data *adequately* (i.e. within a given *tolerance*). Tuning a model for the first time in order to reproduce observed phenomena and measured data is called the *calibration* of the model and has thus been described in Sect. 5.1.

But it may well be that no such tuning is possible or if it is possible, it will reduce the quality of validation achieved in the past considerably. This means that the model leads to $Ver_\tau(\Sigma_0/C_1)$, ..., $Ver_\tau(\Sigma_0/C_{n+1})$, but the verifications of Σ_0 under conditions C_1, ..., C_n are much 'poorer' now. They are still *verifying* cases with regard to the model, but considerably closer to the maximum *level* τ_{max} *of tolerance*, the level up to which we only just accept to support the assumption of verification. In that case we may realize that we will lose more in model-quality by tuning than we may gain from one more verification. Consequently we will have to concede that Σ_0 was *falsified* under condition C_{n+1} (with respect to τ), constituting *one* falsifying case c_f (but *nothing* else!).

As soon as we detect a falsifying case with respect to a model, it suffers a setback. But we have seen that we can still state that Σ_0 is *valid* under the conditions $C_1,...,C_n$ (but invalid under C_{n+1}). For this same reason we had to qualify our proposition 'all swans are white' in Sect. 6.2 and replace it by 'all European swans are white' when it was falsified by the observation of a *Cygnus atratus*. Likewise, we had to limit the range of validity of our algebraic proposition in Ex. 6.4 to a very restricted subset within the domain of the real numbers.

Given C_{n+1}, there has to be a new phase of the set-up of the model, leading to a new and different model Σ_0^+. Nevertheless, C_{n+1} cannot *invalidate* Σ_0 with respect to the conditions C_1, ..., C_n (given a preset degree τ of tolerance), but there is a chance that model Σ_0^+ will be *valid* with respect to all the conditions C_1, ..., C_{n+1} given the predetermined level τ of *tolerance*. Correspondingly, we restrict Ex. 6.6 'all swans are white' to the statement 'all swans except *Cygnus atratus* are white'; maybe, this statement will never be falsified.

This shows that *validating* a model by seeking *verifying cases* is a constant act of comparing the output of the model with data gained from observation and by measurement. Therefore a model cannot substitute measurements carried out in field work, but it may be a means of getting more or less close estimates of parameter values when experiments are not possible, and it can support the analysis of data gained by experiments in the field – this fact has frequently been pointed out, e.g. by Narasimhan (1995) and Philip (1991) with respect to the modelling of environmental scenarios.

9 Tests

As soon as a model has been developed, there is the necessity of testing it. These tests are performed from various points of view. Each of these aspects has been subject of very particular terminologies. Some of the frequently used terms are quite ambiguous, and to some degree the specific interpretations within the various fields of sciences seem to be contradictory; the same also goes for the terminology applied to different stages of testing a model. In order to avoid any confusion based on semantic or pragmatic considerations when testing a model, we give some very formal definitions in order to mark out the scopes of the testing procedures; some of them aim at the objectives of testing and will mostly be derived from the usage of terms in logic. We admit that these definitions will neither be disjunctive nor complete, but they help to avoid some misinterpretations of terms generally applied.

We also emphasize that every test of a model[1] is only *relative*, due to probability considerations as well as to intersubjective judgement and to degrees of tolerance granted to the comparisons between a modelled output and data collected in reality.

9.1 Objectives

The following example will show our point clearly:

Ex. 9.1 We study a sequence of natural numbers, whose first elements are known to us – but not the generic law generating them –, say,

---

[1] Our usage of the generic term 'model' also covers terms like 'hypothesis' and 'theory' subject to being tested.

$$1, 2, 3, 4, \ldots 20, 21, 22, 23, \ldots, \tag{9.1}$$

and we try to 'model' this sequence by a computer programme. We will analyse those known elements of the sequence and develop a programme listing any number of the elements of this sequence. Due to our analysis this programme may, e.g. look like:

1 $i=1$
2 *print i*
3 $i=i+1$
4 *go to* **2**.

This programme looks satisfactory; it meets our demands. First of all the variable i is set to the number '1' in step **1**, and then this number is printed in step **2**. Next the variable i is increased by one in step **3** to the number '2', in step **4** we return to step **2** and the value '2' of i will be printed and so on. This plain programme will thus generate a list of consecutive natural numbers as long as it runs, and we think it is *formally* correct. Of course, we need some reformulations in order to adapt this programme to existing programming languages, but we are convinced that it is correct as long as only its *internal* logic is concerned.

If we now realize this programme and run it on a computer system, it will produce a string of natural numbers, not the complete sequence of all of them, of course, but as many as we wish, say the string 1, 2, ..., 1,000,001. We will now compare this computed string with the 'observed' data, i.e. those given above. Right from the beginning up to '23' our computation will be in concordance with our observation: the strings of observed data and computed data are identical; and there are 999,978 more predictive values 24, ..., 1,000,0001 generated by the programme. Let us now assume the sequence of (9.1) is continued up to the number '31'; in that case our programme will still be in concordance with those data observed in the sequence and thus prove to be extremely dependable in predicting the elements 24, ..., 31 of this sequence. But if the next datum given in 'reality', the 32nd element of the sequence (9.1), is the natural number '1', there will be a problem, as the corresponding computed number is '32'.

This means that we have had a series of 31 cases verifying our programme and one case falsifying it. We may now suspect that the given series is the series of dates of the calendar starting with '1' for Jan 1st up to 31 for Jan 31st, the next number will be '1' again for Feb. 1st, and so on. So if we limit the number of cycles **4→2→4** of our programme to 30 runs, we will get what we want: up to 31 days of the month January. Any further modelled datum will deviate from 'reality'. Nevertheless, the algorithm used for modelling was insofar correct as it generated the string of consecutive natural numbers as we intended; yet in order to match the sequence of observed data, it must be modified accordingly. And even then, the sequence represented by (9.1) may actually be an infinite repetition of the numbers 1, ..., 31, or it may be something like 1, ..., 31, 1, ..., 32, 1, ..., 33, ... only to mention two of the numerous possibilities. ♦

So, there are obviously two different aspects of the quality of a model. There is an *internal* 'correctness' and a degree of belief in its *external* qualities. We show these relations in a schematic representation:

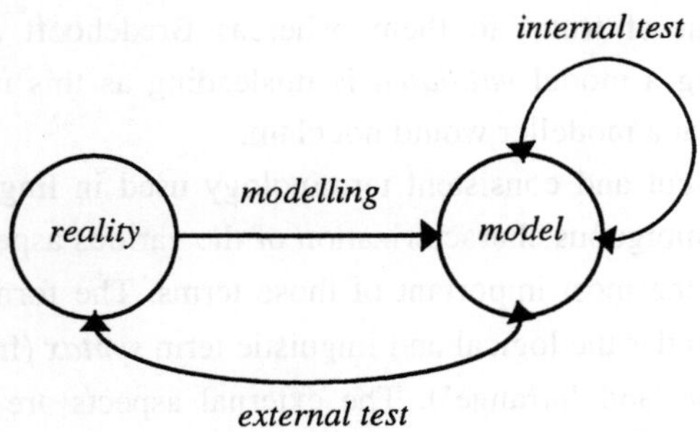

Fig. 9.1. The objectives of testing

We conclude that the 'internal' test is mainly concerned with the formal correctness of the model and largely disregards the relations between the model and its exterior, especially the modelled entities. On the other hand, the 'external' test regards the relations between the performance of a model and its environment, without considering its internal logic.

9.2 Terminology

Before we proceed to discuss the details of 'internal' and 'external' tests we subject a model to, we want to clear up some opposing interpretations applications of the terms 'verification' and 'validation'; they are used interchangeably for several aspects of testing a model.

Schlesinger et al. (1979) recommend to use the term *model verification* to describe the 'substantiation that a *computerized model* represents a *conceptual model* within specified limits of accuracy' whereas – according to them – *model validation* is the 'substantiation that a *computerized model* within its *domain of applicability* possesses a satisfactory *range of accuracy* consistent with the intended application of the model'. Many who model have adopted this terminology, Refsgaard (1996), when evaluating environmental models. Anderson and Woessner (1992) distinguish between the verification of the computer code representing the computerized model and the verification of the model calibration. Oreskes et al. (1994) finally regard a *verified* model as a *true* model, model verification thus being impossible. They propose to restrict the usage of the term verification to the comparison of numerical solutions derived by a computerized model with corresponding analytical solutions. The term *validation*, on the other hand, 'does not necessarily denote an establishment of truth' to them, whereas Bredehoeft and Konikow (1992) think that calling a model *validated* is misleading as this may suggest a degree of correctness that a modeller would not claim.

But there is a clear-cut and consistent terminology used in linguistics and in logic suggesting an unambiguous characterization of the various aspects of testing: we shall try to explain the most important of those terms. The formal aspects of theories are subsumed under the logical and linguistic term *syntax* (from the Greek *sun* and *taxis*: 'together' and 'arrange'). The external aspects are described by *semantics* (Greek *semantikos*: 'significant', 'meaning') and *pragmatics* (Greek *pragmatikos*: 'business').

If we look at the following sentence 'the rock weeps', we realize that it is syntactically correct: it starts with 'the rock', obviously the subject of the sentence, followed by the predicate 'weeps' in the 3rd person singular form; as 'to weep' is an intransitive verb, no object is required. Though it is syntactically (formally) correct, this sentence makes no sense when we interpret it. Interpretations are part of *semantics*, where we ask for the meaning of words and their relations in sentences (or strings of logical symbols). So 'to weep' means 'to shed tears' and describes an emotion a rock as a part of inanimate nature is not capable of.

In addition, there may well be syntactically correct sentences with a proper meaning like 'I will come tomorrow'. If this is the answer you get when you ask a glazier to come and look after a broken window-pane, you may take it literally and hope that he will have repaired it before the next day comes to an end. But, maybe, he just means 'tomorrow' in the sense of 'sometime', and you will have to live with that broken pane for a couple of weeks. This is a *pragmatic* (cf. Subsect. 7.2.3) aspect, also considering the possible intentions of the speaking person. Under pragmatic aspects even 'the rock weeps' may make sense. So, maybe, it is an encoded message only those knowing the code can interpret; it may be a sentence in a fairy-tale; people living in mountainous areas may say 'the rock is weeping' when an avalanche is coming down, or 'Rock' is the nick-name of a boxer knocked-out in the first round and that is the reason why he weeps. So, the interpretation of a proposition depends on the 'business' of the person interpreting it.

When we transfer these terms to the theory of testing models, the 'internal' test of a model performed in order to prove its formal correctness is carried out under syntactic aspects. Therefore we propose to use one of the terms 'syntax check' or 'test of syntactic correctness' to indicate the objective of this test.

The 'external' tests of a model, on the other hand, are part of semantics. When testing the external qualities of a model, we interpret it and try to find verifying or falsifying cases for it. This is done in very much the same way we have applied when we tried to find a sensible interpretation of the sentence 'the rock weeps'. Maybe, there is no solution, or, maybe, we are lucky and even detect the verifying case of the person nick-named 'Rock'.

So, to perform an external test of a model means to look for verifying (or falsifying) cases supporting (or refuting) the model, and to *verify* a model-system Σ_O just means to detect a case c_v verifying it, that means a case c_v supporting Σ_O with respect to a predefined degree τ of tolerance. Likewise, Σ_O is falsified whenever a case c_f conflicting with the results of Σ_O given τ is detected. This *process* of research leading to the consecutive observation of verifying or falsifying cases is the *validation* of the model in question, and it is a relative process always carried out with respect to a preset degree τ of tolerance. We remember that having validated a model *only* means that it has been tested under *those* semantic aspects of interpretation resulting in verifying or falsifying cases; it does *not* mean that the model is *valid* in the sense that there are not and will *never* be any cases falsifying it, it is just well-founded on the basis of these verifications.

We now distinguish the reasons for the objections raised against the application of these terms. One of the arguments is that the lay public may deduce the *validity* of a model – in the sense of an everlasting confirmation – from the fact that it has been validated. But these are the pragmatic aspects of linguistic usage. Among the various scientific communities the implications of the words *verification*, *falsification* and *validation* should be clear. We shall subsequently use these terms as defined above: a verification is the confirmation of a general law – as, for instance, constituted by a model – for a single case, a falsification is the detection of a falsifying case with respect to a general law, and validation is the *process* of verifying or falsifying such a law.

Under pragmatic aspects we may think each of these terms is too expressive and 'carries with it an aura of correctness that we do not believe many of us ... would claim' (Bredehoeft and Konikow 1992). In that case, and especially in view of the discussion with the lay public, we suggest another terminology: validation with the implication of validity should be replaced by *testing* or *evaluation*, words including the possibility of falsification. Furthermore, we may say a case *supports* a model instead of it verifies it; we may even be more cautious and call a verifying case a case *not falsifying* the model or *not invalidating* it. And even the term *falsification* carries with it 'an aura of being *definitely false*' which we do not accept; so we would rather call a falsifying case a case *casting doubt* on the assumption that the model is valid. But these are questions which entirely belong into the linguistic sphere, they are beyond the scope of our considerations, and we shall stick to our terminology described above and talk about verifications, validations and so on in the there mentioned meaning of the words.

9.3 Testing

The characterization of testing procedures under internal and external aspects is not a disjunction. There may well be qualities of a model subject to testing which are internal, but can only be evaluated under considerations regarding their external validity, and *vice versa*. We will study these interdependencies in this section.

The syntactic correctness of a model will be subject to internal testing procedures. The intention is to show that the methods applied for modelling work properly and that they 'do what they are expected to do'. That means, for instance, that computer programmes are debugged and that they transform data input into them correctly into the corresponding output with respect[1] to these programmes. Debugging a programme may be a lengthy and tedious enterprise, but in the end we will know for sure whether we have been successful or not, because a programme will not run unless it is *syntactically* correct.

[1] The additional remark 'with respect to these programmes' emphasizes the fact that the generated output need not necessarily be in concordance with data observed in reality; it is just the correct result of the calculus underlying the given programme.

The question whether a programme 'does what it is expected to do' is more difficult to answer. A programme may be formally correct regarding the syntax of the respective *programming language*, but at the same time it may be *logically* incorrect. This means, the programme runs, but it does not provide the expected results. If, for instance, we change command #3 in the routine of Ex. 9.1 to $i=i+7$ the resulting programme will be syntactically perfect and able to work, but it will result in the sequence 1, 8, 15, 22, ... of natural numbers we did not want to generate. This aspect of correctness of a programme will often be scrutinized by using sets of test data (cf. Subsect. 2.4.3). These data sets contain some significant input data i_1, ..., i_m and also the respective output o_1, ..., o_n. These output data have been gained from the input by methods different from the respective programme, for instance, by being calculated manually. Now i_1, ..., i_m are input into the programme to be tested and its output $*o_1$, ..., $*o_n$ will be compared with o_1, ..., o_n. If there are any differences between $*o_1$, ..., $*o_n$ and o_1, ..., o_n, we conclude that either the test data are incorrect or that the programme does not work properly, though it runs. Having checked the test data, we will be convinced that they are correct, but that the computer programme does not work properly.

On the other hand, if the data sets $*o_1$, ..., $*o_n$ and o_1, ..., o_n output by the programme and by the testing algorithm agree, we should be careful not to jump to conclusions, because this concordance might be pure chance. So we should not be too credulous, but test the programme again with another data set; we must probably repeat this procedure several times. If we always get the same result of data computed by the programme equalling o_1, ..., o_n, then each of these tests is a *verifying case* for the hypothesis that the programme is formally correct. After a series of those tests we may conclude that we may eventually venture to release it to regular runs, even though we know that this programme may still contain some logical errors we did not detect and which will lead to wrong results as soon as the input triggering off these malfunctions occurs. This is a kind of internal test and a case of verification. In the example of computer programmes the results computed from predefined test data are verified. Obviously the formal correctness of a model (cf. Subsect. 5.2.1) includes two facets: let us call them the *syntactic* and the *operative* aspect. The former ensures the technical capabilities of the model, the latter gives evidence of its capability of reproducing the modeller's logical ideas adequately. But both criteria – constituting the internal *test* of the model – are entirely *formal* and have no implications whatsoever for the dependability of a model when it is compared with the part of reality it represents.

When a model has passed these internal tests, its *external* dependability will be probed. As an example we analyse the above-mentioned routine generating natural numbers: it seems to be correct under syntactic and operative aspects. On the condition that our commands can be interpreted by a computer, we will be convinced that we can generate the sequence of natural numbers correctly as soon as this programme is started. But if we run it long enough, the print command #2 will produce a message like 'overflow', because we have reached the limits of capacity of the device. So we must realize that our routine does not produce the *sequence* of natural numbers, but that it will output only a rather long *string* of consecutive natural numbers once it has been started, and under practical aspects this string may be 'long enough' to model any observed 'phenomenon', i.e. any given finite string of natural numbers. In Ex. 9.1 we observed the occurrences of the integers 1, ..., 23; if our programme is tried out and tested, it will supply a long enough string of numbers. So we start it and from its output we venture the prediction that the next 'observed' integer, i.e. appearing in the sequence, will be '24'. If the following number is '24', this observation is in favour of our model; it supports it. We continue predicting 25, ..., 31, and let us assume that these predictions are always in concordance with the observations in 'reality': the sequence is continued with these numbers; these are seven more verifying cases supporting our model. According to the given evidence we may conjecture that this sequence, of which we know the first 31 numbers, is the sequence of positive integers and that our programme used for modelling is dependable, because it never failed in the past. If now something goes wrong, because the next number should be '32', but turns out to be '1' as the sequence shows the dates of the calendar, we have to adjust our model to this new situation.

These purely mathematical examples may look rather farfetched and hypothetical, but they have the advantage of demonstrating what we have to reckon with when comparing a model with its counter-image without bothering about any technical details. So sequences of numbers, for instance, are good for any occurrences of reality. *Any* continuation of a given list of numbers is reasonable, because it may well reflect the law governing the generic sequence. If we are asked to continue the string 1, 2, 3, 4, 5, 6, ..., we may decide for 11, because this is just the law leading to the generic sequence 1, ..., 6, 11, ..., 16, 21, ...26, ..., and any other number but '11' could be used to continue this list beginning with 1, ..., 6, ... with a similar justification. This example comes close to the reality modellers have to cope with, because they never know what the result of the next *validation* of their model will be, just like we do not know what number will be next in a sequence of 1, ..., 6, ...

So we return to the discussion of Ex. 9.1. If we predict the string of numbers 24, ..., 31 to follow the known numbers and if that is correct, then our model has worked successfully for eight times. If there is a set-back like a prediction of '32' computed by the model, but an observation of '1', it has still – given the evidence available prior to these tests – been reasonable to assume that the given sequence is the sequence of natural numbers, where '32' is the successor to '31'.

Now, this disastrous result of the last prediction of '32' in contrast to the observation of '1' will be analysed thoroughly and – if necessary – it will be discussed by experts on this subject. There will probably be different evaluations: some of the experts may suggest the sequence is an infinite repetition of 1, ..., 31, 1, ..., 31, ... Others may *guess* that this sequence represents the calendar. They may also assume that the generic sequence is built according to the law 1, ..., 31, 1, ..., 32, 1, ..., 33, ... or anything else like 1, ..., 31, 1, 1, 1, 1, 1, ... continued with an infinite number of the digit '1'.

This problem calls for experts' ratings. So everybody whose expertise lies in the field of sequences of integers is asked and after some argumentation with 'likely ...', 'more likely ...' and 'most likely ...' the experts will agree upon a certain conjecture like 'this could be the calendar we observe'. They jump to this conclusion because of the *consensus* among the members of their circle, and they reach this consensus on the basis of the evidence available to them and following their 'personal' judgement.

This conjecture will be tested in the future and we will agree that the experts' hypothesis – the sequence represents the calendar – will be supported as long as the observed successive numbers are 2, ..., 28 (or, maybe, 29 if we accept a leap year). On the other hand, any different observation will be a falsifying case with respect to this hypothesis, and the experts will have to come together again and work on a new speculation.

We will now analyse a different *scenario*: we did *not* invent the computer programme described above to model the sequence of observed integers, but we used a generator of integer random numbers. Let us assume that we have started this generator and that its output is a string perfectly matching our observations for some 10,000 elements of the sequence. So, while the experts are still discussing whether the observed sequence is the calendar or obeys some very difficult rules of development, we easily predict the observations with a maximum precision. We must admit that our success has been a stroke of luck, but obviously gambling has proved to be a very effective and *reliable* instrument, possibly much more reliable than the experts' ratings, though their results are scientifically 'justified'. The reliability of a model is an internal *quality*, but it is only detected when the external relations between a model and its counter-image are analysed. In order to be reliable, a model must be syntactically and operatively correct (and objective, as mentioned in Sect. 5.2).

In the previous example of a programme generating random numbers we may have created the impression that this is not the right way to tackle the problem of determining successive elements of a sequence, though this method has turned out to be very reliable as it produced a string of 10,000 correct predictions. As we have already pointed out completing a string of numbers when only some initial elements are known – but not the generic generating law – can be done by any string of numbers. So why not just generate them at random? The only reasonable objection to this method seems to be the limited range of the mentioned generator with an output restricted to a certain range of natural numbers.

But even if we accept that numbers generated at random are a proper means to continue the starting string of the mentioned sequence, we prefer the idea of a generic law leading to this sequence. So we proceed as described above: we will start with the sequence of natural numbers as this is the most *natural* assumption, we will refer it to the calendar and so on, always making use of the available evidence; but each of these suppositions can still be wrong, just like the random generator can be. But who really knows? Maybe, the sequence under research has in fact been generated at random and there is no reasonable way of predicting its elements at all.

Consequently, the question whether a model represents 'reality' adequately very often has to be answered by the *consensus* among the members of a relevant scientific community. They come together, analyse the problem to be modelled, and then they agree on a procedure they trust to be dependable. The reliability of a model is merely a question of precision: the 10,000 data generated at random were very reliable, but even though, we would prefer not to depend on that model.

Objectivity has to be agreed upon in much the same way. As soon as the majority of the scientific community is convinced that modelling has been carried out objectively, we will accept this vote. We may still have some doubts, but in order to get a new estimation of objectivity, we have to falsify this verdict. Falsification in this case is done by delivering evidence supporting the hypothesis that the research was *not* done objectively.

So, the internal test comprises the checks on the syntactic and operative correctness of a model as well as on its objectivity and reliability. The latter tests are carried out under purely syntactic aspects as well as under semantic considerations.

We shall call this part of testing a model its 'test for formal correctness'. Thus, a formally correct model is syntactically and operationally correct, it is objective and reliable. These criteria are relative, which means we will never be definitely sure that a modelling system is truly formally correct; we just do our best and under the aspects of given evidence we have reason to assume formal correctness the way we have described it. If we want to be extremely cautious, we may content ourselves with the statement 'available evidence is *not* inconsistent with our assumption of formal correctness of the model'.

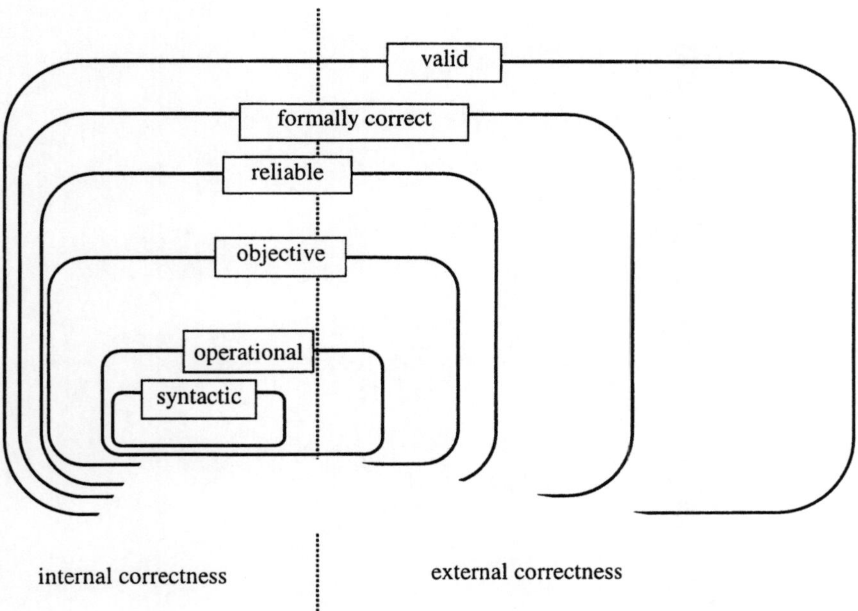

Fig. 9.2. Hierarchy of correctness

The formal correctness of a model thus includes aspects of syntactic and semantic dependability; semantics are part of the operational dependability of the model as far as they are determined by comparing it with its *co*-model. A model must comply with each of these criteria in order to be *valid.* The additional items of the validity of a model are entirely semantic: whether they are fulfilled or not can only be decided by comparing the model and its counter-image. We will give a description of these facets of validation and of the implications we can draw from a statement like ‘this model is valid’ in the following Chap. 10.

These considerations lead to a hierarchy of the criteria used for testing a model. This step-by-step set-up of the dependability of a model is reflected by Fig. 9.2.

10 Validity

We have already touched on the term *validity* of a model in various sections of this book, and we have emphasized that it is a means to measure the degree of dependability up to which a model is *well-founded.* This interpretation is justified by the etymological roots of the word; and if we agree to use the term in that sense it is not an absolute measure, but rather refers to a quality of a model characterized by *degrees* of validity. This is due to the fact that the foundations of scientific propositions (*theories*, *hypotheses* or *models*) are very subject-specific and vary from one field of research to the next, and even within each of those fields they depend on the stage of scientific progress and on the available evidence.

We have also seen that the terminology is far from being standardized as terms like *verification* and *validation* are used interchangeably. In addition, the authors who have a critical look at this problem disagree about the implications of the usage of those words. Some of them suggest banning them altogether for pragmatic reasons, others take the view that those are only words, and 'what's in a name?' as long as a model works. Konikow and Bredehoeft (1992) and de Marsily et al. (1992), for instance, debated about this problem from the point of view of modelling in hydrology. Their arguments caught our attention in particular as they seem to mark out the range of opposite opinions held in the discussion which should not be limited to the analysis of environmental models this controversy was based upon.

So, we will repeat the conclusions we can draw from the *validity* of a model and we shall define the scope of this term. Obviously, there are some serious misunderstandings between the disputing parties, perhaps because of fundamental misinterpretations of epistemological ideas, especially of those of Popper – who was called a witness by Konikow and Bredehoeft (1992) – and of some of his colleagues in the *Vienna Circle* as well as of others working on the subject. These misunderstandings may result from a vague usage of those technical terms mentioned above. Finally, the philosophers' considerations based on the conditional probabilities we described in Chap. 7 allow to develop means suitable for the development of quantifying measures of *degrees of validity*.

10.1 Validation

As those terms *verification* and *validation*, triggering off such a vehement dispute amongst scientists, are of great importance in our book, we think that it may be helpful to give a short survey of this debate and to point out the different positions. The mentioned parties are involved in a controversy about the degree up to which a (hydrological) *model* can be *validated.* This argument was started by Konikow & Bredehoeft (1992) when they cited Popper (1934) who, from an epistemological point of view, states that *under certain conditions* recognition cannot be 'validated'. De Marsily et al. (1992), on the other hand, think that calling on Popper is 'to use a sledgehammer to kill a fly'; their arguments are in short, 'Mr Popper or not, as long as a hydrological *model* is in concordance with our observations it serves its purpose'. In a reply to this comment Bredehoeft & Konikow (1992) describe their position more clearly as the mere – pragmatic – fear of possible misunderstandings between the scientific community and the general public with respect to the term *validation*; this word carrying 'with it an aura of correctness that ... many ..., who model, would (not) claim'. They recommend to abandon the term *validation*, which does not 'imply that model testing and evaluation should be abandoned. ... Models are adjusted until an *adequate* match to some set of historic data is achieved.' They suggest to call this process *history matching*.

If we accept the controversial term *valid* in the meaning of the word as *firm, sturdy* and *well-founded*, to validate a model means nothing more but to look for *any* piece of circumstantial evidence in favour of the assumption that it is 'good' and *dependable.* The judgement of dependability is thus *induced* from a chain of growing evidence, and being inductive implies that we can never be definitely sure about the quality of the model in question as an *absolute* criterion. It is only proved for those cases it has been *verified* for and for those quality criteria tested with the reservation that all those tests were stochastic experiments and that the interpretation of their results may be erroneous.

Accordingly we have to look for all those aspects we have to test in order to validate a model, to make it *well-founded.* As we consider the validity of a model as its generic quality criterion, validity will cover all the other criteria already mentioned and also several additional aspects.

10.2 The Scope

As the meaning of the term *validity* fluctuates between the rejecting interpretation that 'models cannot be validated at all' and numerous definitions of validities in the theory of modelling in, e.g. social sciences, we shall try to modify the definitions used in the latter in order to satisfy the necessities of the former, thus avoiding the rigid and inappropriate verdict of the impossibility of validating a model 'at all'.

Only talking about *validating* a *model* seems to be inappropriate to the attempt to describe what is actually covered by this term, because *validation* is a complex and multilayered process. In Chap. 9 we have seen that the validity of a model is the generic quality criterion covering any other facet of its dependability. Consequently all those aspects subsumed under the term *formal correctness* are part of the validity of a model. As formal correctness covers any item regarded as being part of the internal correctness and also some of the external aspects contained, e.g. in the objectivity and the reliability of a model, there must be further components of the validity which can only be subject to entirely external considerations.

We thus restrict the discussion on the validity of models to these additional external facets. Fig. 10.1 shows in detail the way how the process of *validation* might be subdivided. We have to admit that the individual case may be different and more (or less) complicated with certain concrete models. Fig. 10.1 only and *cum grano salis* outlines the general structure underlying the idea of model-*validation*. The graph is organized vertically by the concept of $I \rightarrow P \rightarrow O$ which we used in Chap. 2 in order to describe systems. The shaded block of Fig. 10.1 contains the fine-structure of that part of *validation* outside the formal correctness of a model. On the left of this block there is the *co*-model represented by a vertical *I-P-O*-system; its input and output are observed phenomena which are made ready for calculations by measurements mapping them into sets of data. The transforming process P of the *co*-model contains the natural laws and processes leading to the phenomena observed in reality. In Ex. 5.1, for example, the input is the amount A_1 of water poured into receptacle R_1 and the meteorological conditions governing the experiment. The output O of the system is the point t_{obs} of time when R_1 runs empty and the amount A_2 of water collected in R_2 at that instant. The transforming process P contains any physical process causing water to disappear from R_1, like percolation and evaporation. Likewise, the columns on the right of the shaded

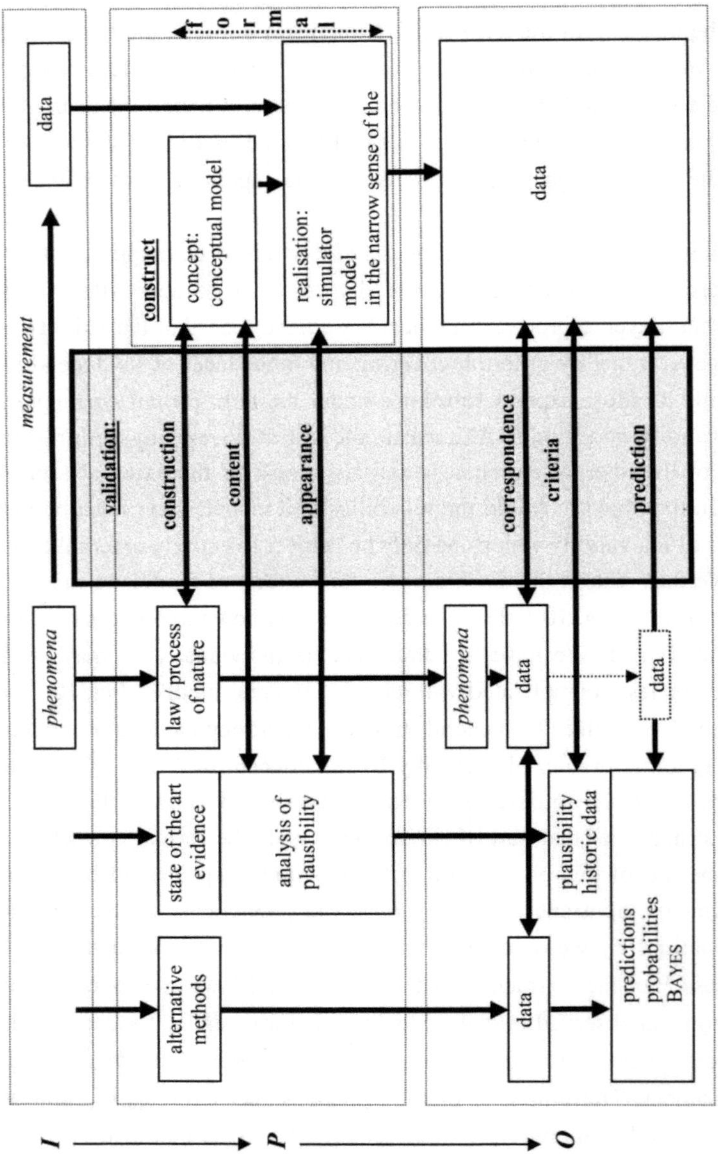

Fig. 10.1. The logic of validating models

block block represent the model, an *I-P-O*-chain of its own; its *process P* is the model in the narrow sense of the word, the simulator reflecting the physical processes of the *co*-model. In Sect. 5.1 we called it the *construct* of the model, containing the conception and its realization, e.g. by a system of computer programmes.

There is an additional entry on the extreme right representing the tests for formal correctness of the construct of the model as a part of its validity. More to the left, there are two supplementary columns; they reflect the evidence based on the experts' knowledge intersubjectively accepted by the relevant scientific community and alternative methods of analysing the problems underlying the model in question.

To *validate* thus means to confirm the internal and external components of formal correctness and to compare the different parts of the (complete) model on the right-hand side of the shaded region with the various items enlisted on its left, *verifying* the model under each of the listed aspects. We thus distinguish validity with respect to the *subordinate* criteria, like syntactic and operative correctness, objectivity and reliability as mentioned in Chap. 9. The further components of validity may be characterized[1] in the following way:

Construction Validity

This facet of validity reflects the similarity between the conceptual model – as part of the construct of a model – and the natural processes governing the real physical system of the *co*-model subject to modelling. The conception is compared with the *co*-image and the kernel of the mapping, and it is analysed under the aspects of not only representing reality appropriately, but also the intentions of the modeller. So, for instance, the physical theories representing the processes observed in the *co*-model must reflect them properly. This criterion is entirely qualitative.

Content Validity

When a model is validated with respect to its content, its acceptation is analysed (by experts) from the viewpoint of *plausibility*. Whereas the construct validity is determined by a direct comparison between the concept of a model and its *co*-model, this scrutiny is based on the evidence available to the scientific community

---

[1] Some of the following terms are frequently used when data models are validated, e.g. in social sciences; we have adopted and modified them in order to make them more generally applicable.

and on the current state of the art. In addition, the representation of the conceptual model must be dependable; this implies that, for instance, the applied physical theories are adequately reflected by state-of-the-art mathematical algorithms. This standard is also qualitative and to some degree it overlaps with aspects of formal correctness.

Appearance Validity

In this stage of validation the realization of the rough concept underlying a model – as part of the construct – is scrutinized under the aspects of plausibility. Whereas the test of formal correctness only deals with the question whether the realization transforms the rough conception adequately into a workable model, its appearance validity guarantees that this realization of the conceptual model also *appears* to be plausible and up to the state of art of the relevant field of research.

Correspondence Validity

Hereby the data output by the *model* are compared with data measured in reality or data gained from alternative scientific methods. This validation describes the way these sets of data match. It examines the extent of concordance between the output from the model and that from the underlying observed physical phenomena.

This aspect of validity covers the quantitative adaptation of the model by calibration, re-calibration and tuning as described in Chap. 5. That means that the correspondence between a model and its co-image cannot be determined in one single test. Correspondence validation is rather an incessant process of scrutiny accompanying a model as long as it is used. If competing models output corresponding series of data, this is in favour of both of them. Often those compared models represent different degrees of insight into the matter; one of them may be historic and serve to evaluate a modern approach to the *co*-model. Correspondence between these models will thus prove that the historic model was not too bad after all and that the recently developed system seems to be dependable.

The determination of correspondences also covers parts of formal correctness. It may be influenced by subjective interpretations as well; we have, for instance, seen in Chap. 5 that the calibration of a model is frequently based on finding best-fitting relations between the input and the output of a system, but the definition of 'best-fitting' is *eo ipso* rather woolly and subjective. Furthermore, Procrustes effects easily occur when outliers within a series of data are eliminated thoughtlessly.

Criterion-related Validity

This aspect particularly covers the aspects of evidence – based on experience and historiography – and plausibility. In this respect criterion-related validity is connected to the aforementioned correspondence. But whereas the correspondence is a quantitative means to evaluate models, this facet of validity is rather focused on qualitative aspects and derived from plausibility and the intersubjective estimation of the relevant scientific community. The criterion-related validity is based on the evidence available to the expert public and their experience and state of knowledge.

Predictive Validity

The predictive validity, finally, is the degree up to which data gained by *modelling* are in concordance with future observations, predictions depending on alternative scientific research methods or on scientifically founded *degrees of rational belief.* If a model is predictive, it satisfies the criterion of correspondence restricted to the comparison of modelled data and future measurements in reality.

Predictive validity is also affected by subjective considerations and the Procrustes effect. This is due to ex-post interpretations leading to statements introduced by the phrase 'Haven't I told you before ... ?'. As we think that the predictive capacity of a model may easily be overestimated, we try to specify our position with the following example:

Ex. 10.1 In Ex. 1.18 we already mentioned a bewildering lecture broadcasted on *TBN* Europe. On Nov. 29th, 2001 we listened to another one dealing with the *WTC* massacre of Sept. 11th, 2001. The lecturer interpreted the original Hebrew version of a passage from the bible and decoded parts of the text as containing words like 'airplane', 'knife', 'Bin Ladin', 'Bush' and 'anthrax', to mention some. He did not clearly say how he came to this *ex-post-facto* interpretation, but he definitely had not been able to *predict* the disaster, derive his exegesis *ex-ante*, otherwise he certainly would have warned the authorities.

This is the trouble with those predictions by, for instance, the lecturer on *TBN* or the interpreters of Nostradamus' visions or whoever they are. There is always the same basic pattern: people are shocked by a disaster and this brings those so-called prophets into the arena. They immediately produce documents handed down on us for generations and begin to interpret the sometimes cryptic words and sentences until they have eventually made out a passage which somehow seems to predict those events, just like Procrustes tortured his captives until they came up to his expectations, i.e. fitted his bed. In a similar way interpretations of works of art often seem to be more reading a special meaning into the object than deriving it from it – especially if the artist is already dead.

On the other hand, there are extremely predictive 'models' of the world we live in. A *normal* number, for instance, is an irrational number containing any possible finite string of consecutive digits. If digits are used to encode the characters of an alphabet, a normal number thus contains any text which has ever been written or will be written and thus predicts any event going to happen and being reported in the future. There is evidence that the numbers e and π may be normal; nevertheless, normal numbers can be constructed, they 'exist' and possess the greatest predictive capacity ever possible; unfortunately we are not able to interpret the predictions contained in these numbers correctly *prior* to the events we want to be warned of. ♦

So the specialist must be aware that evaluating the predictive capacity of a model *ex-post*, i.e. from data gained after the release of the model, but before the validation of this quality criterion, may be done subjectively and may easily lead to false conclusions.

Our classification includes most attempts at finding a consistent terminology covering the process of validation. The facet regarding *correspondences* (and comparisons with predictions as soon as they are historic) comes very close to the above-mentioned ideas of *history-matching* (as proposed by Bredehoeft & Konikow (1992) to be a more appropriate word for what is called the *validation* of a model). The scrutiny of predictive validity of a model is sometimes called *post-audit*; it is the *ex-post* evaluation of a model evaluating its predictive capacity by the comparison of *ex-ante* predictions and *posterior* observations and measurements. We have also read (Leijnse and Hassanizadeh 1994) about the *weak* validation of a model given by *weak* definition – corresponding to our qualitative validation regarding the *construct* and its *content* – and the *strong* validation of a *strongly* defined model, the complete process of *validation* as described above.

Apart from the aforementioned *static* subdivision of the *validation* of a model this process must also be considered under *dynamic* aspects. As we have seen in Sect. 6.2, transcending the borders of the *range of validity* of a model gives rise to improving it without diminishing its value, because at its time it served its purposes. This leads to the well-known cyclic procedure of modelling: systems development followed by periods of evaluation leading to the development of improved systems which must be evaluated again. Consequently, a model can only be *partially* valid, i.e. valid within its *range of validity*, and validation is the process of permanently including verifying cases into this range and excluding falsifying cases from it, thus defining its borderline in an iterative procedure of modelling.

As the discussions on the implications of the term validity have to some extent been based on epistemological theories, e.g. developed by Popper and Carnap, we will give a broad outline of the foundations of these arguments in the following section.

10.3 Epistemic Foundations

Epistemic (Greek *episteme*: 'knowledge') considerations relate to knowledge, its development and its validation. Validation of models deals with the evaluation of the dependability of empirical systems and of the evidence derived from them. This evidence is based on experience produced by experimenting.

Popper (1934) and Carnap (1950, cf. Carnap and Stegmüller 1959, too), both referred to by the disputing parties, worked in the field of the validation of scientific theories. Actually, neither Popper (nor Carnap) can take the credit for the crucial recognition that certain theories cannot be 'validated', as this was already an established fact when they began with their scientific work. This statement reflects that some general laws – formulated as universally quantified propositions – derived by inductive inference cannot be completely verified (cf. Sect. 6.2)[1]: it is impossible to detect each case in an *infinite* scope of the quantifier verifying the proposition. Yet, the general 'law' is easily falsified as soon as a falsifying case is discovered. We thus have to limit the criterion of 'not being completely verifiable' to empirical theories dealing with universally quantified – general – laws whose quantifier possess infinite scopes.

As we have pointed out in Sect. 6.2, Popper identifies empirical theories by their *falsifiability* and thus distinguishes them from others. A quantified statement that is not falsifiable over an infinite universe is either a tautology and true for *logical* reasons (e.g. 'all swans are white or not (white)' or a mathematical theorem like, for instance, Pythagoras'), or it is what he calls metaphysical (e.g. 'there is intelligent life somewhere in the universe'). The mentioned tautologies are universally quantified; the metaphysical statements are existentially quantified. Popper

---

[1] A *complete verification* is tantamount to a *proof*, and, in general, empirical theories *cannot* be proved.

postulates that an *empirical*-scientific system must bear the possibility of *failing* when competing with experience. Still, we think statements like 'there will be a collapse of the (cosmic) universe' are empirical after all, though they are existentially quantified over an infinite (temporal) scope and certainly not falsifiable. According to Sect. 6.2 the borderline between cases *verifying* and *falsifying* a *theory* marks the scope of the range of validity of a theory (a model). Going beyond this borderline by detecting cases falsifying the theory under question leads to new insights and possibly to a new and more comprehensive theory. Hereby the value of the former theory is not belittled; at its time of conception (and the then available evidence) it may have been optimum; but progress is constantly granting us a better understanding of the complex nature of the problems to be investigated.

Popper quite explicitly (and pragmatically) demands the practical applicability of research (Popper 1964). Of course, he knew about the problems connected with validating a hypothesis (or a theory or a model). He therefore subdivides theories to be set up into three classes: *logic-mathematical* (tautologies), *empirical-scientific* and *philosophic* (metaphysical) with different criteria for a proof of the irrefutability of the relevant theorems within each class. Generally speaking, he proposes to validate theories (models) by their problem-solving abilities and he emphasizes:

> '*...wir fragen: Löst die Theorie ihr Problem? Löst sie es besser als andere Theorien? Verschiebt sie es nur? Ist die Lösung einfach? Ist sie fruchtbar? Widerspricht sie vielleicht anderen philosophischen Theorien, die wir zur Lösung anderer Probleme brauchen? ...*'

We could not find an English version of this text, and as Popper was somewhat peculiar about authorizing translations of his works, we try to paraphrase his ideas instead: '... we ask: can the theory (under discussion) solve its (underlying) problem? Is its problem-solving capability better than that of other theories? Does it only postpone the solution of the problem? Is the solution simple? Is it prolific (effective)? Does it conflict with other philosophic theories needed for the solution of other problems (we rely on when solving other problems)?'. In their reply to Konikow & Bredehoeft (1992) de Marsily et al. (1992) touch upon all those central ideas mentioned by Popper and at the same time give an adequate English version:

'Verification will certainly never be perfect, but again we are not striving for certainty and perfection, only to do 'our level best'").

We use Popper's above-mentioned statements to put forward the following *desiderata* - criteria a theory (model) should come up to:

– the *theory* in question must contribute to the solution of the given *problems*,
– the problem-solving capability of the *theory* should not be worse when compared with other *theories*,
– the solution offered by the *theory* should be as simple and fruitful as possible,
– the given *theory* must not be in contradiction to any other *theory* already *tried and tested.*

Yet, the involved philosophers taking these views are far from supporting them unanimously. Popper's 'favourite' opponents seem to have been the members of the *Vienna Circle*. As a matter of fact, Popper never officially belonged to this group, but his ideas – as far as the validation of models is concerned – seem to be closely related to the concepts of its members. Where Popper tries to validate models – i.e. he tries to determine the borderline of its range of validity – by finding falsifying cases and sorting them out, the neo-positivists of the *Vienna Circle* try to fix this boundary by determining verifying cases. A prominent supporter of the latter idea is Carnap who confirms empirical theories on the basis of Bayesian inference which we discussed in Subsect. 7.2.3. Both of them, Popper and Carnap, have developed *probabilistic* means in order to describe *degrees* of the validity of a theory quantitatively; we shall study these measures in the following section.

The discussion proves that Popper certainly did not maintain that empirical theories could not be validated in the sense of the word, i.e. not scrutinized and founded on the recognition gained by testing them, neither did Carnap. We can condense their arguments like this: an empirical theory cannot be proved the way a mathematical theorem is proved; it can only be verified or falsified, and it is valid as long as it is not falsified. In addition, they both regard the validity of a theory (model) as a relative quality depending on the available evidence, and they even recommend means to quantify it by degrees. These degrees are based on probabilities leaving room for tolerances as introduced in Chap. 8. And even the falsification of theories tested with zero-tolerance does not necessarily refute them.

Leinfellner (1965), for instance, concludes that a single falsification of such an (classic) empiric theory does not justify its rejection, but rather calls for transforming it into a proposition employing ideas of *probabilities* thus raising the predefined degree of tolerance.

10.4 Quantifying the Validity of a Model

There are different means to quantify the validity of a model by degrees. Quite naturally, they are all based on probabilities, and these probabilities are always conditional taking the evidence into account that is available to the modeller. The most evident and easily described measure is Carnap's; it is founded on Bayes' theorem (7.7) and (7.9).

10.4.1 Carnap's (Degree of) Confirmation

To Carnap (1950) the *degree of rational belief*, his conception of *probability*$_1$, is a *degree of confirmation* that indicates the degree up to which a *theory* (or *hypothesis* or *model*) is *confirmed* either by the given evidence E – i.e. *a priori* – or on the basis of E, but now enriched by the experience gained from the result R of an experiment – i.e. *a posteriori* with respect to this experiment. These ideas can be traced back to publications by Carnap dating from the 1930s.

This approach is based on a reinterpretation of equation (7.8) leading to the *confirmation of hypotheses by observation*. Event X in (7.8) once more is the *hypothesis* H to be confirmed, E is the *evidence* available to the experimenting observer not knowing whether H can be accepted or not. The term probability p is substituted by *confirmation* c with *prior confirmation* $0 < c(H/E) < 1$ and *posterior confirmation* $c(H/ER)$ of H making use of further information gained from an observation (of the occurrence of a result R of a testing experiment). The transformation of *prior* into *posterior confirmation* follows Bayes' theorem (though Carnap also objects to applying the terms *a priori* and *a posteriori* in this context; he calls his confirmations *prior* and *posterior* with the additional remark that this is meant to be relative with respect to the observation of the result R of the respective experiment[1]). An observation strengthening (weakening) the *confirmation* of a hypothesis is called *positively* (*negatively*) *relevant* to the hypothesis in question[2].

Thus Carnap's *degree of confirmation c* is in fact a *degree of rational belief*, a *probability* measuring the qualities of a hypothesis within the limits $0 \leq c \leq 1$ on the basis of *evidence* and *observation* (testing experiments); its progression is governed by Bayes' theorem and the chain of *learning by experiment* described in Subsect. 7.2.3. The facts about *c* recorded by Carnap are deduced from the axioms of relative (conditional) probabilities (cf. Sect. 7.2).

10.4.2 Popper's Degree of Corroboration

Popper (1934) defines quite different means of 'measuring' the degree up to which a *hypothesis* (*theory*) *H* is *corroborated* by the results *R* of observations made while *testing* it. This *degree C of corroboration* is not a *probability*, but rather a *function of probabilities* reflecting the ability of the underlying theory of being *falsified* (its *falsifiability*). In Popper's works there are different non-equivalent definitions of *C*, for example:

$$C(H/R)=\frac{p(R/H)-p(R)}{p(R/H)+p(R)}(1+p(H)\cdot p(H/R)),\ p(R)\neq 0 \tag{10.1}$$

and

$$C(H/R)=\frac{p(R/H)-p(R)}{p(R/H)-p(RH)+p(R)}, \tag{10.2}$$

both written in the absolute terminology not reflecting *evidence E*, which may easily be introduced by corresponding conditional relations.

This *degree of corroboration* measures the degree up to which *H* and *R* deviate from being *stochastically independent*. This fact will be recognized more easily after a transformation of *e.g.* (10.1) into

$$C(H/R)=\frac{p(RH)-p(R)p(H)}{p(RH)+p(R)p(H)}(1+p(H)\cdot p(H/R)). \tag{10.3}$$

---

[1] We also avoided possible misunderstandings in Subsect. 7.2.3 by adding the additional remark 'with respect to ...'.

[2] Again we realize the guarded description of conclusions by calling verifying or falsifying cases 'strengthening', 'supporting', 'positively relevant' and 'weakening', 'casting doubt' and 'negatively relevant' respectively.

The range of values of this measure is $-1 \leq C = C(H/R) < +1$. The lower limit $C = -1$ will be reached if the result R gained from an observation implies the alternative $\neg H$ to the hypothesis H. On the other hand, the upper limit of $C = C(H/R)$ is $C(H/H) = p(\neg H) < 1$ (if $p(H) > 0$), the degree (probability) up to which hypothesis H can possibly be *falsified*. $C = C(H/R) = 0$ indicates *stochastic independence* between H and R. As to further properties of the *degree of corroboration* we refer to Popper (1934). Equation (10.1) is the latest worked out by Popper; we will concentrate on discussing the consequences arising from Popper's definition by analysing this equation.

Obviously, equations (10.1) and (10.3) are dominated by the *likelihood* $p(R/H)$ of hypothesis H with respect to the result R of the testing experiment. This fact becomes even more evident when (10.3) is transformed into

$$C(H/R) = c_{+} \cdot [p(R/H) - p(R/\neg H)], \tag{10.4}$$

c_{+} being a positive factor and H the *hypothesis* – to be tested – with its alternative $\neg H$. Representation (10.4) of Popper's *degree* $C(H/R)$ *of corroboration* of *hypothesis* H by R shows clearly that the sign of $C(H/R)$ is exclusively determined by the *likelihoods* of H and $\neg H$ given the result R of the testing experiment. $C(H/R)$ will be positive if and only if H is more *likely* (with respect to R) than its alternative $\neg H$. $C(H/R) = 0$ implies that H and $\neg H$ are equally likely.

Consequently, our objections to using *likelihood*-based evaluations when dealing with *single* – or only a small number of – experiments done in order to test the model (Ex. 7.10) also comprise Popper's *degree of corroboration*. Both ideas are based on probabilities defined by *frequency* considerations, unsuitable when evaluating *individual* observations or only a small number of them.

Ex. 10.2 As an example we shall calculate the value of C – according to (10.1) – for the set-up of example Ex. 7.10. We already know that it is more likely to take a white ball out of urn U_4 than from U_6, but because of the relaying experiment we would rather prefer the alternative $\neg H$ to H and guess that it was U_6 we removed the white ball from. With H: $U = U_4$ and $\neg H$: $U = U_6$ Popper's degrees of corroboration will be in favour of H with $C(\neg H/R) < 0 < C(H/R)$.

As a matter of fact, we already know from Ex. 7.10 that $p(R/H) = 0.6$ and $p(R/\neg H) = 0.4$, $p(R) = 0.4002$, $p(H) = 0.001$ and $p(\neg H) = 0.999$, $p(H/R) = 0.0015$ and $p(\neg H/R) = 0.9985$; we thus calculate $C(\neg H/R) = -0.000499 < 0 < C(H/R) = 0.19976$ in favour of hypothesis H. ♦

Nevertheless, Popper's ideas show that he (called a witness for the impossibility of validating models) looked upon these problems from a quite pragmatic point of view; he has even developed means of *quantifying* the criterion in question. Popper assesses the quality of an *empiric* theory as a matter of its *degree of corroboration*, derived from observations 'supporting' or 'casting doubt' upon the theory in question (what we used to call *cases verifying* or *falsifying* the theory). Accordingly Popper does not regard validity as an absolute quality of a model; it is rather a (*relative*) *quantity* varying with the time-dependent *evidence* available to the observer.

10.4.3 Degrees of Validity

In the preceding sections and chapters we have studied means to evaluate recognition referring to the dependability of models. We have presented five different concepts – statistical methods, Bayesian inference, the principle of maximum likelihood, Carnap's degree of confirmation and Popper's degree of corroboration – each of them providing its own method of evaluating the degree up to which a *model* can be *validated* by *verifications* (testing). These 'measures' are all defined by using *stochastic* considerations which are either probabilities – based on frequency considerations or degrees of rational belief –, likelihoods or degrees of confirmation or of corroboration. The first four of the concepts are probabilities, the last one is a function of probabilities.

Degrees of rational belief and Carnap's degrees of confirmation are identical; they both merely interpret Bayes' formula under a more general aspect, or to be more exact they regard only the dependability of theories. Likelihoods and Popper's degree of corroboration are connected with frequency considerations. If there are many experiments carried out in order to evaluate a model, their results can be evaluated by using statistical methods giving rise to trust to a model according to preset levels of significance. If we have the choice between the *Maximum Likelihood Method* and Popper's degree of corroboration, we rather prefer Popper's measure because of the additional information – on the degree of falsifiability of the model – it contains.

But very often the act of *verifying* a model is a *single* event (constituting a minimum sample size $n = 1$ or only very few observations); the appropriate approach under these conditions is based on the concept of *degrees of rational belief*. Within this concept probabilities are defined to determine the *range* within which a model can be verified throughout research. The determination is based on Bayes' theorem (the *principle of inverse probabilities*) identical with Carnap's idea of a *degree of confirmation* of a hypothesis.

To accept a model as dependable is a matter of coming to decisions. We know from decision theory that this process is not only governed by expectation, but also by *risk* and 'emotions' like *joy of winning* and *disappointment* (e.g. cf. Bell 1985), by malicious joy and need for admiration, introduced as measures into the theory of decision making under uncertainty. A detailed discussion of these theories is out of the scope of this book, but we want to outline the basic idea with the following example of 'the two envelopes' which we have slightly modified to make it similar to the goat problem we discussed in Ex. 7.9.

Ex. 10.3 Again a candidate participates in a show. This time he has the choice between two doors, and he is told that there is a cheque for a certain amount of money behind one of the doors and another cheque twice the value behind the other one. The candidate arbitrarily decides for one of the doors, opens it and finds a cheque for the value V; the presenter offers to switch to the other door which is still closed. Now the candidate must decide whether he is ready to risk his winnings by possibly losing half of them or double his stake.

There have been many pros and cons to whether the candidate should switch or not. Obviously, there are – *prima facie* – equal probabilities of finding half the value $0.5{\cdot}V$ or twice the value $2{\cdot}V$ of V behind the door still closed, and the expectation $\in(W)$ of winning is $\in(W)=0.5{\cdot}0.5{\cdot}V+0.5{\cdot}2{\cdot}V=1.25{\cdot}V$. Accordingly it *always* seems to be advisable to switch to the other door; a result which is unreasonable as it will never lead to a decision: whichever door you decide for at the beginning, you will always switch to the other one if you only regard the expected winnings.

Consequently, there must be further aspects leading to decisions. Let us assume the candidate finds a cheque for 1 $ behind the door he has opened: he will probably regard this as disappointing and think there is a good chance of winning 2 $ when switching to the other door. If the value V of the cheque he discovers is $V=100{,}000{,}000$ $, he will probably take them and not risk to lose half this fortune. If the candidate is the *Khan*, we mentioned in Sect. 3.2, and if he is a born gambler, he will perhaps switch doors even if $V=100{,}000{,}000$ $.

On the other hand, if the candidate finds 1$ and this is just the amount of money he needs to survive, e.g to bribe his guards, he will never switch. Likewise, a candidate detecting those 100,000,000 $, but owing 200,000,000 $ to a gang who will kill him unless he pays off his debts within half an hour will eagerly switch to that other door. ♦

So decision making under the aspect of uncertainty depends to a great deal on personal aspects. These decisions cannot be evaluated by only following entirely rational arguments, they must also be pragmatic and take the personal situation of the modeller or of the relevant community into account.

10.5 Evaluating a Model

There is still the question how to determine the values of the terms used to calculate the degrees of confirmation, corroboration or validation or whatever you decide to call them. We have already mentioned this problem in Sect. 7.3 and have seen that it is a matter of experts' ratings.

So experts on the matter and decision makers must come together and evaluate any event supporting the quality of the model in question or casting doubt[1] on it. As we have seen, the validity of a model is its being well-founded, and the panel should take any fact into account supporting this quality criterion or at least not disqualifying it, like, e.g. the expertise of the modellers, the plausibility of the set-up, its being formally correct and properly calibrated and any further piece of circumstantial evidence. Using the means of quantification mentioned in Sect. 7.3 – Delphi Method, the 'wheel of fortune' – may lead to intersubjectively accepted values of those degrees. But quantification is not the factor of crucial importance as long as there is a consensus among the members of the panel that the model is dependable.

Of course, the expertise of the modelling team, the formal correctness and being calibrated should not be overestimated; they just support the assumption that the model is founded – valid –, or, to be more cautious, that these facts do not invalidate the model, they do not cast doubt on it. The quantitative gain may consequently be modest, just leading to a degree of validity slightly more than zero.

But evaluation is a cyclic process lasting over the lifetime of a model. So the panel will come together whenever a new experiment leads to results supporting the model or casting doubt on it, and they will bring their evaluations in line with the then available evidence. This may lead to a different quantification by degrees

[1] We use Popper's terminology.

computed anew or to a qualitative re-evaluation leading to the recognition that this model is now more – or less – dependable than before.

The expression 'casting doubt' on a model – the term used by Popper – seems to be appropriate and much better than falsification, which – to us – carries with it 'an aura of absoluteness that we would not claim' (cf. Bredehoeft and Konikow 1992). The re-evaluation of a model during its lifetime will – in general – be based on tests of its predictive capacity. Imagine a model having correctly predicted phenomena for some thousand times to everyone's satisfaction, but suddenly failing: there is *one* absolutely incorrect prediction. The conclusions drawn from this event can certainly not be that the model was invalid – not well-founded – in the past, because it worked well then, but it turned out to be unsuitable to cope with a new and unforeseen scenario (cf. the example of the *CONCORDE* mentioned in Sect. 2.4) This situation does not really invalidate this model, but casts doubt on it and gives rise to tuning it in order to cover the changed conditions or – if that should prove inadequate – to the development of an entirely new model that considers the new phenomena and can therefore cope with them.

We conclude that the evaluation of models is done under utilitarian aspects, because when validating we ask: does the model (in question) solve its (underlying) problem? Is its problem-solving capability better than that of other models? Does it only postpone the solution of the problem? Is the solution simple? Is it prolific (effective)? Does it conflict with the results gained from alternative methods needed for the solution of other problems (we rely on when solving other problems)? These are the desiderata demanded by Popper, and by repeating them at the end of our book as our own demands we pay respect to him and to de Marsily et al. (1992) who have put forward similar arguments when they argue '.. as long as a model is in concordance with our observations, it serves its purpose...'.

So, eventually, the philosopher and the engineer – and all those modelling outside the fields of sciences – unanimously share the same ideas concerning the validity of a model, its degree of being well-founded:

A model is valid (dependable) as long as it does what it is expected to do: solve its problem and be in concordance with our observations.

11 Suggestions for Further Reading

In addition to the references already quoted in the preceding chapters we would like to mention some further select articles and books dealing with the subject. They facilitate the access to various aspects of modelling and contain thorough bibliographic references.

Many examples of – especially environmental – models are contained in Grütz (1997); general aspects of modelling have been discussed, e.g. by Stachowiak (1973), Anderson and Woessner (1992) and Bossel (1994). Main concern of Bender (1978) or Fowkes and Mahony (1994) is mathematical modelling. The subject is discussed from the epistemic viewpoint by, e.g. Hesse (1966) and by (van) Fraassen (1980), Morris (1938) studies the problems of semiotics and semiosis; (de) Marsily (1994) points out the opposing approaches to modelling by black-boxes and to reflecting the unobservable. We have already quoted some of the articles edited by Freudenthal (1960), by Morgenbesser (1967) and by Abbott and Refsgaard (1996); these volumes contain many more contributions worth reading.

With regard to systems and to systems analysis we also refer to Bertalanffy (1950), Cellier (1991) and Close and Frederick (1995). The 'special effects' described in Chap. 2 and the way they affect stock markets are also described in great detail by Tvede (1999); a general introduction to this problem area is provided by Hull (1997).

For further information on the 'Theory of Science' we recommend Hempel and Oppenheim (1948), Hempel (1965, 1966), Giere (1988) and the papers compiled and edited by Suppe (1974).

Mendelson (1964) wrote a highly recommendable book on deductive inference. Inductive inference governed by frequency considerations has been axiomatically defined by Kolmogoroff (1933); these ideas have been taken up by (von) Mises (1938). 'Degrees of rational belief' is the subject of, e.g. Keynes (1921), Ramsey

(1931), (de) Finetti (1937, 1970), Jeffreys (1939), Savage (1954), Braithwaite (1957), Polanyi (1958), Lindley (1965 a,b), Kyborg and Smokler (1980), Jaynes (1995) and Koch (2000). There have been arguments between the supporters of both theories. Interestingly enough, Kolmogoroff who introduced the axioms of the frequency theory later seems to have favoured degrees of belief (Rényi 1962); Stange (1977) is another example of a determined supporter of frequency considerations, but posthumous publications show that in his old age he took a liking to Bayes' theories. An interesting probabilistic approach to econometrics has been described by Haavelmo (1944); Kendall (1949) shows that both camps can be reconciled. Reichenbach (1935) and Kendall (1940) recapitulate the limited dependability of the 'maximum likelihood method' when only a small number of evaluating experiments is carried out. Procedures used to test statistical data models are described, e.g. by Lamnek (1980) or Rost (1996); a detailed description of the Delphi Method and problems possibly occurring when using it are contained in Wechsler (1978) and in Häder and Häder (2000). We finally want to mention (von) Neumann (1928) and (von) Neumann and Morgenstern (1944) when we have to deal with decision making. Bischoff (1982) contains a wealth of examples of ex-post 'predictions' based on number mysticism.

Further information concerning the scrutiny of models (theories) – their validation, verification, confirmation or whatever it is called – can be found in Hempel (1943, 1945, 1958), Hempel and Oppenheim (1945), Kemeny and Oppenheim (1952), Jeffrey (1956, 1965) or Kamlah and Lorenzen (1967), also cf. Anderson and Bates (2001).

A central theme of our considerations is that the dependability of a model is based upon the *consensus* among the members of a scientific community. We owe this term to Niederer (1990) who pointed out that the 'validity' of a model cannot be proved, but that a model is accepted as being dependable by the experts.

References

Abbott MB, Refsgaard JC (eds) (1996) Distributed Hydrological Modelling. Kluwer Academic Publisher, Dordrecht Boston London

Anderson MG, Bates PD (2001) Model Validation: Perspectives in Hydrological Science. John Wiley & Sons, Chichester New York Weinheim Brisbane Singapore Toronto

Anderson MP, Woessner WW (1992) Applied Groundwater Modeling. Academic Press, San Diego

Antonovsky A (1987) Unraveling the mistery of health. How people manage stress and stay well. Josey Bass, San Francisco

Apostel L (1961) Towards the Formal Study of Models in the Non-Formal Sciences. In: Freudenthal H (ed) The Concept and the Role of the Model in Mathematics and Natural and Social Sciences. D. Reidel Publishing Company, Dordrecht, pp 1-37

Appel K, Haken W (1977) Every planar graph is four colorable. Part I. Discharging. Illinois J. Math. 21: 429-490

Appel K, Haken W, Koch J (1977) Every planar graph is four colorable. Part II. Reducibility. Illinois J. Math. 21: 491-567

Baird JL, Baglive G, Kane JR (1986) The Chronicle of Salimbene de Adam. Center for Medieval and Early Renaissance Studies, University Center at Binghamton, Binghamton, New York

Bell DE (1985) Disappointment in Decision Making Under Uncertainty. Operations Res. 33 (1): 1-27

Bender EA (1978) An Introduction to Mathematical Modeling. John Wiley & Sons, New York Chichester Brisbane Toronto

Bertalanffy L von (1950) The Theory of Open Systems in Physics and Biology. Science 111: 23-29

Bertalanffy L von (1968) General System Theory. George Braziller, New York

Bertalanffy L von, Beier W, Laue R (1977) Biophysik des Fließgleichgewichts. Friedr. Vieweg & Sohn, Braunschweig

Bischoff E (1982) Die Mystik und Magie der Zahlen. Verlag Richard Schikowski, Berlin

Black F, Scholes M (1973) The Pricing of Options and Corporate Liabilities. Journ. Pol. Econ. 81: 637-654

Black M (1962) Models and Metaphors. Cornell University Press, Ithaca London

Bossel H (1994) Modellbildung und Simulation. Friedr. Vieweg & Sohn Verlagsgesellschaft mbH, Braunschweig Wiesbaden

Braithwaite RB (1957) On unknown probabilities. In: Körner S (ed) Observation and interpretation. A symposium of philosophers and physicists. Butterworths Scientific Publications, London

Braithwaite RB (1968) Scientific Explanation. Cambridge University Press, London New York

Braunmühl E von (1975) Antipädagogik. Beltz Verlag, Weinheim Basel

Braunmühl E von (1990) Zur Vernunft kommen. Beltz Verlag, Weinheim Basel

Bredehoeft JD, Konikow LF (1992) Reply to Comment. Adv. Water Resources 15: 371-372

Cannon WB (1932) The Wisdom of the Body. WW Norton&Company, New York London

Carnap R (1950) Logical Foundations of Probability. Routledge & Kekan Paul, The University of Chicago Press, London

Carnap R, Stegmüller W (1959) Induktive Logik und Wahrscheinlichkeit. Springer, Wien

Cellier FE (1991) Continuous System Modeling. Springer-Verlag New York Inc., New York Berlin Heidelberg

Close CM, Frederick DK (1995) Modeling and Analysis of Dynamic Systems. John Wiley & Sons, New York Chichester Brisbane Toronto

Cox JC, Ross SA, Rubinstein M (1979) Options Pricing: A simpilified Approach. Journ. Fin. Econ. 7: 229-264

Dickinson GL (1931) Plato and his Dialogues. Penguin Books, West Drayton Middlesex England New York Melbourne

Finetti B de (1937) La Prévision: ses lois logiques, ses sources subjectives. Annales de l'Institut Henri Poincaré 7: 1-68.

Finetti B de (1970) Teoria delle Probabilità, 2 Volumes. Giulio Einaudi, Torino

Fowkes ND, Mahony JJ (1994) An Introduction to Mathematical Modelling. John Wiley & Sons, New York Chichester Brisbane Toronto

Fraassen B van (1980) The Scientific Image. Clarendon Press, Oxford

Freudenthal H (ed) (1960) The Concept and the Role of the Model in Mathematics and Natural and Social Sciences. D. Reidel Publishing Company, Dordrecht

Frey G (1961) Symbolische und ikonische Modelle. In: Freudenthal H (ed) The Concept and the Role of the Model in Mathematics and Natural and Social Sciences. D. Reidel Publishing Company, Dordrecht, pp 89-97

Giere RN (1988) Explaining Science. The University of Chicago Press, Chicago London

Grütz R (1997) Modellierung und Simulation im Umweltbereich. Friedr. Vieweg & Sohn Verlagsgesellschaft mbH, Braunschweig Wiesbaden

Haavelmo TM (1944) The Probability Approach in Econometrics. Econometrica 12, Supplement: i-viii and 1-118

Häder M, Häder S (2000) Die Delphi-Technik in den Sozialwissenschaften. Westdeutscher Verlag GmbH, Wiesbaden

Hansen W (1996) Phlegon of Tralles' Book of Marvels. University of Exeter Press, Exeter

Hempel CG (1943) A Purely Syntactical Definition of Confirmation. Jour Symb Log 8: 122-143

Hempel CG (1945) Studies in the Logic of Confirmation. Mind 54: 1-26; 97-121

Hempel CG (1958) Empirical Statements and Falsifiability. Philosophy 33: 342-348

Hempel CG (1965) Aspects of Scientific Explanation and other Essays in the Philosophy of Science. The Free Press, New York

Hempel CG (1966) Philosophy of Natural Science. Prentice-Hall, Englewood Cliffs

Hempel CG, Oppenheim P (1945) A Definition of Degree of Confirmation. Phil Sci 12: 98-115

Hempel CG, Oppenheim P (1948) Studies in the Logic of Explanation. Phil Sci 15: 135-175.

Hesse MB (1966) Models and Analogies. University of Notre Dame Press, Notre Dame

Holder-Egger O (ed 1905-1913) Monumenta Germaniae Historica Scriptorum Tomus XXXII, Cronica fratris Salimbene de Adam ordinis minorum, Editio Nova Lucis Ope Espressa 1963. Hahn, Hannover

Hull JC (1997) Options, Futures and other Derivatives. Prentice Hall, Upper Saddle River NJ

Irmscher J (ed 1987) Lexikon der Antike. VEB Bibliographisches Institut, Leipzig

Jaynes ET (1995) Probability Theory: The Logic of Science. http://bayes.wustl.edu/etj.prob.html

Jeffrey RC (1956) Valuation and Acceptance of Scientific Hypotheses. Phil Sci 23: 237-246

Jeffrey RC (1965) The Logic of Decision. Mc Graw-Hill, New York

Jeffreys H (1939) Theory of Probability. Clarendon Press, Oxford

Jung CG (1951) Aion: Untersuchungen zur Symbolgeschichte. Rascher Verlag, Zürich

Kamlah W, Lorenzen P (1967) Logische Propädeutik. Bibliographisches Institut, Mannheim

Kaufmann A (1967) Graphs, Dynamic Programming, and Finite Games. Academic Press, New York London

Kemeny JG, Oppenheim P (1952) Degree of Factual Support. Phil Sci 19: 307-324

Kendall MG (1940) On the Method of Maximum Likelihood. Jour R Statist Soc 103: 388-399

Kendall MG (1949) On the Reconciliation of Theories of Probability. Biometrika 36: 101-116

Kervran LC (1972) Bioligical Transmutations. Swan House Publishing Co., New York

Keynes JM (1921). A Treatise on Probability. Macmillan, London

Knuth DE (1969 The Art of Computer Programming. Addison-Wesley Publishing Company, Inc., Reading Menlo Park London Don Millis

Koch HR (2000) Einführung in die Bayes-Statistik. Springer-Verlag, Berlin Heidelberg

Kolmogoroff AN (1933) Grundbegriffe der Wahrscheinlichkeitsrechnung. Springer, Berlin

Konikow LF, Bredehoeft JD (1992) Ground-water models cannot be validated. Adv. Water Resources 15: 75-83

Kuhn S (1962) The Structure of Scientific Revolutions. The University of Chicago Press, Chicago London

Kyburg Jr. HE, Smokler HE (1980) Studies in Subjective Probability. Krieger Publishing Comp., Huntington New York

Lamnek S (1980) Sozialwissenschaftliche Arbeitsmethoden. edition medizin, Weinheim Deerfield Beach Basel

Le Corbusier (1983) Le Modulor: Essai sur une mesure harmonique a l'echelle humaine applicable universellement a l'architecture et a la mécanique. L'Architecture d'Aujourd'hui, Groupe Expansion, Paris

Lefschetz S (1956) Topology. Chelsea Publishing Company, New York

Leijnse A, Hassanizadeh SM (1994) Model definition and model validation. Adv. Water Resources 17: 197-200

Leinfellner W (1965) Einführung in die Erkenntnis- und Wissenschaftstheorie. Bibliographisaches Institut, Mannheim
Lindley DV (1965) Introduction to Probability and Statistics from a Bayesian Viewpoint, Part 1: Probability. Cambridge University Press, Cambridge
Lindley DV (1965) Introduction to Probability and Statistics from a Bayesian Viewpoint, Part 2: Inference. Cambridge University Press, Cambridge
Liu CL (1968) Introduction to Combinatorial Mathematics. Mc Graw-Hill Book Company, New York St.Louis San Francisco Toronto London Sidney
Mag W (1990) Grundzüge der Entscheidungstheorie.Vahlen, München
Mandelbrot BB (1968) Noah, Joseph and Operational Hydeology. Water Resour. Res. Vol. 4(5): 909-918
Mandelbrot BB (1997) Fractals and Scaling in Finance. Springer-Verlag, New York Berlin Heidelberg
Marsily G de (1994) Quelques réflections sur l'utilisation des modèles en hydrologie. Rev. des Sci. de l'eau 7: 219-234
Marsily G de, Combes P, Goblet P (1992) Comment on 'Ground-water models cannot be validated' by L.F. Konikow & J.D. Bredehoeft. Adv. Water Resources 15: 367-369
Mayor A (2000) The First Fossil Hunters: Paleontology in Greek and Roman Times. Princeton University Press, Princeton
McDougall W (1923) An Outline in Psychology. Methuen & Co. Ltd., London
Mendelson E (1964) Introduction to Mathematical Logic. D. Van Nostrand Comp., Inc., Princeton
Menninger K (1958) Zahlwort und Ziffer. Vandenhoek & Ruprecht, Goettingen
Mises R von (1936) Wahrscheinlichkeit, Statistik und Wahrheit. Springer, Wien
Morgan MS, Morrison M (1999) Models as Mediators. Cambridge University Press, Cambridge New York Oakleigh
Morgenbesser S (ed) (1967) Philosophy of Science today. Basic Books, New York London
Morris CW (1938) Foundations of the Theory of Signs. In: Neurath O, Carnap R, Morris C (eds) International Encyclopedia of Unified Science. The University of Chicago Press, Chicago, Vol. I, No. 2
Narasimhan TN (1995) Models and Modeling of Hydrogeologic Processes. Soil Sci. Soc. Am. J. 59: 300-306
Neill AS (1968) Talking of Summerhill. Gollancz, London
Neill AS (1996) Summerhill school: a new view of childhood. St. Martin's Griffin, New York
Neumann J von (1928) Zur Theorie der Gesellschaftsspiele. Math. Annalen 100: 295-320
Neumann J von, Morgenstern O (1944) Theory of Games and Economic Behavior. Princeton University Press, Princeton
Niederer U (1990) In Search of Truth: The Regulatory Necessity of Validation. In: Proceedings of GEOVAL
Northcott DG (1962) An Introduction to Homological Algebra. Cambridge University Press, Cambridge
Ohsawa G (1971) Biological Transmutation by Louis Kevran, Book Condensation. The George Ohsawa Macrobiotic Foundation, Oroville California
Oreskes N, Shrader-Frechette K, Belitz K (1994) Verification, Validation, and Confirmation of Numerical Models in the Earth Sciences. Science 263: 641-646
Philip JR (1991) Soils, Natural Science, and Models. Soil Sci. 1961: 91-98
Poe EA The Pit and the Pendulum. In: Poe EA Tales, Poems, Essays. Collins

Polanyi M (1958) Personal Knowledge - Towards a Post-Critical Philosophy. Routledge & Kegan Paul, London

Popper KR (1934) Logik der Forschung. Springer, Wien, second commented and modified version. Tübingen: Mohr, 1966.

Popper KR (1964) Über die Unwiderlegbarkeit philosophischer Theorien einschließlich jener, welche falsch sind. In: Szczesny G (ed) Club Voltaire. Szczesny Verlag, München

Pribram KH (1982) Languages of the Brain. Englewood Cliffs, New Jersey

Ramsey FP (1931) The Foundations of Mathematics. Kegan Paul, Trench, Trubner & Co., London

Refsgaard JC (1996) Terminology, Modelling Protocol and Classification of Hydrological Model Codes. In: Abbott MB, Refsgaard JC (eds) (1996) Distributed Hydrological Modelling. Kluwer Academic Publisher, Dordrecht Boston London, pp 17-40

Reichenbach H (1935) Wahrscheinlichkeitslehre. Sijthoff, Leiden

Rényi A (1962) Wahrscheinlichkeitsrechnung. VEB Deutscher Verlag der Wissenschaft, Berlin

Richter H (1956) Wahrscheinlichkeitstheorie. Springer, Heidelberg

Rosenthal R, Jacobson L (1971) Pygmalion im Unterricht. Beltz, Weinheim

Rost J (1996) Lehrbuch Testtheorie Testkonstruktion. Verlag Hans Huber, Bern Göttingen Toronto Seattle

Rousseau J.-J. (1964) Èmile ou De L'Èducation. In: Richard F (ed) Èmile ou De L'Èducation avec une introduction, une bibliographie, des notes et une index analytique par François Richard. Garnier, Paris

Savage LJ (1954) The Foundations of Statistics. John Wiley & Sons, New York

Schlesinger S, Crosbie RE, Gagné RE, Innis GS, Lalwani CS, Loch J, Sylvester RJ, Wright RD, Kheir N, Bartos D (1979) Terminology for model credibility. Simulation 32(3): 103-104

Seidler E (1964) Der Neugeborenenversuch Friedrichs II. von Hohenstaufen. Deutsches Ärzteblatt 39: 2029-2032

Snow CP (1959) The Two Cultures: and a Second Look. Cambridge University Press, London

Spitz R (1954) Genèse des primières relations objectales. Revue Française de Psychanalyse, Presses Universitaires de France

Spitz R (1957) No and Yes. On the Genesis of Human Communication. International Universities Press, New York

Staal JF (1961) Formal Structures in Indian Logic. In: Freudenthal H (ed) The Concept and the Role of the Model in Mathematics and Natural and Social Sciences. D. Reidel Publishing Company, Dordrecht, pp 155-162

Stachowiak H (1973) Allgemeine Modelltheorie. Springer-Verlag, Wien New York

Stange K (1977) Bayes-Verfahren. Springer, Heidelberg

Stegmüller W (1969) Probleme und Resultate der Wissenschaftstheorie und Analytischen Philosophie, Band 1: Wissenschaftliche Erklärung und Begründung. Springer-Verlag, Berlin Heidelberg New York

Suppe F (ed) (1974) The Structure of Scientific Theories. University of Illinois Press, Urbana Chicago London

Taha HA (1976) Operations Research, an Introduction, Second Edition. Macmillan Publishing Co., Inc., New York

Troelstra AS (1977) Aspects of Constructive Mathematics. In: Barwise J (ed) Handbook of Mathematical Logic. North-Holland Publishing Company, Amsterdam New York Oxford

Tvede L (1999) The Psychology of Finance. John Wiley & Sons, Chichester New York Brisbane Singapore Toronto

Ubbink JB (1961) Model, Description and Knowledge. In: Freudenthal H (ed) The Concept and the Role of the Model in Mathematics and Natural and Social Sciences. D. Reidel Publishing Company, Dordrecht, pp 178-194

Webb JK, Murphy MT, Flambaum VV, Dzuba VA, Barrow JD, Churchill CW, Prochaska JX, Wolfe AM (2001) Further Evidence for Cosmological Evolution of the Fine Structure Constant. Physical Review Letters 87-9: 091301 (1-4)

Wechsler W (1978) Delphi-Methode: Gestaltung und Potential für betriebliche Prognoseprozesse. In: Aschoff C, Müller-Bader P Wirtschaftwissenschaftliche Forschung und Entwicklung. Florentz GmbH, München

Wiles A (1995) Modular elliptic curves and Fermat's Last Theorem. Ann. Math. 141: 443-551

Appendix

In the introductory first chapter of this book we have discussed models in the broadest possible sense of the word. According to our definitions anything produced by man, any paper and any work of art is a model.

We now want to conclude our considerations by presenting some concrete examples of *scientific* models. Of course, any article in any journal of any branch of science is an example of modelling in sciences. Yet, we have selected eight manuscripts illustrating the ideas discussed in the previous chapters, thus giving a practical touch to our sometimes rather theoretical explanations. Among the contributions we received, we have confined ourselves to the field of natural sciences. All of those articles are outstanding confirmations of our ideas, and it was extremely difficult to arrange them.

We finally decided to classify them chronologically: so the first four models are – retrospectively – directed towards the past; they give explanations of phenomena we observe. We start with C. Theis' model of the evolution of galaxies, which is an excellent example of the attempt to *ex post facto* explain the fact that there are galaxies. The validity of this model is entirely founded on the evidence available to the modeller and is based on the consensus among the members of the relevant scientific circle. As it cannot be improved by experimenting, this model entirely depends on the growing insight into the astrophysical actuality.

The second model, presented by S. Kempe and J. Kazmierczak, deals with the reconstruction of the genesis of the earth and the generation of environmental conditions favourable for the development of life. These conclusions have been drawn from the close scrutiny of some modern Crater Lakes which seem to be governed by conditions similar to those having enabled the development of life in Precambrian 'Soda Oceans'. The validity of this model is based on plausibility and arguments of analogy.

C. Hertling, F. Schrenk and T. Bromage have developed a model elucidating the evolution of the genus *Homo*. It has been induced from finds by the paleoanthropologists and serves the purpose of reconstruction and explanation. The dependability of this model is entirely founded on the evidence gained from the evaluation of fossils. It is accepted as valid by the experts on this subject and is thus scientifically based on a consensus among the members of this community.

Developing a steadily growing demand for *Lebensraum* the human race is inevitably endangering the quality of its environment. M. Ostrowski, M. Lempert and S. Heusch have developed a 'Water Balance Model' in order to identify human impacts on floods. Their rainfall - runoff model shows that urbanization leading to changes in land use and the development of engineering infrastructures have a considerable effect on floods. On the one hand, these investigations explain observed phenomena; on the other hand, the model is used for the prediction of the development of floods.

We now turn our attention to phenomena occurring in the time being, but hardly ever observable.

M. Kelm's model of a glass melter describes the immobilization of nuclear waste by vitrification. It is related to mass data evaluated in experiments using a physical scale model; but it is only a rather limited evidence of the actual distribution of the waste particles in the melter, because they cannot be observed. The experts just have to trust to this model without being able to find cases verifying or falsifying it in any way.

The model described by J. Lützenkirchen and B. Kienzler likewise gives account of the potential underground transport of nuclear wastes and of the interactions between these hazardous contaminants and the matrix material. The political leaders who have to decide whether to license underground nuclear disposals bear a great responsibility as the measure will have serious consequences. They entirely depend on the evaluation of these models when weighing up the risks of the mobilization of radionuclides caused by groundwater intruding into the disposals and thus spoiling the biosphere.

Both systems model the unobservable and at the same time serve the purpose of predicting consequences arising under certain conditions. They share this aspect with the model developed by Ostrowski et al.

The interest is now focused on entirely predictive models: the first problem to be dealt with is the question 'what will the weather be like tomorrow and on the following days?'. G. Adrian and D. Majewski describe an extremely sophisticated approach to numerical weather modelling. Observations of the weather give daily rise to falsifications or verifications of the forecasts of the past few days, but not of the model. This is due to the effects of multifinality and the impossibility of parametrizing these models correctly.

U. Bielert, W. Breitung and A. Kotchourko finally present a model simulating the aftermath of a severe accident in a nuclear power plant. Unlikely as they may be these disasters possess an extremely high damage potential. The scenario can be simulated and the results of these simulations can be evaluated in order to gain estimations of what might happen if one of these catastrophes should occur; but, of course, nobody will ever carry out a 'real' experiment in order to determine the range of validity of these models.

And if we want to look further into the future and try to find out more about catastrophes lying ahead of us, we may return to the beginning, to Theis' model which also describes – applying analogous physical theories – the interactions between galaxies, possibly leading to their extinction.

Modelling the Evolution of Galaxies

Christian Theis

Introduction

If one compares astrophysics with other branches of applied physics, the adopted theoretical approaches (i.e. fundamental equations or numerical techniques) are quite similar, whereas on the experimental side usually big differences exist: astrophysicists are rarely able to perform experiments with the objects of their research. We have mainly to rely on the information coming from electromagnetic radiation at different spectral ranges and – to a far lesser extent – on cosmic rays or other high energy particles (I neglect here the close vicinity of the Sun which is accessible by spacecrafts or can be probed by meteorites). Moreover, due to the long timescales of most cosmic processes we see only snapshots of their evolution. Additionally we observe them only from our position in space, i.e. at a distance of about 7.5 kpc from the centre of a spiral galaxy which is separated by about 16 Mpc from the centre of the next galaxy cluster. Therefore, the classification of observations has become a very important tool and guide for our understanding of celestial phenomena. However, it is not an easy task to recover a whole movie from having only single snapshots, especially if the snapshots might belong to different movies!

Therefore, the rise of computational power during the last decades which gave birth to the field of Computational Physics became also very essential to astrophysics. The complexity of many astrophysical systems usually prevents a quantitative investigation on the basis of simple analytic or semi-analytic models. E.g. for the analysis of stellar atmospheres millions of line transitions might be important. Another example is the Newtonian dynamics of even small open star cluster which is described by hundreds of coupled non-linear equations of motion. Many of these astrophysical systems can now be studied to some level of sophistication by numerical calculations. Even more important is the fact that astrophysical objects can be investigated in great detail by *numerical experiments*. Simulations allow to follow different evolutionary steps which can be 'observed' by a 'numerical telescope' and compared with real observations. Last, but not least, numerical experiments allow us to modify the fundamental laws of nature and, by this, to investigate the implications of such changes. Therefore, numerical simulations became an invaluable tool for astrophysics.

The term *Modelling* related to (astro)physics usually means that we describe complex systems by reducing them to a few fundamental components and interaction processes from which a variety of observed phenomena is derived or *modelled.* 'Fundamental components' might mean real fundamental building blocks, e.g. when describing the dynamics of nucleons by quarks and gluons as done in the framework of quantum chromodynamics, or just basic, but still very macroscopic subunits like stars or whole galaxies, e.g. when solving the stellardynamical equations for the evolution of a star cluster or a cluster of galaxies, respectively. The latter are far from being fundamental, because they are obviously composed of billions of atoms or even larger objects like stars or very massive gaseous clouds.

A second aspect of the term *Modelling* deals with its goals. One goal might be to understand the qualitative behaviour of a system. For example we might be interested whether an ecosystem can reach a stable configuration or whether the climate will change in the next millennium. For both we might need a description of the detailed processes, e.g. the number of predatory animals or the weather in a special area. However, no one (probably!) would believe the exact numbers of these quantities, e.g. the weather forecast for March, 14th 2154.

Similarly, most of the large scale simulations in astrophysics are not used for describing special systems in very detail, but for providing a qualitative insight into their behaviour. This does not exclude to extract quantitative relations from these simulations. E.g. most state-of-the-art cosmological simulations on structure formation in the universe are not performed to match exactly the local universe, i.e. the distribution of galaxies in the Local Group. On the other hand, statistical quantities like the mass and angular momentum distribution in the formed structures at a given epoch can be derived and compared with observations giving hard tests for the underlying model assumptions. Another example are high-accuracy N-body simulations for the evolution of globular clusters which consist of about 10^5 stars. Though a tremendous numerical and computational effort is required to determine the trajectories of the stars, no one is interested in their exact positions or velocities at some time.

Another possible goal of *modelling* is an as exact as possible description of a real system. E.g. when an engineer investigates the flight characteristics of a new type of airplane, (s)he is not interested in demonstrating that aircrafts in general can fly, but that this new airplane will do the job. Similarly, in astrophysics special objects have been modelled in great detail. Examples are the structure and atmosphere of the Sun or the structure of our Milky Way galaxy. The main problem of this detailed modelling in astrophysics is usually the high level of complexity of many astrophysical systems and the vast parameter space which has to be investigated. Both often requires very high computational efforts. Therefore, different to studies of the qualitative behaviour of prototypical systems for which only a few calculations are required, thousands of model calculations can be necessary for fitting real observations in detail.

A typical example for this kind of problems is the formation and evolution of galaxies. In this article I will describe both aspects of modelling, i.e. the *qualitative modelling* and the detailed or *quantitative modelling*, with respect to galaxy evolution. After an overview about main observed properties of galaxies in Sect. 1 I introduce the *chemo-dynamical modelling* for galaxies as an example for a state-of-the-art, qualitative model (Sect. 2). In Sect. 3 I discuss the interaction of galaxies as an example (and challenge) for detailed modelling. Finally, a summary is given.

1 Galaxies

1.1 Isolated Galaxies

Though diffuse spiral-like nebulae have been known for centuries (e.g. some members of the famous catalogue published by Messier 1784 belong to this kind of nebulae), their real nature as extragalactic stellar systems became widely accepted during the 1920s. It was Edwin Hubble who derived in 1924/25 the distance to the spiral nebula NGC 6822 from a period measurement of Cepheid stars. From that it became clear that NGC 6822 is not a member of the Milky Way, but a remote stellar system. Similar results for M33 and M31 followed soon (Hubble 1926a, 1929), by this corroborating the hypothesis that spiral nebulae are *island universes*. Already one year after the results on NGC 6822 Hubble (1926b) introduced a phenomenological classification scheme for galaxies, the 'tuning fork diagram'. Though it has been slightly modified since its first publication, it is still used. The main division is between elliptical galaxies (E) and spirals (S). The latter group is split in those spirals with (SB) or without (S) a central bar. Additionally, the spirals of each group are sorted by the relative contribution of the mass of the central spheroid (the bulge) to the mass of the disk as well as by the shape of the spiral arms. Systems with the most massive bulges are called Sa (or SBa) galaxies, whereas those with small bulges are denoted as Sc (or SBc). Irregular galaxies like the Magellanic clouds belong to the group of Sd galaxies. In the next paragraphs some characteristics of galaxies are summarized (for details see Binney and Tremaine 1987 and Binney and Merrifield 1998).

Elliptical Galaxies

Elliptical galaxies are almost devoid of cold gas. However, they can be immersed in an extended halo of tenuous, hot (10^5 - 10^6 K) gas. The stars in ellipticals frequently seem to be formed in one instant of strong star formation (a *star burst*). Morphologically, the isophotes have an elliptical structure with an axis ratio ranging from 1 down to a minimum of 0.3. Elliptical galaxies show almost no rotation, but a large amount of irregular motion, i.e. a high velocity dispersion. Therefore, ellipticals are pressure-supported stellar systems. At the low-mass end the flattening of the ellipticals can be explained by their (slow) rotation, whereas the very massive ellipticals are flattened by an anisotropic velocity dispersion. The intensity distribution I(R) of stars follows nicely the empirical de Vaucouleurs- or $R^{1/4}$-law

$$I(R) = I_0 \cdot exp\left\{-7.67\left[\left(\frac{R}{R_{eff}}\right)^{1/4} - 1\right]\right\}$$

A typical effective radius R_{eff} is about 3 kpc.

Spiral Galaxies

Morphologically spiral galaxies are composed of different components. Most impressive is the flat disk which mainly consists of stars and gas. The latter amounts to about 10% of the stellar mass. Additionally, there is an admixture of dust which is clearly visible due to absorption, but its mass is usually less than 1% of the mass of the gaseous phases. The spiral structure found in many disks is strongly pronounced by young and massive stars which are very luminous compared to old less massive stars. The stellar density enhancement, however, is typically only of the order of 10% of the azimuthally averaged stellar density, whereas the ratio of arm to inter-arm gas densities can be much larger. The azimuthally averaged stellar mass distribution in disks follows an exponential profile with a scale length of about 3-5 kpc. The scale height above the plane of the disk is about several 100 pc (depending on which galactic component is considered). Contrary to elliptical galaxies, the disk is rotationally supported. Most of the disk stars move on almost circular orbits with typical speeds of about 200 km s^{-1}. Deviations from circular orbits result in a velocity dispersion of the order of several 10 km s^{-1}. E.g. the rotation speed of the Sun is about 220 km s^{-1}. Assuming a distance of 8.5 kpc from the Galactic centre, a galactic revolution takes about $2.4 \cdot 10^8$ years.

In addition to the disk, spiral galaxies have spheroidal components, too. In the centre a rotationally flattened bulge well described by a de Vaucouleurs profile can be found. The mass of the bulge can reach about 50% of the disk's mass (for Sa galaxies) declining to 10% for Sc galaxies. In at least half of the spiral galaxies the centre has the form of a large bar. A second spheroidal component is the extended, old and almost non-rotating stellar halo. Its mass is much smaller (~10%) than the mass of the disk.

Special interest has been attracted by the rotation curves of spiral galaxies. Many disk galaxies show a rapid increase of the rotational speed within a few kpc to about 200 km s^{-1} and then the rotation curves remain almost constant. Best measurements are obtained from radio observations of neutral hydrogen (HI) which can be traced far outside the optically visible regions of galaxies. Many of these HI data show no decline of the rotational speed. The usual interpretation is that galaxies are embedded in a halo of *dark matter* which only interacts by gravity with normal baryonic (and leptonic) matter (this dark matter halo should not be confused with the stellar halo!). The nature of the dark matter is still unclear, but different candidates ranging from exotic states of baryonic matter to special elementary particles have been suggested. Discoveries like the non-vanishing rest mass of the neutrino seem to be a hint that particle physics might provide substantial input to this field. Independently from the composition of the dark matter, the total dynamical mass in a disk galaxy can be derived easily from the rotation curve. As a result the mass ratio of dark to baryonic matter in a disk galaxy is about 5-10:1. However, inside 1-2 disk scale lengths usually no or only a small amount of dark matter is required in order to explain the observed rotation curves. Alternatively, a modification of the law of gravity could be invoked to explain the rotational velocities. However, it is still unclear how to embed such kind of ad-hoc prescriptions into fundamental theories of gravity.

Dwarf Galaxies

Almost all galaxies of the Hubble scheme are so-called 'normal galaxies', i.e. their stellar masses are about several 10^{10} $M_{\odot}$ to some 10^{11} $M_{\odot}$. However, the world of galaxies shows much more variety: in the mass range below the normal galaxies, a wealth of 'dwarf galaxies' has been found. Some of them seem to be low-mass copies of large elliptical galaxies (dE), whereas others look rather irregular and contain a large fraction of gas (dIrr). At the upper mass limit we find galaxies like the Magellanic Clouds. They are satellites of the Milky Way with an actual galactocentric distance of about 50 kpc (LMC) and 65 kpc (SMC), respectively. They are disk-like, gas-rich galaxies and show an ongoing star-formation activity. On the lower end of the mass scale we find diffuse galaxies which have masses of only 10^6 $M_{\odot}$, i.e. they have masses similar to large globular clusters. Some of the dwarf spheroidal galaxies (dSph) of the Local Group belong to this galaxy type.

1.2 Interacting Galaxies

During the last decades of the 20th century the classical picture of galaxies being *Weltinseln* (e.g. Kant 1755, v. Humboldt 1850), i.e. island universes, has completely changed. In the 1920s, galaxies were thought to be mainly isolated stellar systems (*islands*) with just a few obvious exceptions like M51. Systems not fitting Hubble's classification scheme have been neglected – by definition – as peculiar. Interest in these systems has increased when the galaxy catalogues by Vorontsov-Velyaminov (1959) and Arp (1966) became available. They demonstrated that many peculiar objects have common features which are described as bridges, tails, or rings. Moreover, better observational instruments allowing for deeper images and new wavelength ranges revealed many new details. From that many galaxies formerly classified as non-interacting seem to have had some gravitational interaction in the past which is still reflected in e.g. warps, flares, or thick disks. Especially useful are observations of HI which probe a much larger radial range than the optical data. Therefore, HI is not only important for the determination of the dynamical mass at large galactocentric distances, but it is also a very good indicator for tidal interactions. An excellent example is NGC 4449, a prototype of a Magellanic type dwarf galaxy: on optical images which cover a galactocentric distance to about 5 kpc no signs for a tidal interaction are visible, whereas the extended HI shows a distinct structure made out of several connected streamers.

These streamers are a clear hint for an interaction with the small companion DDO 125 (Theis and Kohle 2001).

Today it is believed that interactions between galaxies are very common in the history of a galaxy. In the standard cosmological scenario the formation of galaxies starts with small subunits which merge during their evolution. By this larger aggregates – like galaxies – are created in a hierarchical manner. From cosmological simulations we expect furthermore that each normal galaxy has had at least one encounter with a galaxy of the size of the Large Magellanic Cloud. We are witnessing such an event right now in the Milky Way which 'swallows' the Sagittarius dwarf galaxy. As a result of such a minor merger the thin disk might be heated giving rise to a thick disk component which is seen in many disk galaxies. Another process related to galaxy interaction is the merging of normal galaxies, provided their orbital parameters allow for a sufficient loss of orbital energy and angular momentum. A possible result is the formation of an elliptical galaxy out of two spirals or the creation of a very massive elliptical galaxy in the centre of a cluster of galaxies.

Besides the structural changes induced by an interaction, the interstellar medium is also strongly perturbed. As a result the star formation activity can be substantially enhanced leading to a star burst. Different stellar populations might be tracers of such events even if the interaction occured long ago and did not leave any visible 'damage'. Another aspect of interaction-induced star formation is the formation of massive young star clusters in tidal tails (as seen e.g. in the Antennae galaxies) or the formation of dwarf galaxies in tidal tails.

The details of the dynamics of an interaction strongly depend on the kinematical and structural properties of the interaction partners. Therefore, understanding galaxy encounters or mergers gives us an additional key for the investigation of the matter distribution in galaxies. Especially the dark matter might be studied by the structure and kinematics of extended tidal tails. Moreover, the knowledge of exact timings and strengths of interactions also gives tight constraints on timescales and conditions for star formation in the universe.

2 Modelling of Galaxies

In this section theoretical approaches for modelling the evolution of galaxies are described. First, I will present in Sect. 2.1 a list of necessary ingredients for a complete model of galactic evolution. This will include a brief description of both, the different galactic components and their interactions. Then, in Sect. 2.2 I will discuss the backbone of all galaxy simulations, the treatment of the dynamics. This will be split into stellar dynamics and gas dynamics.

2.1 The Grand Picture

2.1.1 Galactic Components

The first question when modelling a dynamically evolving system concerns its constituents: a galaxy consists of dark matter and 'normal' baryonic matter (including the leptons!). The baryonic matter mainly exists as a coupled star-gas-system. Its mass ranges from 10^6 $M_\odot$ to 10^{11} $M_\odot$. In the central region the mass ratio of the dark matter to baryons is small, but increases quickly with galactocentric distance. Already outside one or a few scale lengths of the baryonic mass distribution the contribution of the dark matter becomes equal to that of the baryons. In the outskirts of galaxies which can be traced by neutral hydrogen, by globular clusters or by small satellite galaxies (dSph) the mass of the dark matter continues to increase without a clear limit or an outer edge (Zaritsky 1999). This is also in agreement with theoretical models in the framework of CDM cosmology (Navarro et al. 1996).

The baryonic component in galaxies is usually dominated by stars. The main parameter of a star with respect to its impact on galactic evolution is its mass: low mass stars (0.08 - 1 $M_\odot$) are very long-lived and very numerous, by this acting mainly as mass sinks in galaxies. Intermediate mass stars (1-8 $M_\odot$) have a significantly shorter lifetime. In their final stages they undergo planetary nebulae events during which processed matter is inserted into the ISM on a typical timescale of 10^9 years. Their remnants are white dwarfs with an average mass of 0.6 $M_\odot$. Finally, there are massive stars (>8 $M_\odot$) which have a very short lifetime of only a few up to 20 Myrs. At the end of their life they explode as type II supernovae, by this ejecting large amounts of energy and metal enriched matter into the interstellar medium. Stellar remnants of this group are either neutron stars or stellar mass black holes. Below a mass of 0.08 $M_\odot$ (normal) hydrogen burning is not possible. Objects in this mass range (brown dwarfs or, at the very low mass end, planets) behave – from a galactic evolutionary point of view – like low mass stars, i.e. they are long-lived mass sinks.

A special group of compact baryonic objects are massive black holes (BH) for which evidence has been found in many galaxies (Kormendy and Richstone 1995). Their masses are in the range of 10^6 $M_\odot$ (e.g. the black hole in the centre of the

Milky Way, see Melia and Falcke 2001) up to $10^{9.5}$ $M_{\odot}$ (e.g. in M87). Their Schwarzschild radius is – even for the most massive BHs – only of the size of the Solar system. Therefore, their influence on galactic evolution is sufficiently described by Newtonian dynamics of a single, very massive particle, though they are relativistic objects.

The class of non-compact baryonic matter consists of the components of the interstellar medium: gas and dust. The gas mass fraction is about 10% in normal spirals like the Milky Way. It can increase to 20-30% in dwarf galaxies. The gas component can be split into several groups. The classical 3-phase model of McKee and Ostriker (1977) assumes a clumpy component and several diffuse gaseous phases. 'Clumps' denote basically cold molecular clouds. They have a temperature between 10 and 100 K and they are the sites of star formation. The second group contains both, the warm interstellar medium which has a temperature of about 8000 K and the hot intercloud medium which is heated by supernovae to about 10^6 K. Additionally, a substantial amount of dust can be found in galaxies.

2.1.2 Interactions between Galactic Components

The main interaction coupling all galactic components is gravity. Besides gravity, magnetic forces might act on different components of the interstellar medium (stars and dark matter are believed to be unaffected by magnetic fields). Additionally, there are several phase transitions and interactions like star formation, stellar winds or stellar death which couple gas and stars. Dense clouds and the surrounding diffuse gas exchange matter and energy by condensation and evaporation processes. Moving clumps experience a drag force due to the ram pressure of the ambient diffuse gas.

Dynamically the stars and the dark matter can be treated as collisionless and non-dissipative, whereas cooling in the interstellar medium and/or transfer of kinetic energy into inner degrees of freedom (e.g. in cloud-cloud collisions) result in energy dissipation. Especially during galaxy formation when almost all baryonic matter was in gaseous form, the dynamics of a galaxy is strongly affected by dissipative processes. E.g. it influences the formation timescale of a thin disk (Burkert et al. 1992, Samland et al. 1997) or the density profile created in a collapse (Hjorth and Madsen 1991, Theis and Hensler 1995). Additionally, energy dissipation is very important for the stability properties of galactic disks (Romeo 1994, Orlova et al. 2002) or the matter redistribution in galaxies, e.g. the mass transport to the galactic centre (Noguchi 1988). Moreover, star formation and the related energy feedback sensitively depend on the stability of the gaseous components. This results in a complex reaction network which can exhibit many properties of non-linear systems like self-regulation, run-away evolution, periodic and chaotic behaviour. An example for such a network is shown in Fig. 1.

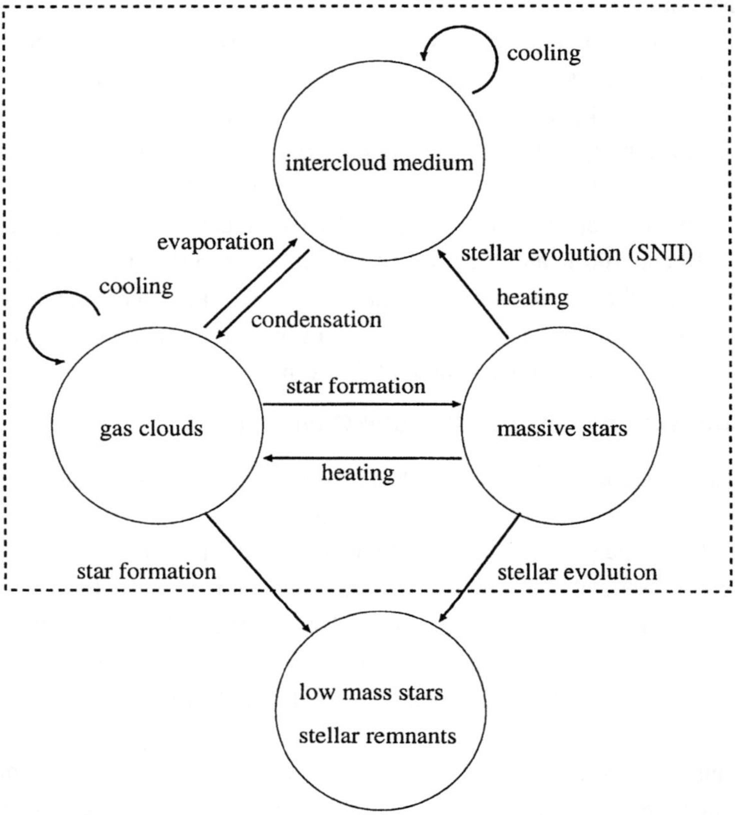

Fig. 1. A typical chemo-dynamical interaction network (adopted from Theis et al. 1992). Shown are the relations between gaseous and stellar components. The components within the box are mutually coupled, by this allowing for feedback mechanisms, whereas the low-mass stars (and stellar remnants) mainly act as mass sinks.

2.1.3 Chemo-Dynamical Models

Summarizing the mentioned characteristics of galaxies a complete model must include the dynamical evolution as well as the interactions between the galactic components. For such kind of models the term *chemo-dynamical model* has been coined (Burkert and Hensler 1987, Theis et al. 1992). It means that the coupling between chemical and dynamical evolution has to be taken into account. A prerequisite is a multi-phase treatment of the interstellar medium, because different phases of the ISM usually have different chemical compositions. Moreover, the dynamical status of the components of the ISM might differ, too. Both affects the chemical composition of newly born stars in a galaxy and makes the necessity of a consistent coupling between chemical and dynamical evolution of galactic components obvious.

How to treat the dynamics of the galactic components? A natural ansatz for stars and dark matter is the description as an N-body system, i.e. as particles moving in a common gravitational potential created by all masses in a galaxy. The equation of motion for particle *i* reads then

$$\frac{d^2\vec{r}}{dt^2} = -\nabla\Phi$$

The potential Φ is derived from the Poisson equation

$$\Delta\Phi = 4\pi\, G\rho\,(\vec{r}\,).$$

$\rho\,(\vec{r}\,)$ is the mass density of all galactic components at position $\vec{r}$ and G is the constant of gravity.

The diffuse components of the interstellar medium are naturally described by the hydrodynamical equations, e.g. the mass transport of a gas phase is given by the continuity equation

$$\frac{\partial\rho}{\partial t} + \nabla\cdot(\rho\,\vec{v}\,) = S(\rho)$$

$\vec{v}$ is the mean velocity of the gas component and $S(\rho)$ is the source function describing phase transitions like star formation or condensation and evaporation processes between different gas phases. Similar equations are required for the momentum and energy transport.

The cloudy component of the ISM is also diffuse on small scales. Therefore, the best treatment would be a high-resolution gas dynamical simulation, too. Unfortunately, the computational requirements necessary for simulating millions of objects covering many orders of magnitude in size and density are by far out of reach. Thus, simplified descriptions based either on a statistical model for the clumps (within a hydrodynamical approach) or based on an identification of individual clumps with so-called *sticky* particles (within an N-body approach) are used.

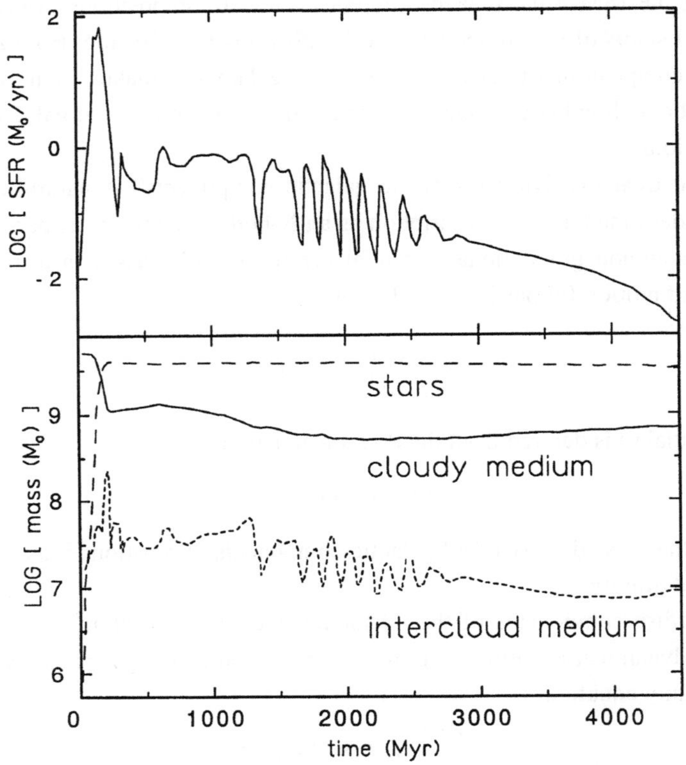

Fig. 2. Temporal evolution of the star formation rate (upper diagram) and the masses of the galactic components (lower diagram) for a $5 \cdot 10^9\,M_\odot$ galaxy.

Even with these simplifications in mind, 3-dimensional chemo-dynamical models became only possible recently (e.g. Berczik et al. 2002, Harfst et al. 2002, Semelin and Combes 2002). Historically, the first chemo-dynamical models were performed in spherical symmetry using a gas-dynamical approach for the components of the ISM. The stars have been treated by the moment equations of the Boltzmann equation up to second order (e.g. Theis et al. 1992). The latter is very similar to the hydrodynamical equations, except for the inclusion of an anisotropic 'pressure', i.e. an anisotropy of the stellar velocity dispersion. The price for this simplification of stellar dynamics is that orbit dependent effects are completely neglected, e.g. violent relaxation will not occur. On the other hand, a gas dynamical description has the advantage of an easy implementation of interactions between galactic components by a proper choice of sink and source terms.

As an example for some aspects of chemo-dynamical simulations, the evolution of different components of a dwarf galaxy is shown in Fig.2. The star formation rate exhibits a burst-like event within the first 300 Myrs. Afterwards a phase of almost constant star-formation activity follows which ends in a highly variable phase lasting for about 1 Gyr after which the star formation becomes negligible. This evolution is also reflected in the masses confined in different galactic components: in the first star burst 80% of the cold clouds are converted into stars. Later the cloud mass remains almost constant due to the replenishment by intermediate mass stars. The high variability of the star formation rate is mainly reflected by the amount of hot gas being ejected from type II supernovae. Adding up all baryonic masses in the simulation, a drop of the total mass is found: gas is lost as a galactic wind due to the strong heating in the first star burst phase.

The main disadvantage of the first chemo-dynamical models is their restriction to spherical symmetry. Hence, the next generation of chemo-dynamical simulations were 2d-dimensional axisymmetric models (e.g. Samland et al. 1997). These calculations were applied to the formation of disk galaxies. Similarly to the 1d-models, the spatio-temporal evolution of the ISM and the stars has been followed. E.g. these models demonstrated a delayed formation of the disk and they reproduced for the first time the distribution of elements in different regions of the Milky Way as a result of a universal stellar nucleosynthesis model and a global star formation law. This result nicely demonstrated the close coupling between chemical and dynamical evolution as well as the necessity of a multi-phase treatment of the interstellar medium.

However, the 2d-models also suffered from their geometrical restrictions. E.g. no spiral arms or bars can be formed and thus the mass and angular momentum transport related to non-axisymmetric structures are neglected. Moreover, different galactic components like the disk and dark matter halo need not to be aligned, by this breaking the adopted axisymmetry. Last, but not least, except for head-on collisions, no galaxy interactions can be modelled by a 2d-code. A direct extension of the old 2d-chemodynamical code based on the Boltzmann moment equations to three dimensions, however, is not the best way to go. First, these calculations became very CPU-time-consuming. A single simulation might need about 1 yr CPU-time on a modern vector machine. Second, though the distribution function described by the Boltzmann equation contains all information about the stellar system, the moment equations (which are derived by integrating over velocity space) reduce the distribution in velocity space to a mean velocity and the velocity dispersion. This loss of knowledge is naturally avoided when using N-body simulations instead of the Boltzmann moment equations. Additionally, the computational power of modern computers has reached now a level allowing for hybrid codes combining N-body simulations and hydrodynamical calculations. Most 3d-implementations of chemo-dynamical schemes are based on N-body calculations (Berczik et al. 2002, Harfst et al. 2002, Semelin and Combes 2002).

2.2 Dynamics of Galaxies

In the previous section the "grand picture" of chemo-dynamical models has been described. The backbone of these state-of-the-art chemo-dynamical models, but also of less ambitious models is the proper treatment of the dynamics, i.e. of stellar and gas dynamics. The classical gravitational N-body problem is easily formulated and solved: the formal solution for the gravitational potential (and, by this, for the related force) at the position $\vec{r}_i$ of particle i with mass m_i is

$$\Phi(\vec{r}_i) = -\sum_{j=1, j\neq i}^{N} \frac{Gm_j}{\left|\vec{r}_i - \vec{r}_j\right|}.$$

Computationally, this sum is very expensive, because it scales with $O(N^2)$. Thus, practically it is restricted to particle numbers N up to $\sim 10^{3\text{-}4}$.

In order to avoid this difficulty several approximations depending on the application have been made. The strongest simplification is the assumption of stationarity which might hold over some time for isolated galaxies. Then, no simulations are necessary and the analysis of potential-density pairs allows for the investigation of the dynamical properties (including stability) of these stellar systems.

When time-dependent effects like the onset and growth of spiral instabilities are investigated, the stellar system might be approximated successfully by a fluid, using the Boltzmann equation or the gasdynamical equations. This approach is especially useful, when the system is highly symmetric, but it becomes computationally expensive, when real 3d problems have to be tackled.

Both mentioned approximations just circumvent the classical N-body problem. However, there are also ways to simplify the galactic N-body problem by means of approximative methods. The next two sections will briefly describe these methods focussing on the development of N-body simulations.

2.2.1 Stellar Dynamics

The theoretical understanding of galactic dynamics, especially of interacting systems, suffered for a long time from the lack of computational power allowing for a numerical solution of the gravitational N-body problem. After a remarkable treatment by Holmberg (1941) who built an analogue computer (consisting of light bulbs and photo cells) for the determination of the gravitational force, it took 20 years until N-body simulations were performed on a general purpose computer (Pfleiderer and Siedentopf 1961). The basic idea of their *restricted N-body* simulations of interacting galaxies is the assumption that the potential of interacting galaxies can be adequately modelled by two particles (representing the two galaxies) moving under their mutual gravitation, i.e. moving on Keplerian orbits. With these assumptions all the other particles are just test particles, and the complete N-body problem is reduced to N single body problems for a time-dependent potential. In a remarkable series of simulations Toomre and Toomre (1972) applied this technique and determined the parameters of several well-studied interacting systems including M 51 + NGC 5195 and NGC 4038/39.

Motivated by these simulations and by a remarkable coincidence of the estimated fraction of strong interactions with the fraction of elliptical and S0 galaxies Toomre (1977) put forward the idea of a merger origin of elliptical galaxies. According to this suggestion a whole class of galaxies would have been formed mainly by strong interactions. However, restricted N-body simulations were not able to check this idea, because they do not include the self-gravity of galactic disks. Thus, they do not treat self-consistently processes like dynamical friction or transfer of orbital angular momentum of the galaxies into galactic spin. Moreover, the assumption of unaffected point-like galaxies becomes rather poor for merging systems. Thus, computationally much more expensive self-consistent calculations were required. Already first models using N=500 particles proved the idea of merging of galaxies (White 1978, Gerhard 1981).

As a next step new N-body techniques have been developed which increased the accessible particle number by many orders of magnitude. The TREE-method determines the gravitational potential by superposing the potentials of different groups of particles. The particles are organized in such a way that the more distant particles are combined in larger groups. The contributions of the individual groups to the common field of gravity are then considered by a multipole expansion of their potentials (Barnes and Hut 1986, Hernquist 1987). The computational effort scales like ~ $O(N \log N)$ which is also superior to direct simulations.

Using his TREE code, Barnes (1988) was able to simulate encounters of disk galaxies including all dynamical components, i.e. disk, bulge, and halo as N-body systems. Compared to faster grid-based methods (e.g. Sellwood 1980) or expansion methods (e.g. Hernquist and Ostriker 1992) direct N-body simulations (or semi-direct methods like TREEs) are more flexible with respect to strongly varying geometries and scale lengths. This makes them the preferred tool for simulations of interacting galaxies or galaxy clusters.

An alternative to these techniques are special-purpose computers (like the machines of the GRAPE project (Sugimoto et al. 1990). They implemented Newton's law of gravity (modified for gravitational softening) in the hardware by constructing a special chip. This allows for a very fast direct determination of the gravitational forces, though there still exists the N^2 -bottleneck. E.g. for simulations with N=10^5 particles a GRAPE3af (with 8 GRAPE processors) is competitive with a TREE-code running on a CRAY T90.

2.2.2 Gas Dynamics

Gasdynamical simulations can be split into Eulerian or Lagrangian calculations. Eulerian methods use a fixed (or adaptive) grid on which the hydrodynamical equations are solved. The mentioned 1d- and 2d-chemo-dynamical models belong to this group.

An alternative approach are Lagrangian codes which solve the hydrodynamical equations on a comoving fluid element. In three dimensions a natural realization of this idea relies on particles probing the fluid. In this *smoothed particle hydrodynamics* scheme (SPH) the equations of motion of a particle include not only gravity, but also hydrodynamical forces related to pressure gradients or viscosity (e.g. Hernquist and Katz 1989). The corresponding terms are calculated by averaging ("smoothing") over the close vicinity of a particle. Such a description of the ISM emphasizes its diffuse nature.

An alternative ansatz focuses on the clumpiness of the ISM treating the clouds as *sticky particles*: Without physical collisions or close encounters the clouds move on ballistic orbits like stars. However, in the case of a collision the clouds might merge or lose kinetic energy depending on the adopted microphysics (e.g. Palouš et al. 1993, Theis and Hensler 1993).

3 Modelling Interacting Galaxies

In the previous section physical ingredients and numerical methods of state-of-the-art simulations for galactic evolution have been discussed. Because such calculations are computationally very expensive, they are mainly used for improving our basic knowledge about fundamental physical processes related to galaxy evolution. Thus, they belong to the class of *qualitative* models.

In this section I will describe the problems related to *quantitative* modelling of galaxies, i.e. an as-close-as-possible match of observations. As an example interacting galaxies will be discussed.

3.1 Approaches and Problems

Relatively few papers exist on modelling special objects. This deficiency is the result of two factors. First, high resolution data in configuration and velocity space are required, covering a large fraction of the space between the interacting galaxies. In principle, HI would be suitable, however, there are just a few observational sites which give data of sufficient quality.

The second problem is the large parameter space for a galactic interaction resulting in two connected difficulties: finding a good fit and determining its uniqueness (or other acceptable parameter sets). Observationally, only three kinematical quantities – the projected position on the sky and the line-of-sight velocity – can be measured. The determination of the galactic mass depends on the availability of velocity data, the determination of the distance and the inclination of a galaxy, as well as on the reliability of the conversion from velocities to masses. In case of a two-body problem and if one neglects the centre-of-mass data of the interacting system the 14 parameters reduce to 7 parameters containing the relative positions, relative velocities and the mass ratio of both galaxies. From these 7 parameters 3(-4) values might be determined observationally. Moreover, these 7 values just fix the orbits of the galaxies (or more exact their centres). One still has to specify the parameters that characterize both stellar systems, e.g. characteristic scales, orientation, or rotation. The final result is a high-dimensional parameter space which is in general too large for a standard search method. For instance, the interaction of a galactic disk with a point-mass galaxy is described by at least 7 parameters. A regular grid with a poor coverage of 10 grid points per dimension demands 10^7 models or 3400 years of GRAPE3 simulation time (assuming 3 CPU-hours for a single simulation) or still about a year with the faster, but cruder restricted N-body method (assuming 3 CPU-seconds per simulation).

3.2 Modelling as an Optimization Problem

Summarizing the requirements of modelling observations of interacting galaxies, one has to solve an optimization problem in a high-dimensional parameter space. The aim is to find one or several regions in the parameter space which give an absolute maximum of the quality function. The quality function itself (also denoted later as *fitness*) describes the goodness of fit. Historically, the first models of galaxy interactions were not systematic scans of the parameter space, but they were derived from a trial-and-error strategy. Though their results (e.g. Toomre and Toomre 1972) are very remarkable they suffer from the ‘human bias’: it is not clear, what is quantitatively a good fit (‘fit by eye’); the optimization path is not reproducible (or just completely unknown); psychological effects might influence the estimate of the goodness of a fit.

More systematic searches suffer from the problems mentioned in Sect. 3.1. Therefore, very efficient simulations techniques as well as efficient search methods are required. Additionally, a quality function has to be specified explicitly. Since the dependence of the quality function on the parameters (the ‘fitness landscape’) is almost unknown *a priori*, the search method should be insensitive to its starting point. Moreover, it should have the capability to leave local optima in order to avoid trapping. Genetic algorithms fulfill these conditions. In the next sections their concept is described and applications to interacting galaxies are presented.

3.3 A Genetic Algorithm Approach

The idea of applying models of organic evolution for optimization problems dates back to the 1960s and 1970s (e.g. Rechenberg 1965). Unlike standard deterministic gradient techniques for optimization (e.g. the downhill simplex method (Press et al. 1992) Rechenberg's *Evolutionsstrategie* is probabilistic: starting with a more or less random *parent*, i.e. a single point in parameter space, a *child* is generated by a random mutation of the parameter set characterizing the parent. The quality of both individuals with respect to the optimization problem (i.e. their *fitness*) determines the parent of the next generation. Repeating this process of mutation and selection improves the quality of the individual monotonously.

An alternative evolutionary concept are genetic algorithms which operate on a population instead of a single individual (Holland 1975, Goldberg 1989). Each member of a population represents a single point in parameter space, i.e. a N-body simulation with a given set of parameters. The members are characterized by their fitness which quantifies the agreement between the simulations and the reference map (observation or numerical model). In order to determine the 'parents' two individuals are selected according to their fitness. These parents are two points in parameter space. The parameters of each individual are converted to a "universal" alphabet (here 4-digit numbers) and then combined to a single string, the *chromosome.* This chromosome is subject to a *cross-over*[1] and a *mutation* operation resulting in a new individual which is a member of the next generation. Such a breeding is repeated until the next generation has been formed.

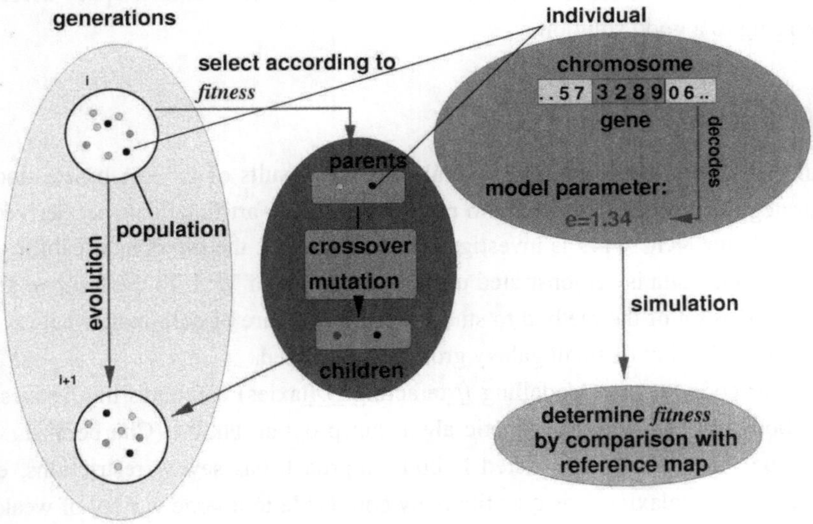

Fig. 3. Schematic diagram of genetic algorithm approach. For details see text.

Finally, the whole process of creating new generations is repeated iteratively until the population confines one or several regions of sufficiently high fitness in pa-

[1] Cross-over e.g. is realized by swapping the ends of two chromosomes at a randomly chosen cross-over position.

rameter space. The scheme of such a GA is shown in Fig. 3. For more details see Theis (1999) or Theis and Kohle (2001).

Although genetic algorithms (GA) have been used in many branches of science, there are just a few applications in astrophysics, e.g. for fitting rotation curves or analysis of Doppler velocities in δ Scuti stars (for a review see Charbonneau 1995). Recently, Wahde (1998) and Theis (1999) demonstrated the ability of genetic algorithms to recover orbital parameters for artificial observations generated from N-body simulations of interacting galaxies. The application of GAs has several advantages: compared to complete grids in parameter space, the probabilistic but oriented nature of the evolutionary search strategy allows for an efficient check of a high-dimensional parameter space. Compared to gradient methods which are very fast near the optimum, GAs do not need any gradient information which is computationally expensive. They depend only weakly on the starting point and – most important – they are able to leave local optima. The price for these features is a large number of fitness evaluations or test points in parameter space before converging to a good solution.

3.4 Some Results

In this section I will present several tests and results of the GA-based modelling strategy. Afterwards its ability to reproduce a given artificial data set derived from models for NGC 4449 is investigated. In Sect. 3.4.2 the direct applicability to observational data is demonstrated using the example of M51. In the last two sections the extension of the method to studies of the structure of dark matter haloes and to the investigation of small galaxy groups is discussed.

Our code MINGA (*M*odelling *IN*teracting *GA*laxies) combines the fast restricted N-body method with the genetic algorithm program PIKAIA (Charbonneau 1995). Though the classical restricted N-body approach has severe restrictions, e.g. no merging of galaxies, it is quantitatively applicable to a large variety of weak to intermediate encounters of galaxies as comparisons with self-consistent calculations proved. Even in merging systems it might be applicable over short periods of the evolution. Its main advantage is the low computational cost. A complete simulation can be done in about 1 CPU second on a modern PC which allows for 10^4 to 10^5 simulations in a single day. A GA run usually needs at least about 10^4 evaluations of the fitness function, i.e. 10^4 N-body simulations. Because self-consistent N-body simulations are about a factor of 1000 more expensive, the restricted N-body approach is the only approach allowing for a sufficient fast computation.

3.4.1 Tests of the Method

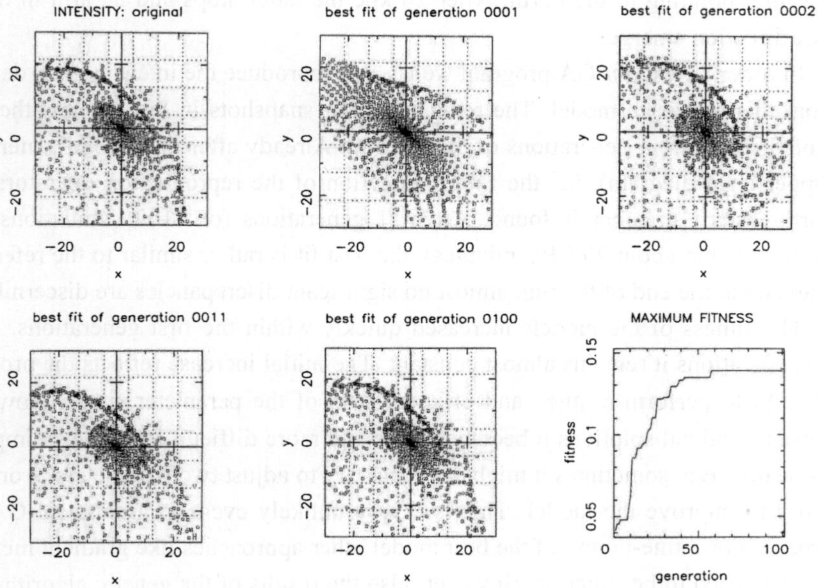

Fig.4. Best fit model during the course of a GA fitting procedure.

The projection of the particles on the x-y-plane and the corresponding grid for the intensity evaluation are displayed: the original data (upper left), the best fit of the GA after initialization (upper middle), after the first breeding (upper right), after 11 generations (lower left) and at the end of the fitting procedure after 100 generations (lower middle). The evolution of the maximum fitness is shown in the lower right diagram. The number of test particles is 900.

As a first example we studied HI data of NGC 4449 suggesting an interaction with the dwarf companion DDO 125. NGC 4449 is very similar to the Large Magellanic Cloud, but it has no normal galaxy in its close vicinity. Therefore, it became a prototype of this class of objects (for details see Kohle 1999). On optical images it shows a bar and strong star formation activity, but no hints for an interaction. Already in the 1970s a very extended HI halo was detected (van Woerden et al. 1975), however only recent observations revealed its detailed structure (Hunter et al. 1998). Very remarkable are the streamers with their abrupt changes in the direction. The upper left diagram in Fig. 4 shows a snapshot of a model qualitatively reproducing the streamer structure. Visible are the three streamers: a weak feature starts south of the centre to the south west which abruptly turns to a vertical streamer pointing to the north. After 20 kpc the latter stops and an arm in north-east direction emerges.

In order to test our GA program we tried to reproduce the intensity map derived from this reference model. The remaining four snapshots in Fig. 4 show the best model of different generations of the GA run. Already after the second generation (upper right diagram), i.e. the first application of the reproduction operators, the north-eastern streamer is found. After 11 generations (or 1100 simulations performed within about 20 CPU minutes), the best fit is rather similar to the reference map and at the end of the run, almost no significant discrepancies are discernible.

The fitness of the models increased quickly within the first generations. After 50 generations it remains almost constant. The initial increase reflects the property of GAs to perform a quick and efficient scan of the parameter space. However, close to optimal solutions it becomes more and more difficult for a GA to improve its results. E.g. sometimes it might be necessary to adjust two parameters at once in order to improve the model. This is a very unlikely event in a classical GA approach. For a fine-tuning of the best model other approaches like gradient methods are supposed to be superior. However, also the results of the genetic algorithm are quite accurate as the comparison between the parameters of the reference map and those of the best model of a generation demonstrate (Fig. 5). All parameters are reproduced to better than 10% relative accuracy, some of them even better than 1%. Such a high level of accuracy is much better than the observational uncertainties, e.g. in case of the mass ratio of the galaxies. Additionally, the systematic errors introduced by the uncertainties of the (non-accessible) initial conditions of both galaxies is probably not smaller than 10%. From that the accuracy of the GA models should be sufficient in all cases when restricted N-body simulations are applicable at all.

Last, but not least, the reproduction of the parameters of the reference model in all our GA runs for NGC 4449 gives a clear hint that its HI structure seems to be produced by a unique region in parameter space. Though this is not a mathematical proof, the investigation of millions of models is a much better basis for such a claim than the usual few models investigated in a search ‘by hand’.

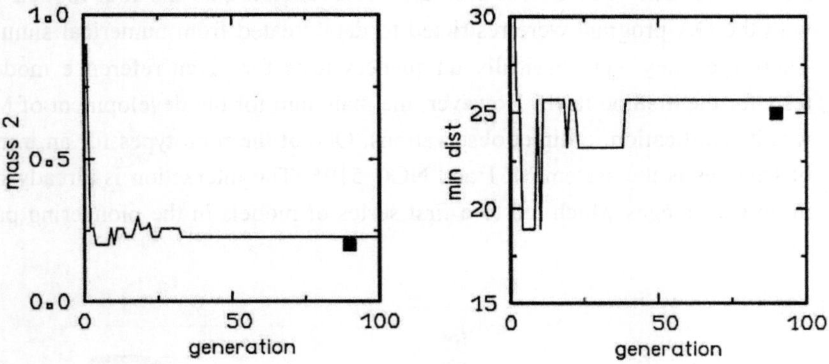

Fig.5. Development of two parameters of the best fit model during a GA run. The parameters are the mass of the secondary galaxy (left) and the minimum distance (right). The filled squares show the parameters of the fitted artificial reference model.

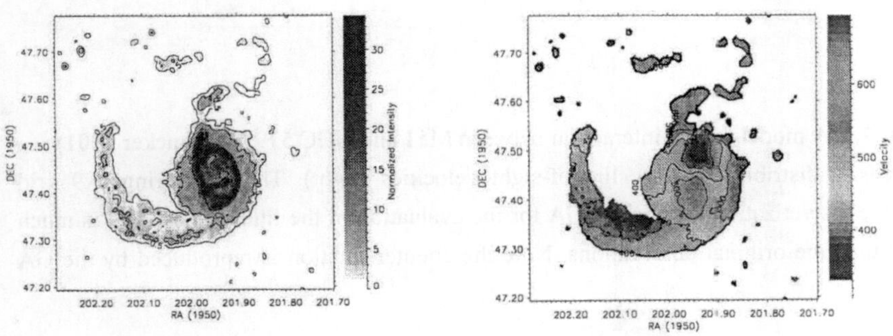

Fig. 6.
Left: HI intensity distribution in M51. The optical visible region including the interaction partner NGC 5195 is roughly identical with the dark shaded central region. Note that the HI lumps north of the centre seem to be part of a large-scale coherent structure. Right: Line-of-sight velocities derived from HI observations of M51. Note the counter-rotation seen in the identical velocities north of the centre of M51 and in parts of the tidal tail (data from Rots et al. 1990).

3.4.2 Automatic Fit: The Case of M51

Due to the lack of sufficient accurate observational data the first models studied with the GA program were restricted to data created from numerical simulations. Therefore, they were basically uniqueness tests for given reference models and tests for the method itself. However, the main aim for the development of MINGA was its application to direct observations. One of the prototypes for an interaction of galaxies is the system M51 and NGC 5195. The interaction is already obvious in optical images which led to a first series of models in the pioneering paper by

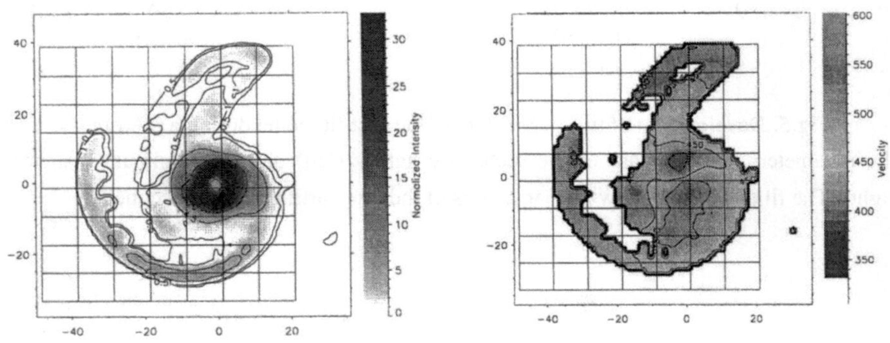

Fig. 7. GA model for the interaction between M51 and NGC 5195 (Spinneker 2001) HI intensity distribution (left), line-of-sight velocities (right). The underlying 7x9 grid shows the discretization used in the GA for the evaluation of the fitness. This grid is much coarser than the original observations. Note the counter-rotation is reproduced by the GA model.

Toomre and Toomre (1972). However, again HI observations revealed a long extended tidal arm which is not visible in the optical (Fig. 6 left). Especially, the classical models from Toomre and Toomre are not able to reproduce the extended HI structure. A second problem stems from the dynamics: there is a counter-rotation between parts of the tidal arm and the central region of M51 (Fig. 6 right) which is not reproduced by the classical scenario.

Using real data immediately adds the requirement of preparing the observational data. The first step is the reduction of the observational data which includes all steps from transferring the raw data into 'astronomer's data' (e.g. it might include calibrations, flat field corrections etc.). The next step is to extract the information related to the object of interest. E.g. data can include unwanted fore- or background objects. Depending on the nature of these "perturbers" one can try to remove them from the data (e.g. in case of point-like objects) or to incorporate them in the modelling procedure. This step always enters some kind of arbitrariness, because especially for extended features it is not always *a priori* clear, if they belong to the investigated object. Practically, it turned out that the usage of masks adopting different weights for its cells is very helpful. Unreliable regions can get a low weight or they might be even completely neglected.

Additionally, a mask allows to emphasize regions which contain important information about the interaction. For example, the bright centres of galaxies are normally much less affected by an interaction (if they are affected at all), because the tidal radii usually exceed the central region. Therefore, centres are not very helpful in constraining an interaction scenario. On the other hand, the typically weak, but very extended outer regions (e.g. the HI arm seen in M51) contain much more information about the interaction history. So, matching them is more important than modelling the central areas. Their special importance can be expressed by a higher weight of the corresponding positions in the mask.

Fig. 7 shows a GA fit for the M51 system. From this model we expect two passages of the companion through the plane of M51's disk within the last 500 Myr. The eccentricity of the orbit is 0.46, i.e. the galaxies form a gravitationally bound system. The perigalactic passage happened 320 Myr ago at a distance of about 12 kpc. Our model is qualitatively similar to one of the models suggested by Salo and Laurikainen (2000). Both models reproduce the counter-rotation between the centre and parts of the tidal arm. Despite the success of reproducing the large-scale intensity distribution as well as the gross dynamics, there are still unclear features: none of our models could reproduce the exceedingly high velocities seen in the spur north of NGC 5195. Is that gas lump really a part of the M51 system? Is its velocity related to special initial conditions? Is there some hidden physics not included in our models so far?

3.4.3 Recovering Dark Matter Halos?

An important motivation for studying interacting galaxies is to learn more about the invisible galactic components, i.e. about the dark matter distribution of galaxies. Dubinski et al. (1999) demonstrated that the properties of dark matter halos have a strong influence on the formation of tidal tails. However, this is just a necessary, but not a sufficient condition for constraining the dark matter profiles by modelling tidal features. E.g. systematic degeneracies between halo and other galactic parameters might also exist.

A more technical problem is the applicability of restricted N-body calculations to the interaction of extended objects. A prerequisite is that the dark matter halo keeps its properties during the interaction (or varies only in a "simple" way). Qualitatively, this assumption seems to hold for many interactions, because dynamically hot systems (like halos) couple much weaker to perturbations than cold systems (like disks). A nice example is NGC 4449 (Theis and Kohle 2001). Another difficulty is related to the calculation of the accelerations: the forces between two extended, overlapping spheres are not identical to the forces between two point-like objects of the same mass. Therefore, the orbits of the galaxies became non-Keplerian. This roughly doubles the computational time for a single simulation.

Combining such a modified restricted N-body scheme with a genetic algorithm and applying it to artificial 'observations" gives a good reproducibility of the halo parameters. In case of a run for NGC 4449 all parameter except one have been reproduced to better than 10% (Gerds 2001). The only larger discrepancy (a factor of 2) was the size of the companion's halo which was already in the reference model too small to affect the interaction significantly. Additionally, the GA runs showed that the fits are less sensitive to the exact adopted dark matter profiles than to the gross properties of the halo like spatial size or total mass.

3.4.4 Small Groups of Galaxies

It is natural to apply the GA not only to systems of two galaxies, but also to small galaxy groups like some of the Hickson groups. Fig. 8 demonstrates the excellent result for a system of three galaxies. Though the enlarged parameter space requires larger populations and more generations, the match between the intensity maps seems to be almost perfect. Mass ratios are reproduced to better than 2% (!), whereas z -positions and proper motions are fixed on average to about 10-25%. It should be noted that the results have been derived from the intensity map only.

Eventually available information about the line-of-sight velocities will probably improve the fits even more. Thus, the enhanced parameter space seems not to prohibit the application of the GA to small groups of galaxies, provided sufficient tidal signals are visible.

Summary

In this paper different techniques for modelling galaxies have been described. *Qualitative* modelling, i.e. calculations performed to improve our basic knowledge on fundamental processes related to galactic evolution, have been discussed in the framework of state-of-the-art chemodynamical models. Different to such complex models, the direct analysis of observed galaxies, i.e. the *quantitative* modelling, has to start with much simpler models in order to cope with the very extended parameter space. It has been demonstrated that in the case of interacting galaxies, a combination of a fast, but approximative N-body solver (restricted N-body) and a

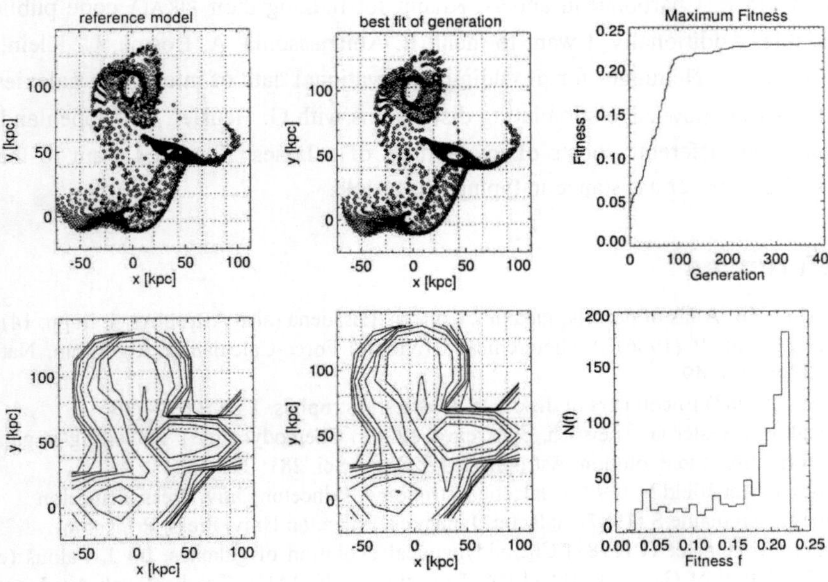

Fig. 8. GA run for a small group of galaxies (Gerds 2001).
Left column: reference model (upper left: positions of particles, lower left: intensity map used in the GA code), middle column: best fit after 400 generations with 899 individuals, right column: evolution of fitness (upper right) and final fitness distribution (lower right).

genetic algorithm allows for a reliable and very efficient scan of the parameter space. If results are found at all, their relative accuracy is usually of the order of 10%. They allow to restrict the properties of dark matter haloes as well as modelling of small galaxy groups. The example of M51 proved the applicability to direct observations. Moreover, even if sufficient observational data are missing, the GA code can be used for uniqueness tests of preferred interaction scenarios.

On the other hand, GA-based models are just a first step. A necessary second step is a check of the applicability of the restricted N-body method for the system under investigation. Therefore, the results have to be compared with self-consistent stellardynamical and gasdynamical simulations. And finally the regions of interest in parameter space should be studied with very detailed state-of-the-art methods like the chemo-dynamical calculations.

Acknowledgements

It is a pleasure to thank my students S. Harfst, Ch. Spinneker and Ch. Gerds for discussions and engaged improvements of the GA-code MINGA. I am especially grateful to P. Charbonneau and B. Knapp for making their PIKAIA code publicly available. Additionally, I want to thank E. Athanassoula, A. Bosma, U. Klein, S. Kohle and N. Neininger for providing observational data of interacting galaxies. I gratefully acknowledge stimulating discussions with G. Hensler, J. Köppen and P. Kroupa on different aspects of interactions of galaxies. Finally, I want to thank Mrs. Kuhr for her assistance in typing this article.

References

Arp H (1966) Atlas of peculiar galaxies. Caltech, Pasadena (also Astrophys. J. Supp. 14)

Barnes J, Hut P (1986) A Hierarchical O(NlogN) Force-Calculation Algorithm. Nature 324: 446-449

Barnes J (1988) Encounters of disk/halo galaxies. Astrophys. J. 331: 699-717

Berczik P, Hensler G, Theis Ch, Spurzem R (2002) Chemodynamical modelling of galaxy formation and evolution. Astrophysics & Space Sci. 281: 297-300

Binney J, Merrifield M (1998) Galactic Astronomy. Princeton Univ. Press, Princeton

Binney J, Tremaine S (1987) Galactic Dynamics. Princeton Univ. Press, Princeton

Burkert A, Hensler G (1987) Chemodynamical evolution of galaxies. In: J. Palouš (ed.) Evolution of Galaxies (10th ERAM meeting of the IAU). Czechoslovak Academy of Sciences, Ondreov, pp 275-277

Burkert A, Truran JSW, Hensler G (1992) The collapse of our Galaxy and the formation of the Galactic disk. Astrophys. J. 391: 651-658

Charbonneau P (1995) Genetic Algorithms in Astronomy and Astrophysics. Astrophys. J. Supp. 101: 309-334

Dubinski J, Mihos JC, Hernquist L (1999) Constraining Dark Halo Potentials with Tidal Tails. Astrophys. J. 526: 607-622

Gerds Ch (2001) Signaturen dunkler Materie bei der Wechselwirkung von Galaxien. diploma thesis (in German), Univ. of Kiel, Germany

Gerhard O (1981) N-body simulations of disc-halo galaxies – Isolated systems, tidal interactions and merging. Monthly Not. of the Royal Astron. Soc. 197: 179-208

Goldberg DE (1989) Genetic Algorithms in Search, Optimization, & Machine Learning. Addison-Wesley, Reading

Harfst S, Theis Ch, Hensler G (2002) Exchange processes in a multi-phase ISM. Astrophysics & Space Sci. 281: 313-314

Hernquist L (1987) Performance characteristics of tree codes. Astrophys. J. Supp. 64: 715-734

Hernquist L, Katz N (1989) TREESPH – A unification of SPH with the hierarchical tree method. Astrophys.J. Supp. 70: 419-446

Hernquist L, Ostriker JP (1992) A self-consistent field method for galactic dynamics. Astrophys. J. 386: 375-397

Hjorth J, Madsen J (1991) Violent relaxation and the R exp ¼ law. Monthly Not. of the Royal Astron. Soc. 253: 703-709

Holland J (1975) Adaptation in natural and artificial systems. Univ. of Michigan Press, Ann Arbor

Holmberg E (1941) On the Clustering Tendencies among the Nebulae.II. A Study of Encounters Between Laboratory Models of Stellar Systems by a New Integration Procedure. Astrophys. J. 94: 385-395

Hubble E (1925) NGC 6822, a remote stellar system. Astrophys. J. 62: 409-433

Hubble E (1926a) A spiral nebula as a stellar system: Messier 33. Astrophys J. 63: 236-274

Hubble E.(1926b) Extragalactic Nebulae. Astrophys. J. 64: 321-369

Hubble E (1929) A spiral nebula as a stellar system, M31. Astrophys. J. 69: 103-158

von Humboldt A (1850) Kosmos, Vol.III. Cotta'scher Verlag, Stuttgart

Hunter DA, Wilcots EM, van Woerden H, Gallagher JS, Kohle S (1998) The Nature of the Extended HI Gas around NGC 4449: The Dr. Jekyll/Mr. Hyde of Irregular Galaxies. Astrophys. J. 495: L47-L50

Kant I (1755) Allgemeine Naturgeschichte und Theorie des Himmels oder Versuch von der Verfassung und dem mechanischen Ursprunge des ganzen Weltgebäudes nach Newtonischen Grundsätzen abgehandelt. Ebert H (ed.), 1890, Verlag Wilhelm Engelmann, Leipzig

Kohle S (1999) Ph. D. thesis, University of Bonn

Kormendy J, Richstone D (1995) Inward Bound --- The Search for Supermassive Black Holes In Galactic Nuclei. Ann. Rev. Astron. & Astrophys. 33: 581-624

McKee CF, Ostriker JP (1977) A theory of the interstellar medium – Three components regulated by supernova explosions in an inhomogeneous substrate. Astrophys. J. 218: 148-169

Melia F, Falcke H (2001) The Supermassive Black Hole at the Galactic Center. Ann. Rev. Astron. & Astrophys. 39: 309-352

Navarro JF, Frenk CS, White SDM (1996) The Structure of Cold Dark Matter Halos. Astrophys. J. 462: 563-575

Noguchi M (1988) Gas dynamics in interacting disc galaxies – Fuelling of nuclei by induced bars. Astron. & Astrophys. 203: 259-272

Orlova N, Korchagin V, Theis Ch (2002) Global spiral modes in multi-component disks. Astron. & Astrophys. 384: 872-878

Palouš J, Jungwiert B, Kopecký J (1993) Formation of Rings in Weak Bars – Inelastic Collisions and Star Formation. Astron. & Astrophys. 274: 189-202

Pfleiderer J, Siedentopf H (1961) Spiralstrukturen durch Gezeiteneffekte bei der Begegnung zweier Galaxien. (in German) Zeitschrift. f. Astrophysik 51: 201-205

Press WH, Teukolsky SA, Vetterling WT, Flannery BP (1992) Numerical Recipes in FORTRAN – The Art of Scientific Computing. Cambridge Univ. Press, Cambridge

Rechenberg I (1965) Cybernetic solution path of an experimental problem. Royal Aircraft Establishment, Library Translation No. 1122, Farnborough

Romeo A (1994) How faithful are N-body simulations of disc galaxies? Astron. & Astrophys. 286: 799-806

Rots AH, Bosma A, van der Hulst JM, Athanassoula E, Crane PC (1990) High-resolution HI observations of the Whirlpool galaxy M51. Astron. J. 100: 387-393

Salo H, Laurikainen E (2000) N-body model for M51 – I. Multiple encounter versus single passage? Monthly Not. of the Royal Astron. Soc. 319: 377-392

Samland M, Hensler G, Theis Ch (1997) Modeling the Evolution of Disk Galaxies. I. The Chemodynamical Method and the Galaxy Model. Astrophys. J. 476: 544-559

Sellwood JA (1980) Galaxy models with live halos. Astron. & Astrophys. 89: 296-307

Semelin B, Combes F (2002) Formation and evolution of galactic disks with a multiphase numerical model. Astron. & Astrophys. 388, 826-841

Spinneker Ch. (2001) Die Modellierung wechselwirkender Galaxien am Beispiel von M51. diploma thesis (in German), Univ. of Kiel, Germany

Sugimoto D, Chikada Y, Makino J, Ito T, Ebisuzaki T, Umemura M (1990) A Special-Purpose Computer for Gravitational Many-Body Problems. Nature 345: 33-35

Theis Ch, Burkert A, Hensler G (1992) Chemo-dynamical evolution of massive spherical galaxies. Astron. & Astrophys. 265: 465-477

Theis Ch, Hensler G (1993) Dynamical evolution of dissipative cloud systems. Astron. & Astrophys. 280: 85-104

Theis Ch, Hensler G (1995) Dissipative Collapse of a Non-Rotating System. In: H. Hippelein, K. Meisenheimer, H-J. Röser (eds.) Galaxies in the Young Universe, Ringberg, pp. 201-205

Theis Ch. (1999) Modeling Encounters of Galaxies: The Case of NGC 4449. Rev. of Modern Astronomy 12: 309-345

Theis Ch, Kohle S (2001) Multi-method-modeling of interacting galaxies. I. A unique scenario for NGC 4449? Astron. & Astrophys. 370: 365-383

Toomre A, Toomre J (1972) Galactic Bridges and Tails. Astrophys. J.178: 623-666

Vorontsov-Velyaminov BA (1959) Atlas and Catalogue of Interacting Galaxies. Sternberg Institute, Moscow

Wahde M (1998) Determination of orbital parameters of interacting galaxies using a genetic algorithm. Description of the method and application to artificial data. Astron. & Astrophys. Supp. 132: 417-429

White SDM (1978) Simulations of merging galaxies. Monthly Not. of the Royal Astron. Soc. 184: 185-203

van Woerden H, Bosma A, Mebold U (1975) Distribution and motions of neutral hydrogen in the giant irregular galaxy NGC 4449. In: Weliachew L (ed.) La Dynamique des Galaxies Spirales, pp 483-488

Zaritsky D (1999) The Mass and Extent of the Galactic Halo. In: Gibson B et al. (eds.) The Galactic Halo, ASP Conf. Ser. 165: 34-45

Modern Soda Lakes
Model Environments for an Early Alkaline Ocean

Stephan Kempe and Jozef Kazmierczak

Introduction

The reconstruction of Earth's history relies on evidence recovered from its rock record. From this record we know that life originated prior to 3.8 Ga (10^9 years) (e.g. Schidlowsky et al. 1979, Mojzis et al. 1996), that multicellular life arose at around 0.6 Ga, and that biologically controlled (enzymatic) biomineralization started with the beginning of the Phanerozoic eon, at around 0.54 Ga ago (e.g. Lowenstam and Margulis 1980, Simkiss 1989). Geochemical forcing most probably triggered these events. The nature of this forcing, however, remains unknown and one can only formulate working hypotheses. The Earth system is governed by too many parameters, therefore excluding the possibility of building models, which, by representing all the physical interactions, would create life. Today we know that Earth must be a unique planet in a special planetary system to have not only made biogenesis possible but also to sustain life's long-term evolution towards multicellularity and self recognition. In fact so many preconditions have been met that a biologist might ask, "would God have invented life if he were a physicist?" Therefore we rather have to ask, what are the necessary environmental conditions under which life could have been initiated, and, can we find such environments even today in order to study these primordial conditions?

Most researchers agree that life arose in an aqueous solution, most probably in the Hadean ocean, even though special niches such as hydrothermal systems, lagoons or even fresh water or volcanic lakes cannot be excluded *a priori.* For an overview of the different hypotheses of origin of life see for example Davis & McKay (1996). Also life could not have arisen in an oxygenated environment, organic matter would have been oxidized too quickly. We furthermore suggest that life cannot have arisen in a solution rich in Ca^{2+} such as in the present ocean (10 mmol Ca/l) (e.g. Kempe and Kazmierczak 1994). Appreciable concentrations of Ca^{2+} or other divalent ions in the cell-fluid prohibit the proper functioning of proteins. In fact, Ca^{2+} concentration cannot rise to higher than 10^{-6} M without damaging the cell proteins irreversibly. For the proper functioning of cells a level of $<10^{-7}$ M needs to be maintained (e.g. Borle 1981, Campbell 1983, Marmé 1985, Carafoli 1987, Berridge et al. 1998). Otherwise the very low solubility product of $[Ca^{2+}]$ and $[PO_4^{3-}]$ would lead to the precipitation of calcium phosphate which would be deleterious to cell functioning functioning (Lowenstam and Margulis 1980, Simkiss and Wilbur 1989). Today cells regulate their internal Ca-concentration by complicated proteins, called Ca-pumps, situated in the phospholipid cell membranes. These certainly have not been available *ab initio*, excluding, to our understanding, environments high in Ca from the possible choice of biogenesis solutions. In search for likely places for the creation of life, not only on Earth but possibly also on Mars and on the Jupiter moon Europa (Kempe and Kazmierczak 1997, 2002), we therefore must look for environments which have very low Ca-concentrations by nature.

Such environments exist: alkaline lakes (soda lakes) rich in HCO_3^- and CO_3^{2-} (e.g. Eugster and Hardie 1978). Due to the relatively low solubility product of Ca, Mg and Fe carbonates, alkaline solutions are low in those ions. Furthermore, in the absence of silica-utilizing (metabolizing) algae and animals, high silica concentrations, as they are typical for alkaline solutions, should have reduced the Ca^{2+} concentrations even further. It is therefore very interesting to study modern soda lakes as analogues of a Precambrian "Soda Ocean" (Kempe and Degens 1985). Such lakes are found today mostly in areas were fresh volcanic rocks are exposed to weahtering, i.e. along the volcanic plate boundaries or at hotspot volcanic sites (Fig. 1). Here we will present data from two sodium-carbonate lakes, one of them containing only a dilute solution, illustrating the mode and rapidity with which alkaline lakes can form, and one containing a highly alkaline solution illustrating the long-term stability of such lakes. These lakes are the Crater Lake of the Nemrut Volcano and Lake Van, both of them situated in Eastern Anatolia, Turkey (Fig. 2A). Data there were collected during the three expeditions of 1974, 1989 and 1990 (Kempe 1977, Reimer 1995, Landmann 1996).

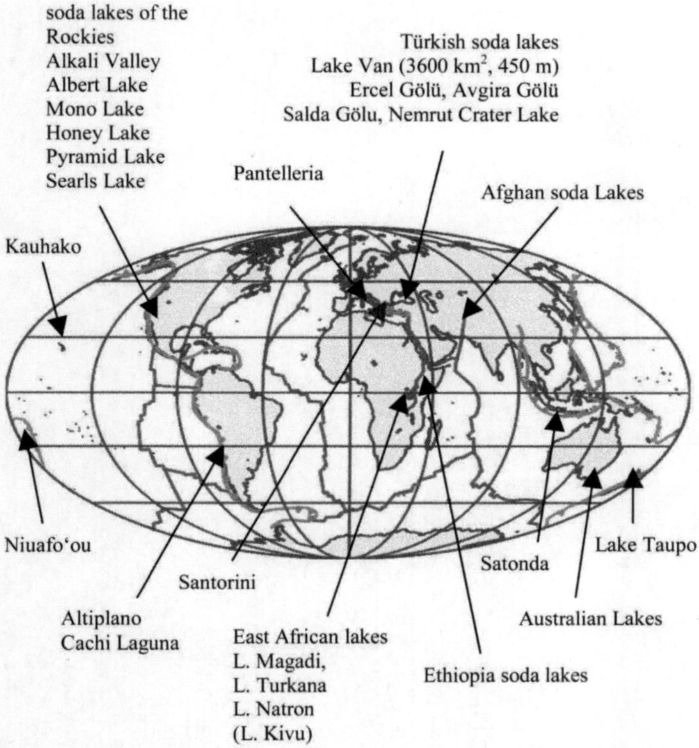

Fig. 1. Locations of soda lakes with relation to plate boundaries

1 Chemistry of Recent Soda Lakes

2.1 Nemrut Crater Lake, Turkey

The Nemrut is a volcano in eastern Turkey (Fig. 2a). It has a very large summit caldera. In this caldera several lakes exist, the largest of which, Nemrut Lake, has a depth of over 150 m. Temperature measurements in 1990 showed a minimum of 3.61°C at 68 m and a significant temperature increase of up to 4.60°C below, indicating the thermal influence of the volcano on the lake. This volcanic influence is also seen with regard to the PCO_2 (internal CO_2-pressure), which increases from 1,200 ppmv at the surface to ca. 30,000 ppmv un the lower water column.

Fig. 2. Lake Van, Turkey

A – NASA Challenger photograph (red arrow indicates the Nemrut Volcano); B – Whitings along western shore at tributary of a Ca-rich stream; C – Diver taking water sample at outlet of a calcareous pinnacle; D – Submerged (water depth 17 m) cyanobacterial tufa-towers with white caps of freshly precipitated inorganic calcium carbonate around groundwater orifices (scale bar = 1 m); E – Submerged (water depth 5 m) bush-like calcareous structures precipitating inorganically at Ca-rich groundwater seeps on the lake bottom (chemical garden) (scale bar = 10 cm).

Table 1. Comparison of lake water chemistry for Lake Nemrut and Lake Van with modern seawater.

Lake	Alkalinity meq/l	Cl meq/l	SO_4 meq/l	Na meq/l	Mg meq/l	Ca meq/l
Nemrut 10 m	4.61	0.43	0.22	3.67	0.52	0.96
Nemrut 150 m	4.77	0.45	0.25	3.80	0.53	0.97
Van 3 m	151.2	160.2	48.86	337.9	8.870	0.210
Van 440 m	155.8	166.5	50.9	347.5	9.072	0.174
Seawater at surface	2.33	546	56.6	468	106.2	20.6
Lake	Cond µS/cm.	pH	PCO_2 ppmv	SI_{Cc}	SI_{Ara}	SI_{Dol}
Nemrut 10 m	456	8.27	1240	0.37	0.22	0.43
Nemrut 150 m	469	6.89	28,600	-1.11	-1.27	-2.59
Van 3 m	26000	9.74	305	1.126	0.977	4.069
Van 440 m	26700	9.69	252	1.05	0.895	3.780
Seawater at surface	-	8.24	363	0.59	0.44	1.96

Data from Reimer (1995), saturations calculated by PHREEQUE (Parkhurst et al. 1990). Saturation index of minerals is defined by = -log (Ion activity product/$K_{mineral}$), i.e. zero denotes saturation, positive values denote supersaturation, negative values undersaturation.

Table 1 and Fig. 3 give the composition of the waters in the crater lake. The main cations liberated during the weathering of the volcanic rocks of Nemrut are sodium, followed by calcium, magnesium and potassium (ca. 0.2 meq/l). The main anion is HCO_3, measured as alkalinity, followed by chloride (factor of ten less than alkalinity), and sulphate. Since the main ions in the lake waters are Na and HCO_3, sodium carbonates such as natron (synnonym "soda") ($Na_2CO_3*10H_2O$; monoclinic), trona ($Na_3(CO_3)(HCO_3)*2H_2O$; monoclinic) or thermonatrite ($Na_2CO_3*H_2O$; orthorhombic) would precipitate upon severe evaporation. Were it not for the dilute conditions and the high PCO_2 in the lake, the waters would thus become highly alkaline, as is typical for soda lakes, Lake Nemrut is thus a soda lake in the making. Since no rivers enter the lake and since it must be geologically young, being located in a caldera of a volcano active throughout the Holocene, as shown by multiple ash layers in the Lake Van sediment record (Kempe 1977,

Landmann 1996), all solutes must have derived from the internal weathering of the volcanic rocks under the influence of volcanic gases, mostly CO_2. The lake apparently degasses CO_2, leaving the surface layer slightly supersaturated with regard to calcite (Cc), aragonite (Cc) and dolomite (Dol), while the lake is highly undersaturated with regard to these minerals at towards the bottom. Further degassing would certainly increase their saturation and, in the geological future, as the lake increases in alkalinity it will finally start to precipitate calcium carbonate minerals.

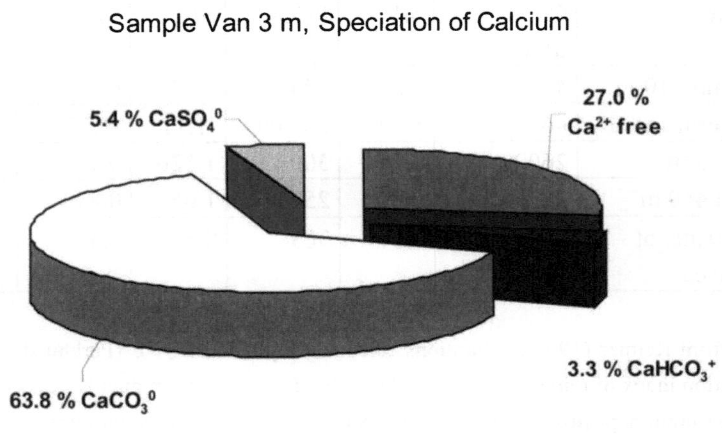

Fig. 3. Plot of the relative composition of the five analyses given in Table 1. Note that potassium has been dropped to simplify the graph and that therefore the cation/anion balance is slightly off. Note the dominance of chloride in seawater in comparison to the other samples.

1.2 Chemistry of Lake Van, Turkey

In contrast to the still dilute situation in Lake Nemrut, Lake Van is a true highly alkaline soda lake (Table 1, Fig. 3). It is situated to the east of the Nemrut volcano. In fact, the Nemrut volcano caused the formation of Lake Van. The volcano blocked the Euphrates valley, cutting off the headwaters now tributary to Lake Van. The date of this blockage is not known as yet, but must predate the Last Glacial. Lake Van is by far the largest soda lake of today (morphometric data in Table 2). It is at the same time the fourth largest (by volume) closed lake on Earth after the Caspian Sea, Lake Aral, and Lake Issykul.

Table 2. Lake Van morphometric data compiled from Reimer (1995)

Total Basin Area	16,096 km^2
Total Tributary Area	12,574 km^2
Lake Surface	3,522 km^2
Lake Volume	576 km^3
Lake max. Depth	451 m
Annual Water Exchange	3.78 km^3
Altitude of Lake Level	1648 m a.s.l.

Being a closed lake, the water level fluctuates according to the annual water balance. This fluctuation has an amplitude of ca. 50 cm (mean of 29 years of lake level record) corresponding to a seasonal difference in water balance of 1.76 km^3 between river inputs (2.1 km^3/a), precipitation over the lake (1.68 km^3/a), and evaporation (3.78 km^3/a) (Kempe 1977, Reimer 1995). For long years the lake surface fluctuated at around 1646 m a.s.l. but it increased between 1986 and 1995 by 2.2 m, most likely due to long-term climatic change. Terraces around the lake mark high stands, the highest occuring during the Last Glacial Maximum at 20,300a BP (calibrated calendar years; BP = before present, i.e. years before A.D. 1950) (Kempe et al. 2002). Then the lake level dropped and by ca. 15,000a BP the lake had dried completely (Landmann et al. 2002) due to severe shifts in the climatic zones during the Postglacial.

Today Lake Van has acquired a salt content of 22 g/l and is highly alkaline with a pH of around 9.7 and an alkalinity of around 153 meq/l (Table 1). There is only a small gradient between the surface layer, which is seasonally diluted by river input, and the bottom waters. Below 60 m the lake has, with regard to the main cation and anions, a very stable composition. The lake appears to be mixed every few years. The PCO_2 is rather low throughout the water column; apparently no major volcanic input exists. All three main carbonate alkaline earth minerals are highly supersaturated. Supersaturation decreases somewhat down column because of a small drop in the Ca concentration with depth. This is due to the continued input of Ca from hardwater rivers and submersed springs to the lake's surface layer.

The source of the sodium carbonate in the lake is thought to be the weathering of the volcanic rocks which are predominantly to the east (Nemrut Volcano) and north (Süphan Volcano) and from the ash layers these volcanoes deposit on the landscape from time to time. The lake was already alkaline before and during the Last Glacial since the sediments of the terraces contain calcite pseudomorphs after wheat grain-shaped aragonite, which is a habitus typical for crystallites precipitated in the water column in the lake today (whiting) (Kempe et al. 2002). When the lake dried around 15,000a BP the alkaline solution sank into the sediments of the lake bottom from where it now diffuses upward. This process is documented in the gradient of the pore water alkalinity: it more than doubles in concentration down core to over 400 meq/l in the central lake basin (not in those from higher positions!). This can only be caused by a source highly concentrated in sodium carbonate salts which exists below the reach of our coring devices (the longest core recovered measures 890 cm). The chloride content of the pore waters, interestingly, does not increase at the same rate as the alkalinity down core. This suggests that the chloride in the present lakes comes from a different source. Such a source could be a Miocene salt deposit outcropping to the east of the lake, where it is recovered today in salt pans.

Due to the very high supersaturation of alkaline earth carbonate minerals in the lake, precipitation of calcite and aragonite continuously occurs in the lake. Four modes must be discerned, two associated with submersed seepage of Ca-rich ground waters, and two occuring in the form of surface water whitings. Where Ca-rich ground water seeps from carbonate aquifers into the lake, white, organic carbon-poor, soft, inorganic globular and microcrystalline calcite and aragonite fabrics are precipitated often in the form of delicately branched bushes or coral-like structures as in chemical gardens (Fig. 2E). If this occurs below the wave action, the surfaces of these apparently inorganic precipitates are quickly settled by coccoid cyanobacteria, diatoms, and peritrichid ciliates forming a microbial mat. Within the sheaths of the cyanobacterial colonies a mixture of microgranular (micritic) aragonite and authigenic silicates with a much higher organic carbon content is precipitated forming hard microstromatolitic crusts, which can grow to form sublacustrine tuffa towers up to 40 m high (comp. Figs. 2C, D; see also Kempe et al. 1991).

In contrast to these localized precipitates, two types of whitings also precipitate carbonates regionally in the lake. The first type is associated with the plumes of Ca-rich rivers (Fig. 2B). Here precipitation seems to occur usually in spring when the river flow is high. The second type of whiting is caused by the strong evaporation of the surface layer in summer. These precipitates form annually laminated sediments, i.e. varves, which have been used to establish one of the longest uninterrupted Postglacial and Holocene chronologies to date (Kempe 1977, Landmann 1996, Landmann et al. 1996 a, b).

2 Models for an Alkaline Ocean.

The interest in alkaline environments rests, among others, on the fact that in such environment the free Ca concentration, so detrimental to the functioning of proteins, and the spontaneous formation of phospholipid encased vesicles (e.g. Irion 2002) can be depressed to almost cytosolic levels. This is due to the fact that in solutions high in bicarbonate and carbonate ions, ionpairs are formed. The same happens in the presence of sulphate. Lake Van serves here as a good example: The total Ca concentration in Lake Van is only 102 µmol/l (10^{-4}M) (= 0.204 meq/l). Due to the formation of ion pairs, $CaCO_3^0$, $CaHCO_3^-$ and $CaSO_4^0$, the free Ca^{2+} is only 22 µmol/l ($2*10^{-5}$M) (for calculations see Reimer 1995) (Fig. 4).

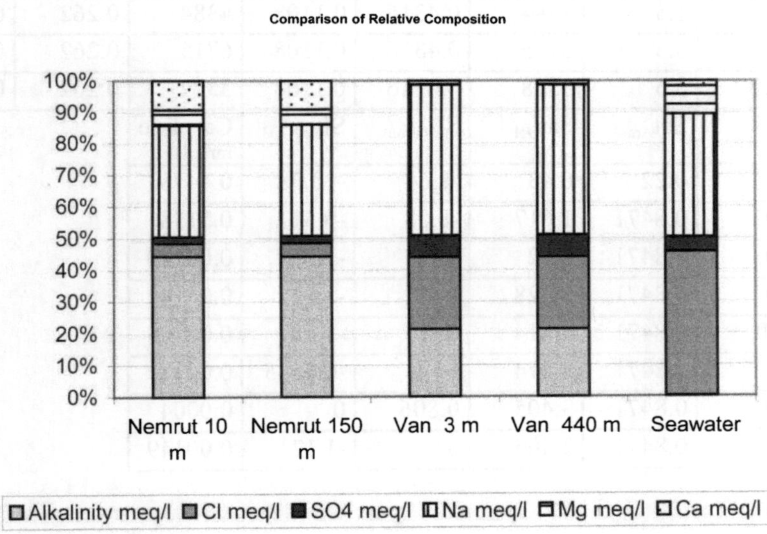

Fig. 4. Distribution of ion pairs in Lake Van, 3 m depth (data after Reimer, 1995).

The numerical model PHREEQUE (Parkhurst et al. 1990) allows us to calculate the thermodynamics of aqueous solutions. In Table 3 various model runs are given based on the 10 m water sample from the Nemrut Crater Lake. This sample was selected because it has the typical characteristics of a soda lake water, i.e. its alkalinity is much higher than its concentrations of alkaline earth elements (i.e. HCO_3 >> Ca+Mg) or, phrased differently, its Na+K << $Cl+SO_4$. These conditions are precisely those, which lead upon evaporation to highly alkaline solutions (see the basic paper of Garrels and Mackenzie 1967). The concentrations are given in millimolalities (millimols of chemical/kg of solvent) as this is the output of the numerical model.

Table 3. Numerical simulations of the Nemrut Crater Lake water

line	pH	$pPCO_2$	Alkal. mequ	Cl mmolal	SO_4 mmolal	Na mmolal	Mg mmolal	Ca mmolal
1	8.27	2.906	4.61	0.4316	0.1108	3.628	0.262	0.479
2	8.85	3.5	4.83	0.4316	0.1108	3.628	0.262	0.59
3	10.29	3.5	3379	0.4316	0.1108	3380	0.262	0.0306
4	10.29	3.5	3379	0.4316	0	3380	0.262	0.0306
5	9.77	2.5	3619	0.4316	0.1108	3619	0.262	0.0318
6	9.22	1.5	4384	0.4316	0.1108	4384	0.262	0.037
7	10.41	3.5	6715	0.4316	0.1108	6715	0.262	0.0167
8	10.29	3.5	3378	0.4316	0.1108	3379	0.262	0.0312
	Sat_{Cc}	Sat_{Ara}	Sat_{Dol}	SI_{Natron}	SI_{Na2CO3}	Ca free mmolal		
1	0.37	0.22	0.43	-8.64	-10.08	0.4478		
2	1.00	0.8471	1.617	-8.08	-9.51	0.5158		
3	1.00	0.8471	3.12	0	-1.12	0.00149		
4	1.00	0.8471	3.128	0	-1.12	0.00149		
5	1.00	0.8471	3.114	0	-1.08	0.00159		
6	1.00	0.8471	3.074	0	-0.9515	0.0018		
7	1.00	0.8471	3.405	0.806	0	0.0004		
8	1.00	0.8471	3.405	0	-1.12	0.00149		

Sample 10 m depth, 12.8°C; K = 0,1985 mmolal; PHREEQUE, using extended Debey-Hückel equations as provided by the program); $pPCO_2$ = negative logarithm of the CO_2 partial pressure; SI = Saturation index = log ($IAP/K_{mineral}$). Line 8: Saturation of hydroxylapatite at PO_4 mmolal = 1.511.

Line 1 repeats the original analysis but gives the new concentrations. Due to the relatively low concentration of alkalinity the free Ca concentrations is a little less than that of the total Ca. Line 2 brings the solution to a calcite supersaturation of 1.0 and equilibrates it to a $pPCO_2$ of 3.5 roughly representative of ambient PCO_2 ($pPCO_2$ = 3.5, i.e. 316 ppmv). In result the total and the free Ca concentrations rise, as well as the alkalinity and the pH. Now aragonite and/or calcite could spontaneously precipitate in the water column and/or the supersaturation could trigger the permineralization of algal mats. These processes would keep the water at this saturation state, not allowing a significant further rise in the Ca concentration. Other constituents in the water column may, however, rise further. This is illustrated in line 3, where the solution is allowed to acquire also the saturation with regard to natron, yielding a pH of 10.29. Now the Ca-concentration is depressed to $3*10^{-5}$ and the free Ca-concentrations drops to $1.5*10^{-6}$. If the sulphate is taken out of the calculation (line 4), changes in the composition are so small that they do not show. However, if one chooses Na_2CO_3 as the limiting mineral for the Na and alkalinity concentrations, then (line 7) the free Ca concentration can be as low as $4*10^{-7}$, i.e. within the range of cytosolic levels. The pH rises to 10.41.

The increase in the PCO_2 of the atmosphere has a relatively small effect on the Ca of these highly buffered carbonate solutions as illustrated by the calculation of the simultaneously natron and calcite saturated sample for a $pPCO_2$ of 2.5 and 1.5 (lines 5 and 6), i.e. for an atmosphere ten and a hundred times as rich in CO_2 as present. Also the addition of phosphate up to the saturation of hydroxylapatite (line 8) does not alter the Ca-concentrations significantly.

This approach of thermodynamic modelling is not answering the question as to the location and composition of the environment where life came about since there are many other geochemical parameters which may have played a large role in the early weathering of Earth. However, recent soda lakes and mass balance calculations are reasons to explore such a model in more detail. The degassing of the mantle and the arrival of water by cometary impacts would have led to the immediate onset of weathering reactions in early Earth. We roughly know the mass balance of the inorganic acids H_2CO_3, HCl and H_2SO_4 that have been consumed during weathering throughout Earth history, i.e. ca. $65.5*10^{21}$ g of C or $5.5*10^{21}$ mol, $52*10^{21}$ g Cl or $1.6*10^{21}$ mol, and $5.2*10^{21}$ g S or $0.16*10^{21}$ mol (Kempe and Degens 1985). 3.7 times more carbonic acid than hydrochloric acid and 34 times more carbonic acid than sulphuric acid were available. The primordial komatiitic crust had a [Ca+Mg]/[Na+K] of 1.6. Therefore not enough Cl and SO_4 were available to balance Na and K, i.e. some of the alkaline metals must have been balanced by carbonate. Early weathering solutions, just as weathering solutions of to-day, must have had a predominance of carbonate ions rather than that of chloride as in today's ocean. Garrels and Mackenzie (1967) have shown what happens if solutions of various compositions are subjected to evaporation: As long as there is a small surplus of carbonates over Ca + Mg and a surplus of Na + K over Cl + SO_4, solutions will become alkaline quickly, i.e. of a high carbonate-alkalinity and pH. How quickly that can happen is illustrated by today's soda lakes which all have histories not significantly longer than several 10,000 years, considering the tremendous glacial and interglacial climate changes in the recent Earth history.

One can illustrate the transition of a possible early soda ocean to the present halite and gypsum dominated ocean by numerical mixing modern seawater with an example soda solution such as that in line 3, Table 3. The result is given in Fig. 5. The program mixes the endmember solutions linearily and then adjusts them to a supersaturation of calcit (SI calcite = 1.0) and a $pPCO_2$ of 3.5, ca. ambient pressure of today. The resulting mixtures show that a small increase in alkalinity causes a large decrease in the total and even greater decrease in the free calcium concentration. Interestingly enough the rather high sulphate concentration in the seawater end of the mixing range does not influence the calcium concentration significantly, i.e. the Ca concentration of the ocean appears to be entirely alkalinity driven. This is in accordance with the fact that the ionpair $CaCO_3^0$ (see Fig. 4) plays the largest role for the Ca ionic species.

The model therefore not only illustrates the thermodynamic feasability of a highly alkaline ocean, but it also shows that relative to soda dominated solutions small changes in alkalinity can shift the oceanic Ca concentration significantly. The model also shows that highly supersaturated conditions are not linked to high Ca-concentrations, but to high alkalinities. Therefore, even in the Phanerozoic (i.e. in the time since the onset of the Cambrian) supersaturation changes could occur in times when excess alkalinity is produced in sulfate reducing anaerobic marine basins (Kempe 1990, Kempe and Kazmierczak 1994, Kempe 2002). As the example of Lake Van shows, such levels of supersaturation would lead to the advent of microbialite deposits like those that dominated the entire Precambrian and like those that re-occured in certain times and regions in the Phanerozoic (Kazmierczak et al. 1985, 1996). Alkalinity changes inducing Ca-concentration change in the early ocean could also provide the geochemical forcing behind major biological evolutionary events such as eukariogenesis, the development of multicellularity, and finally the onset of biomineralization at 0.54 Ga ago (see Kempe and Kazmierczak1994 for more information).

Acknowledgements

The Lake Van studies were supported by the Deutsche Froschungsgemeinschaft (S.K. and J.K.), support was also obtained from the Foundation for Polish Science (to J.K). This paper was written during a sabbatical leave at the residence of Carol and Dr. Marlin Spike Werner, Hilo, Hawaii. Whithout their enduring care this would never have finished in time. Mahalo nui Carol and Spike!

References

Berridge MJ, Bootman MD, Lipp P (1998) Calcium – a life and death signalling. Nature 395: 645-648

Borle AB (1981) Control, modulation and regulation on cell calcium. Rev. Physiol., Biochem. Pharmacol. 90: 14-153

Cambell. AK (1983) Intracellular Calcium; Its Universal Role as a Regulator. J. Wiley & Sons Ltd., Chichester

Carafoli E (1987) Intracellular calcium homeostasis. Annual Rev. Biochem. 56: 395-433

Davis WL, McKay CP (1996) Origins of life: a comparison of theories and application to Mars. Origins of life and Evolution of the Biosphere 26: 61-73

Eugster HP, Hardie LA (1978) Saline lakes. In: Lerman A (ed) Lakes - Chemistry, Geology, Physics. Springer, New York, pp 237-293

Garrels RM, Mackenzie FT (1967) Origin of the chemical composition of some springs and lakes. In: Equilibrium Concepts of Natural Water Systems", American Chemical Society Advances in Chemistry, 67: 222-242

Irion R (2002) Astrobiologists try to 'follow the water of life'. Science 296: 64748

Kazmierczak J, Ittekkot V, Degens ET (1985) Biocalcification through time: environmental challenge and cellular response. Paläont. Zschr. 59: 15-33

Kazmierczak, J, Coleman ML., Gruszczynski M, Kempe S (1996) Cyanobacterial key to the genesis of micrite and peloidal limestones in ancient seas. Acta Palaeontologica Polonica 41: 319-338

Kempe S (1977) Hydrographie, Warvenchronologie und organische Geochemie des Van Sees, Osttürkei. Dissertation, Mitt. Geol.-Paläont. Inst. Univ. Hamburg 47: 125-228

Kempe S (1990) Alkalinity: the links between anaerobic basins and shallow water carbonates? Naturwissenschaften 77: 426-427

Kempe S (2002) Alkalinity pump and carbonate precipitation: A comparison of the crater lakes of Kauhako, Hawai'i, Satonda, Indonesia and Niuafo'ou, Tonga. – 6th Intern. Symp. Geochem. Earth's Surface, 20-24 May, 2002, Honolulu, Hawaii, Abstract Vol: 136-137

Kempe S,. Degens ET (1985) An early soda ocean? - Chem. Geol. 53: 95-108

Kempe S, Kazmierczak J (1994) The role of alkalinity in the evolution of ocean chemistry, organization of living systems and biocalcification processes. In: Past and Present Biomineralization Processes. Considerations about the Carbonate Cycle (edited by F. Doumenge), Bull. Inst. Oceanogr, Monaco, no. spec. 13, 61-117

Kempe S, Kazmierczak J (1997) A terrestrial model for an alkaline martian hydrosphere. Planet. Space Sci., 45(11): 1493-1499

Kempe S, Kazmierczak J (2002) Biogenesis and early life on Earth and Europa: Favored by an alkaline ocean?, Astrobiology 2(1): 123-130

Kempe S, Kazmierczak J, Landmann G, Konuk T, Reimer A, Lipp A (1991) Largest known microbialites discovered in Lake Van, Turkey. Nature 349: 605-608

Kempe S, Landmann G, Müller G (2002) A floating varve chronology from the Last Glacial Maximum terrace of Lake Van/Turkey. Zeitschr. Geomorphol., Supplem. 126 Research in Mountians and Deserts of Afrika and Central Asia: 97-114

Landmann G (1996) Van See/Türkei: Sedimentologie, Warvenchronologie und Paläoklima der letzten 15 000 Jahre. Dissertation, Facult. of Geosciences, Univ. Hamburg, 137pp, unpublished

Landmann G, Reimer A, Lemcke G, Kempe S (1996a) Dating Late Glacial abrupt climate changes in the 14,570 years long continuous varve record of Lake Van/Turkey. Paleogeography, Paleoclimatology, Paleoecology, 122: 107-118

Landmann G, Reimer A, Kempe S. (1996b) Climatic induced lake level changes of Lake Van/Turkey during the transition Pleistocene/ Holocene. Global Biogeochemical Cycles 10(4): 797-808

Lowenstam H, Margulis L. (1980) Evolutionary prerequisites for early Phanerozoic calcareous skeletons. BioSystems 12: 27-41

Marmé D (1985) Calcium and Cell Physiology. Springer-Verlag, Berlin

Mojzisis SJ, Arrhenius G, McKeegan KD, Harrison TM, Nutman AP, Friend CRL (1996) Evidence for life on Earth before 3,800 million years ago. Nature 384: 55-59

Parkhurst DL, Thorstenson DC, Plummer LN (1990) PHREEQE: A computer program for geochemical calculation. Conversion and upgrade of the prime version of PHREEQE to IBM PC-compatible systems by Tirisanni, J.V. Glynn, P.D., U.S: Geol. Surv. Wat. Res. Rep. 80-96

Reimer A (1995) Hydrochemie und Geochemie der Sedimente und Porenwässer des hochalkalinen Van Sees in der Osttürkei. Dissertation, Facult. of Geosciences, Univ. Hamburg, unpublished

Schidlowski M, Appel PW, Eichmann R, Junge CE (1979) Carbon isotope geochemistry of the $3.7*10^9$) yr old Isua sediments, West Greenland: Implications for the Archean carbon and oxygen cycles. Geochim. Cosmochim. Acta 43: 189-199

Simkiss K (1989) Biomineralization in the context of geological time. Trans. Royal Soc. Edinburgh: Earth Sciences 80: 193-199

Simkiss K, Wilbur KM (1989) Biomineralization, Cell Biology and Mineral Deposition. Academic Press, San Diego

Modelling in Palaeoanthropology: Reconstructing the development of paleospecies

Christine Hertler, Friedemann Schrenk and Timothy G. Bromage,

Introduction

Models play a crucial role in palaeoanthropological research. Since it is not possible to observe macro-evolutionary processes directly we must reconstruct what has happened. For this purpose evolutionary research is based on actualistic models developed from recent biological and geological contexts applied to interpretations of fossil hominids and their environments.

Thus models help us to reconstruct the life ways of our fossil human ancestors, we apply them in systematic classifications and judgments concerning the evolutionary level reached by fossil taxa, and they are essential to our reconstructions of evolutionary settings and processes.

In spite of their importance there is no general agreement about how to construct models, or about how they should be used. However, we consider this not a reason to refrain from using models, rather we feel they must be explained and the range of their explanatory value must be stated.

With this in mind, let us take a close look at the procedure followed in the description of one particular fossil, and how this description contributed to its classification and a reconstruction of its role in early human evolution.

1 Describing a Fossil: HCRP-UR 501

In 1991 and 1992 the Hominid Corridor Research Project (HCRP) discovered the remains of a nearly 2.5 million years old lower jaw at Uraha in Northern Malawi

(Bromage et al., 1995a, Schrenk et al., 1993). This lower jaw with the catalogue number HCRP-UR 501 was clearly hominoid. The teeth of apes and humans are formed in a quite characteristic way and they can be easily distinguished from those of other primates.

This jaw was broken in two pieces, almost in the middle, where the two halves of the jaw join to form a structure called the symphysis. In the description provided after lengthy preparation, a decision had to be made concerning which measurements to take. As a convention the initial description of a fossil should not only contain photographs or drawings of several normal views, but it should also provide qualitative descriptions and quantitative measurements.

Although there are systems for taking measurements and how exactly this should be done, it is not immediately clear which system to choose. We decided to follow a particular system (Bromage et al., 1995a) developed and introduced by Wood (1991) that was originally developed for the description of early hominids of Koobi Fora, Northern Kenya. The sediments there are a bit younger than those in Malawi, but the Koobi Fora hominids cover a range of variability, larger than at any other hominid site in eastern or southern Africa. Moreover, Wood's system focuses particularly on cranial and dental remains, thus providing the essential elements for our description of HCRP-UR 501.

The system proposed by Wood (1991) also allows one to distinguish between different species of hominids. It is based primarily on anatomical and morphological data and allows one, particularly, to differentiate between robust and gracile forms in eastern and southern Africa, respectively. Wood chose to attribute systematic meaning to identify hominid groups based primarily on anatomical features. This is why it serves as a typological model, allowing one to distinguish two different steps in a description, which can be easily confused: the morphological description on the one hand, and the systematic interpretation on the other.

Wood proposed to distinguish four different types of hominids in the African PlioPleistocene: gracile and robust forms in southern Africa, abbreviated SAFGRA SAFROB respectively, and such forms in eastern (EA) Africa as well, abbreviated EAFROB and EAFHOM respectively. Specimen HCRP-UR 501 displays a unique mixture of features: Its overall size is quite large, as well as the crown areas of molars and the dimensions of the mandibular corpus. It therefore somewhat resembles the robust forms, SAFROB and EAFROB alike.

There are, however, other uniquely derived characteristics such as the particular shape of the molars and the thickness of the enamel on the premolars which clearly distinguish this fossil from the robust forms. Moreover, this particular combination of features occurs in another fossil which has been found in Koobi Fora, East Africa, and is considered by Wood as belonging to the category EAFHOM (eastern African *Homo* sp.) For this reason it was decided to attribute this fossil hominid mandible to the latter category.

In this particular context the system provided by Wood (1991) was used as a framework and reference for the anatomical and morphological description of HCRP-UR 501. The system provided several categories allowing one to distinguish between different morphological types occurring among hominids of the Plio-Pleistocene in Africa. As this system represents a standard for comparisons it serves also as a model for the description and reconstruction of fossil hominids.

In this case the attribution to a particular category is based mainly on dental characters, but since the categories include more than that, it also allows inferences about other cranial and post-cranial features which usually go together with this particular set of dental features. We can therefore make predictions about parts of the body we have yet to recover from Uraha.

2 Reconstructing a Species: *Homo rudolfensis*

The next problem to address is how we can determine to which species the fossil belongs. The eastern African early *Homo* record is quite diverse. Yet, a very close morphological relationship exists between the HCRP-UR 501 mandible from Malawi and the mandible KNM-ER 1802 from Koobi Fora which was attributed to the species *Homo rudolfensis* by Wood (1992). But how did Wood arrive at this conclusion, giving the fact that *Homo rudolfensis* is only one among many different early hominid species, some of which themselves are still in dispute? A good example is provided by early hominid finds at Olduvai Gorge, Tanzania.

The first hominid remains discovered at Olduvai in 1959 were attributed to the Genus *Zinjanthropus*, popularly known as nutcracker man (Leakey, 1959). Their skulls and jaws were quite robust and the molars were huge. Today they are included in the East African robust genus *Paranthropus*. Yet, the very same fossil beds in Olduvai Gorge yielded simple stone tools and handaxes. Few people felt comfortable to attribute the capability of producing stone tools to such ape-like-looking creatures.

When in 1964 the remains of another hominid species were found at Olduvai, far more gracile and possessing a comparatively large brain case, it was felt that they were the likely manufacturers of the tools, and not the ape-like *Zinjanthropus*. The new remains were therefore reasonably attributed to the genus *Homo*, which was thereby partly defined by their large brain size and their capability of manufacturing tools. Its specific name, *habilis*, meaning "handy" was created for this particular reason (Leakey et al., 1964) The postulation of this new species seemed to be necessary to accommodate the distinctive features exhibited by these fossils. The next younger species in time, *Homo erectus*, had an even larger brain case and several other features which allowed one to distinguish clearly between them.

Since that time practically all early *Homo* fossils of a certain age from Africa where simply attributed to *Homo habilis*. After several decades and the discoveries of new sites in East Africa, especially in the vicinity of Koobi Fora in Northern Kenya, the material included in *Homo habilis* gradually came to contain a large range of variability. There were fairly large skulls reaching almost the cranial capacity of *Homo erectus*, but also very small ones with a dentition similar to that of the australopithecines.

It was therefore necessary to determine whether all of these fossils could still be reasonably included in one single species or if this inclusion represented an unjustified lumping of more than one species. This has been done by a statistical procedure in which a matrix generated by a number of different standard measurements was tested against the null hypothesis of a single species (Wood, 1992). A comparison between the fossils originating from two different sites in East Africa, Olduvai and Koobi Fora, showed that the variability occurring among the Koobi Fora material attributed to *Homo habilis* was much larger than the variability among the Olduvai material. The Olduvai material may indeed refer to a single species *Homo habilis* as originally described by Leakey et al. (1964), but not the material from Koobi Fora.

The situation in Koobi Fora seemed much more complex. There were indeed fossil remains that fell into the range of variability known from the Olduvai material. Therefore, fossils belonging to *Homo habilis* occurred in the Koobi Fora sample. But there were also larger forms similar to *Homo erectus* which were attributed to a new species, *Homo ergaster*, perhaps even conspecific with *Homo erectus*. However, there was still early *Homo* material left unassigned with a slightly bigger brain case than *Homo habilis*, but more primitive in almost every other respect. For this group, a new species, *Homo rudolfensis*, was created. The fossil forms representing this species seem to show a quite interesting mixture of *Paranthropus*-like and *Homo*-like features.

Since the complications surrounding the origin of the species name *Homo rudolfensis* are quite interesting from a historical as well as technical point of view, and were discussed extensively by Kennedy (1999) and Wood (1999), we should elaborate on this for a moment. The full name of the species is *Homo rudolfensis* Alexeev 1986, meaning that the specific name '*Homo rudolfensis*' has been given by Alexeev in 1986 for the first time. However, the publication in which Alexeev published his description was not a proper scientific publication but a popular book on human evolution. Moreover, he called the fossil in question, the skull KNM-ER 1470 from Koobi Fora, "*Pithecanthropus rudolfensis*", refering to the old genus name *Pithecanthropus*, which was formerly used for fossils nowadays attributed to *Homo erectus* (as proposed by Mayr, 1950).

The procedure for how a new specific name can be given, where and in which form this has to be published, and which name has priority if more than one name is given to one and the same species, is laid out in the International Code of Zoological Nomenclature (ICZN 2000). Therefore, the full and correct name of the species is *Homo rudolfensis* (Alexeev 1986) Wood 1992.

In his inaugural paper introducing *Homo rudolfensis*, Wood (1992) investigated metrical and non-metrical characters of a number of different fossils from Koobi Fora. Among them was not only the skull KNM-ER 1470, the type specimen of the newly named species, but also the mandible KNM ER 1802 which was characterized by a marked relief on the external surface of the mandibular corpus, an everted base, a relatively large third molar, a comparatively large premolar talonid (a specific cusp area on the tooth crown), and large plate-like tooth roots.

This is the very set of characters which we can as well observe on HCRP-UR 501 from Malawi, forming the basis of referring the mandible to *Homo rudolfensis*. The model used here was a statistical approach particularly designed for the distinction of different hominid species on the basis of selected anatomical features.

But what does it tell us about human evolution? Reconstruction of the early evolution of the genus *Homo* is based on the species described. The distribution of fossils which were formerly attributed to one single species, *Homo habilis*, to at least three different ones, *Homo habilis*, *H. ergaster*, and *H. rudolfensis*, has considerably changed the basis of the reconstruction. If there is only one species, *Homo habilis*, there is only one single path which evolution can follow. It leads from an australopithecine ancestor through one single larger brained and tool producing form of early *Homo* to *Homo erectus*, who more or less dominated the Early Pleistocene environment even beyond Africa. However, if there is more than one contemporaneous taxon, the situation becomes much more complex. According to Wood (1992), who refers to a cladistic analysis in this context, *Homo habilis* and *Homo rudolfensis* share a common ancestor which neither shares with any other hominid group, i.e. they are sister groups among a bunch of early hominid species. This also means that neither of them gave rise to the particular lineage leading to other species of the genus *Homo*. Yet, these cladistic analyses have the problem that they are merely based on statistical procedures judging the evolutionary valence of different anatomical (or genetic) features. They do not allow a test of whether such distinguished species have been forming a reproductive community, i.e. whether they were species in a biological sense or whether we are dealing with subspecific arrangements of characters which do not necessarily identify a species.

Living species demonstrate that the amount of variability which may occur within a species or a population cannot be easily standardized. There are so called sibling species, which resemble each other so closely that we are unable to distinguish between them by anatomical or behavioural means. However, they do not interbreed and must thus belong to different species. On the other hand, animals as different as lions and tigers are still able to interbreed when they are artificially kept together. As long as we do not have a model which allows us to explain the extent and direction of occurring variations we cannot reasonably base a judgement of specific affiliation solely on this variation.

3 Species and Evolution

Given all of the difficulties and obstacles in the way of characterizing *Homo rudolfensis*, why is it nevertheless important to know to which species a particular fossil has to be attributed? Is it really necessary to know one's name as long as we know the fossil itself, its age and place of origin and can easily determine all of the data necessary for understanding its structure? It is not. According to major approaches in evolutionary theory, species – and not the fossil individual preserved through millions of years - have to be understood as the result of evolutionary the process. As evolution proceeds, new species develop and older ones disappear, i.e. they die out. Evolution consists of processes occurring in populations composed of conspecifics. The species is understood as a reproductive community. It is defined as a one or more actually or potentially interbreeding natural populations which are reproductively isolated from other such groups (Mayr, 1942, 1950). When new species arise, reproductive barriers will be established dividing a single population into two or more. This is a gradual model at first hand, since reproductive barriers can be established in a number of ways: The simple spatial separation of part of a population by a river, a steep valley etc., excludes reproduction between members of the separated population parts. If this separation continues for many generations, there will be enough time to establish behavioural and/or morphological reproductive barriers within the reproductive system itself. This process is called allopatric speciation.

Although the biological species concept appears to be operational in most cases of living organisms (excluding e.g. organisms which propagate asexually, and many plants possessing different kinds of reproductive systems), it cannot be applied to fossil species. We cannot conceive any direct test for interbreeding amongst fossil groups. When we are observing them in their boxes, drawers and vaults, they definitely do not interbreed. Even more so, we can readily assume that most of them could not have interbred when their bearers were alive, since the sites where they were found are often separated by ten thousands of years in time and by thousands of kilometres in space.

However, the biological species concept leaves open the possibility that they could interbreed, provided they met. We therefore have to check this possibility by indirect hints and clues. Another proposal put forward by the geneticist Theodosius Dobzhansky consists of the morphological species concept. When individuals look similar to each other and are separated from other such groups by a marked morphological gap, we may count them as members of one species, belonging to one and the same species. Dobzhansky proposed to infer morphological species from "genetic" (i.e. biological) ones actually existing in nature by judging morphological similarity or difference between individuals (Dobzhansky, 1941, 1942, 1944).

However, in practice this sounds far easier than it is. We do know from observations of populations of living organisms, that members of a population are not identical replicas of each other. They differ according to their individual age, due to sex differences and individual variation. The individuals of one particular population may even show a certain degree of variability, which we might otherwise have interpreted as representing more than one species in the fossil record. The amount of variability occurring in one population is difficult to standardize independently.

Currently, statistical procedures are used to overcome this difficulty, but we can see how difficult it still is. This procedure depends ultimately on the precise definition of characters and the statistical means which are chosen for an analysis of variation. Most of these analyses are not based on biological criteria and they do not reflect the historical dimension. They are only suitable for delimiting contemporary species in a taxonomic sense, which is the value of Wood's analysis of eastern and southern African hominids (Wood, 1991). However, they allow one to formulate a hypothesis which can be tested in a scenario which is based on a certain biological setting.

4 Ecological Species Concept and Habitat Theory

The ecological species concept was first introduced by the botanist van Valen as follows: "A species is a lineage (or a closely related set of lineages) which occupies an adaptive zone minimally different from that of any other lineage in its range and which evolves separately from all lineages outside its range" (van Valen, 1976). At first sight this looks only as shifting the delimiting criteria for species differentiation from the organisms to their environment, and this has in fact been one of the major criticisms of van Valen's proposal. It depends on how adaptive zones are defined. An adaptive zone is much more than just a resource space of a group of organisms existing independently from them: there is a particular relationship between an organism in a population and its particular environment – this is its adaptation (or function and/or biological role in a non-evolutionary sense). If the environment changes, or if populations are migrating to an area with other environmental conditions, they may react and restructure their life styles accordingly. They may need, for instance, to develop new strategies of food acquisition, to change their diet, and/or to establish other modes of locomotion to move through a habitat with a different structure etc. If the changed conditions last with a certain degree of stability, they will have to change their life style in an appropriate way and establish a new kind of balance with their environment. Adaptation is not a process in which organisms, populations or species are suffering passively until selective pressures push them over the edge to extermination. They contribute actively by establishing new relations to their particular environment.

Habitat theory focuses on the structure of these relations and their changing circumstances in evolution (Vrba, 1988, 1992, 1999). According to this approach evolution, and thus speciation, is determined by three different factors: the inherited properties of lineages and the physical and biotic contexts at a given place or region within a given time interval. Interactions between these three factors directly determine the composition of the habitats, i.e. the particular environments in which populations and thus species exist. The relations between populations and their environments constitute a complex network which can be studied by relating the dynamics of the geographic distribution of species to changes in different environmental variables, such as the effects of climatic cycles and tectonics on temperature or humidity. Within a certain area these changes will lead to a corresponding shift in the distribution of habitats of particular populations, thereby resulting either in expansion or shrinking and fragmentation (vicariance) of these populations. Species can respond in three different ways to their changing environment: they may either speciate, thereby adapting to new environmental conditions; they may maintain familiar relations to the environment as long as this is possible, but have to migrate, i.e. 'drift' accordingly; or they may go extinct. This general approach can be applied to interpretations of species distributions in a certain area and time interval.

Vrba derived a series of hypotheses and predictions from this general concept and used it for the interpretation of the fossil record in Africa during the Plio-Pleistocene. Let us look at some of the hypotheses in more detail.

Vrba puts strong emphasis on changes in the physical conditions caused by climatic cycles and/or tectonism. These changes operate as initiating factors on speciation and changes in the geographic distribution of species, i.e. distribution drift. "Strong habitat changes are causally associated with significant evolutionary changes, and more nearly constant habitats are associated with stable evolutionary lineages" (Vrba, 1999). It has therefore to be demonstrated, as a first step, which environmental changes occurred in the relevant range and these changes have to be related to global climatic cycles and/or more regional tectonic events. If the environment does not change, the evolutionary lineages, populations and species are in stable equilibrium and their relations remain more or less constant. The assumption of a direct and necessary link between environmental change and speciation and/or change in geographic distribution may be one of the weaknesses of Vrba's Habitat Theory, since this reference is not absolutely essential for her argument, as the environmental changes may not necessarily act as initiating causes.

What happens to an organism or a population, when the climate changes? Most populations persist by following their habitat, shifting their distribution area accordingly; i.e. populations migrate together with their habitat. However, if their particular environment becomes isolated, e.g. extensive forests during a period in which the climate becomes more arid, formerly uniform populations will be isolated as well. If this situation continues, speciation may occur, new species arise and old ones disappear. This general model can be used for the interpretation of the fossil record at different sites and the geographic distribution of certain species, thereby providing a reconstruction of the distributional range of different species and the ecosystems which are represented at different fossil sites. This allows one to reconstruct the evolutionary processes leading to the development of populations and species showing certain adaptations as a response to changes in their habitat. This overall model is called a scenario.

5 Reconstructing a Scenario: *Homo rudolfensis* in its Habitat

Habitat theory has been applied to interpret the ecological communities represented at several eastern and southern African fossil sites, including Uraha in Malawi, and to reconstruct the role *Homo rudolfensis* plays in early hominid evolution (Bromage et al. 1995b, Bromage & Schrenk 1995, Schrenk et. al 1995).

About 10 million years ago a global climatic trend towards more cool conditions began. This process reached its maximum about 2.5 million years ago after which the climate trend reversed. This global process is paralleled in Africa by the formation of the African Rift valley leading to an effective rain barrier in regions east of the valley itself. The climatic changes along the rift valley lead to cooler and more arid conditions, conditions under which australopithecines developed in equatorial Africa and migrated southwards along the rift valley until they reached southern Africa about 3 million years ago. As the climatic trend continued, southern and eastern African species were more or less isolated by distance during this time. New species arose in eastern Africa in particular. Since the tropics generally show a larger amount of diversity, it can be assumed in accordance with Habitat Theory that faunal turnover will occur with higher rates of speciation, extinction and migration in lower latitudes than it does in temperate regions. Responding to these environmental changes, eastern African forested areas were partitioned and food plants adapted to this climate change by developing a better protection against loss of water, thereby becoming mechanically tougher. This process is reflected by the development of a more pronounced masticatory apparatus of hominids during this period, particularly well represented by *Paranthropus*, the large toothed and jawed australopithecine in eastern Africa. However, at least some hominid populations started to develop a material culture using simple tools for food processing. As we can observe in the *Homo rudolfensis* mandible, HCRP-UR 501 shows large molars with relatively large crown areas compared to modern humans, though this feature is surpassed by the eastern African robust australopithecines. In the case of *Homo rudolfensis*, functions normally performed by the masticatory apparatus alone were now performed with the aid of tools.

About 2.5 million years ago the climatic trend reversed. Southern and eastern African habitats were connected again through the Rift valley. This is reflected by the faunal assemblage in Uraha. Numerous species which were restricted to eastern

Africa inhabited the environments around Lake Malawi, among them *Homo rudolfensis*. However, some southern African endemic species of bovids migrated to this area as well. This demonstrates that a connection between southern African habitats and the Rift valley had been established, along which some species migrated into the southern reaches of the Rift valley.

This scenario represents an application of Habitat Theory providing a model for the reconstruction of habitats and speciation, extinction and migration of different species. This allows us to understand the evolutionary development of particular species. It has been applied mainly to interpret the fossil record of bovids (Vrba 1985, 1987) and hominids (Bromage & Schrenk 1995). However, comparable processes and developments are expected to be reflected in other taxonomic groups as well.

Every study contributes in testing the validity of the scenario proposed. Moreover, in exploring Habitat Theory there are more ecological hypotheses related to evolutionary processes in populations like speciation, extinction and migration which can be derived.

The reconstruction of the relations between organisms and their environments and of the changes and shifts which occurred during evolution, therefore, is a significant and extensive field of scientific modelling.

References

Alexeev VP (1986) The Origin of the Human Race. Progress Publishers, Moscow

Bromage TG, Schrenk F, Zonneveld FE (1995a) Paleoanthropology of the Malawi Rift: An early hominid mandible from the Chiwondo Beds, northern Malawi. J Hum Evol 28:71-108

Bromage TG & Schrenk F (1995) Biogeographic and climatic basis for a narrative of early hominid evolution. J Hum Evol 28:109-114

Bromage TG, Schrenk F, Juwayeyi YM (1995b) Paleobiogeography of the Malawi Rift: Age and vertebrate paleontology of the Chiwondo Beds, northern Malawi. J Hum Evol 28:37-58

Bromage TG & Schrenk F (1999) (eds) African Biogeography, Climate Change, and Early Hominid Evolution. Oxford University Press, New York

Dobzhansky T (1941) Genetics and the origin of species. Columbia Press, New York.

Dobzhansky T (1942) Races and methods of their study. Transact New York Acad Sci Ser II 4(4):115-123

Dobzhansky T (1944) On Species and Races of Living and Fossil Man. Am J Phys Anthropol 2 N Ser. 2(3):251-265

ICZN (International Commission on Zoological Nomenclature) (ed) (2000) International Code of Zoological Nomenclature. 4th edition. University of California Press, Berkeley

Kennedy GE (1999) Is "*Homo rudolfensis*" a valid species? J Hum Evol 36:119-121

Leakey, LSB. (1959) A new fossil skull from Olduvai. - Nature 184:491-493.

Leakey, LSB, Tobias PV. & Napier JR. (1964) A new species of the genus Homo from Olduvai Gorge. - Nature 202:7-9.
Mayr E (1942) Systematics and the Origin of Species. xxx Publisher, New York
Mayr E (1950) Taxonomic Categories in Fossil Hominids. Cold Spring Harbour Symposia on quantitative Biology 15:109-118
Schrenk F, Bromage TG, Betzler CG, Ring U, Juwayeyi Y (1993) Oldest Homo and Pleiocene biogeography of the Malawi Rift. Nature 365:833-836
Schrenk F, Bromage TG, Gorthner A & Sandrock O (1995) Palaeoecology of the Malawi Rift: Vertebrate and invertebrate faunal contexts of the Chiwondo Beds, northern Malawi. - J. Hum. Evol. 28 (1): 59-70.
Valen L van (1976) Ecological Species, Multispecies, and Oaks. Taxon 25(2/3):233-239
Vrba ES (1985) Ecological and adaptive changes associated with early hominid evolution. In: Delson E (ed.) Ancestors: The Hard Evidence. Alan R. Liss, New York, pp. 63-71
Vrba, E.S. (1987): Ecology in relation to speciation rates: Some case histories of millioncene-Recent mammal clades. - Evol. Ecol. 1:283-300.
Vrba ES (1988) Late Pliocene Climatic Events and Hominid Evolution. In: Grine FE (ed) Evolutionary History of the "Robust" Australopithecines. de Gruyter, Hawthorne New York, pp 405-426
Vrba ES (1992) Mammals as key to evolutionary theory. J Mamm. 73:1-28
Vrba ES (1999) Habitat Theory in Relation to the Evolution in African Neogene Biota and Hominids. In: Bromage T, Schrenk F (eds) African Biogeography, Climate Change, and Human Evolution. Oxford University Press, Oxford, New York, pp 19-34
Wood BA (1991) Koobi Fora Research Project : Vol. 4. Hominid Cranial Remains. Clarendon Press, Oxford
Wood B (1992) Origin and evolution of the genus *Homo*. Nature 355:783-790
Wood B (1999) '*Homo rudolfensis*' Alexeev, 1986 – fact or phantom? J Hum Evol 36:115-118

A Mathematical Model for the Identification of Human Impacts on Floods: the Modau Case Study

Manfred Ostrowski, Marcus Lempert and Steffen Heusch

The model presented WBM (Water Balance Model) is a spatially distributed conceptual, yet physically based continuous hydrological / water resources systems model. The model is generic, it has been developed for the simulation of arbitrary water resources systems. Elements within such systems can be natural river basins, urban catchments including components of urban drainage systems, channel networks and reservoirs. It is a scale flexible model and provides various approaches to calculate the water fluxes within the different stages of the rainfall – runoff process as part of the overall hydrological cycle. Simulation can be performed both for single storm events and continuously over several years. The model has a clear modular structure taking into account all important hydrological subprocesses and their interaction with various components of a basin. This makes it a comfortable tool to investigate effects of changes in the basin and along as well as in the river. Recent severe floods throughout the world show that such models are urgently needed.

The model was applied to the Modau River in Germany which is a tributary to the River Rhine in the southern part of the Federal State of Hesse, Germany. The Modau basin covers an area of 90 km² up to the flow gauge used for model parameter estimation and verification. The river originates from a hilly, rural area where land use is dominated by forests and agriculture into a flatter region with intensive urban development. The data base for the simulation of the basin was satisfying. From four stations in the basin rainfall was available over 30 years, temperature and potential evaporation data was given for one climatological station. With this data it was possible to calculate long term continuous water balances as well as single flood hydrographs. To compare different basin conditions, air photographs and land use statistics were analysed for several historic states of the basin, leading to an answer to the question whether and to what extent human activities have had an impact on floods. The River Modau was chosen as its basin has been heavily modified during the 20th century.

1 Simulation Approach

WBM is a physically oriented but still conceptual rainfall-runoff model designed to simulate various elements of a river basin including reservoirs and urban drainage systems. Hence, it is a practical tool for analysing different system states and types of changes in a river basin. For calculation of water fluxes different approaches are implemented for runoff generation and runoff concentration. Runoff generation describes, how much rainfall will infiltrate, the remainder becomes direct runoff. Runoff concentration describes how water will move on the surface towards a selected downstream control section in the drainage system.

1.1 Direct Runoff Formation

Hydrologic processes within the basin can be described by means of Hydrologic Similar Units (HSU). In each HSU a detailed simulation of the Soil – Vegetation – Atmosphere – Transfer – System (SVAT) considering all relevant transport and storage processes in a vertical column is applied to determine runoff generation. For an HSU it is assumed that basin charcteristics can be averaged due to negligible inhomogeneity.

An alternative to the HSU approach is a grid based simulation option, based on a high degree of spatial resolution and interaction between all elements of the catchment. Again the SVAT-simulation generates the runoff components.

Within the SVAT-simulation each element is represented by a maximum of five layers. The vegetation layer on top is the major link between atmosphere and soil. It is described by an interception model to account for storage of rainfall on biomass and its subsequent evaporation.

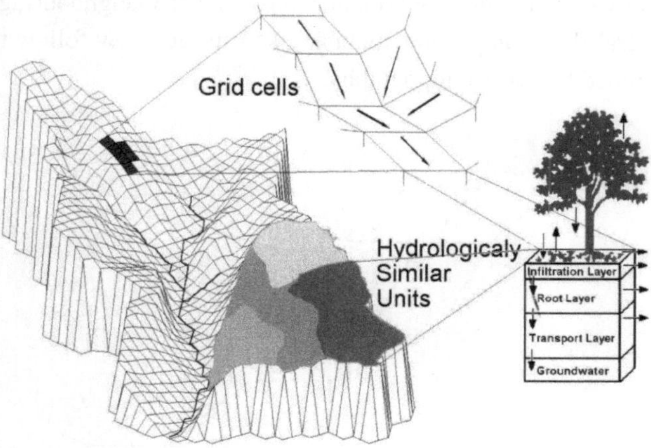

Fig. 1. Options of catchment representation in WBM

The soil below has three horizons according to figure 1. Horizon 1 accounts for infiltration and evaporation processes, horizon 2 mainly determines transpiration and percolation, horizon 3 is the transient transport zone to transform percolation to groundwater recharge. The lowest layer of the column is the groundwater zone. Mathematically, each soil column is represented by a series of storages. All processes are unidirectional vertical and lateral fluxes between these storages, they are defined as non-linear functions depending on storage content, e.g. soil moisture. Piecewise linearisation of these process functions led to computationally efficient analytical solutions at a high degree of accuracy. This approach is still conceptual, however, it is closely connected to the non-linear set of differential equations of the physical approach.

1.2 Runoff Concentration

Two approaches are used alternatively for runoff concentration. A conceptual approach describes surface flow by a conceptual cascade of three linear reservoirs, which assumes that outflow from the surface storage is proportional to its content. For the more physically based approach all transport processes are calculated between adjacent elements. Thus, this approach is a real cascade system representing the natural flow pattern in the catchment according to actual topography of the basin. The relationship between elements is derived by means of geometric information using digital elevation maps (DEM). A search algorithms determines the cascade flow system with fixed relationships between neighbouring units (Figure 2). All lateral flow components including sub surface flow follow this cascade system determined by surface topography.

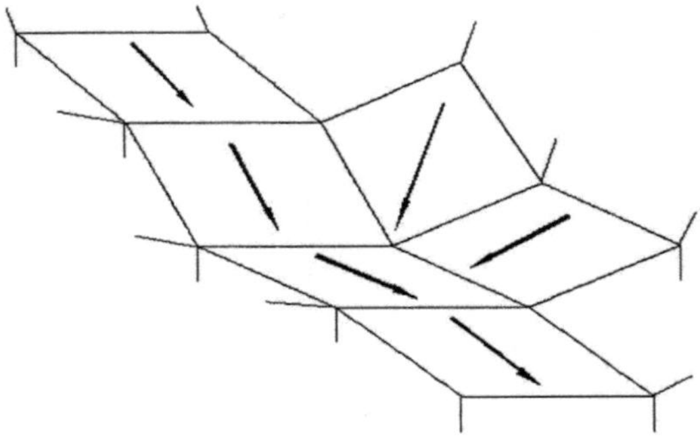

Fig. 2. Overland cascade system

1.3 Flood Routing

Channel network elements (transport elements) are open channel (river) segments described by a hydrologic flood routing method, being again a cascade of linear reservoirs. This model describes, how a flood hydrograph is dampened and decelerated while passing a river valley. The parameters involved, which are the number of storage reaches and the retention constant, can be derived from river geometry and roughness conditions of river banks.

1.4 Urban Catchments

Urban catchments show a most different hydrologic bahaviour from natural catchments. Surfaces are mostly smooth and impervious, the drainage system is artificial. For urban catchments again two alternative modules are implemented. A simple approach represents flow from impervious areas by a storage cascade, flood routing through the drainage network included. For more detailed considerations urban drainage systems are described by several elements, which can be connected in any order.

The elements are impervious areas, sewers, overflows, separators and stormwater tanks with overflows.

A further important element within the model is a multipurpose reservoir. Since this component permits the use of different release rules, a wide range of applications with regard to reservoir system operation is possible. The model was designed to simulate arbitrary water resources systems. Therefore, all modules described can be connected in any order, according to the physical flux of water.

2 Basin Case Study - River Modau

The model was applied to the River Modau Basin. The catchment is located in the southern part of the Federal State of Hesse, Germany. The Modau river is a direct tributary of River Rhine, it is 44 km long and its total catchment area is about 204 km². The area investigated up to the gauging station of Eberstadt ends 27,7 km downstream from its source covering 90 km². Some characteristics of the basin are given in Table 1.

Table 1: Characteristics of the Modau catchment

Modau catchment									
Area	Elevation [m]			Precipitation	Temperature	Discharge [m^3/s]			Stream length
	Aver	Low	High	Mean	mean	Mean	Low	High	
90 km^2	272	110	590	750 mm	7.5 - 8.5 °C	0.809	0.107	21.2	27.7 km

The lower part of the catchment is highly urbanised. The headwater areas are dominated by forest, agricultural land, grassland and small villages.

2.1 Data Base

The data base comprises spatial information on topography, soils and landuse as well as time series of precipitation, discharge and climate with different resolution from 24 hour down to 5 minute time step. The data base is shown in Table 2.

Table 2: Data base

Data source	Scale / Resolution / Station	Time period
Topographic map	1-25000	-
Air photos / air photographic maps	1-25000 - 1- 10000	1935/ 1953/ 1977/ 1988/ 1997
Soil maps	1-25000 , 1- 50000	-
Digital elevation model (DEM)	40m x 40m	1995
Land use statistics	based on the communities	1950/ 1960/ 1970/ 1974/ 1977/ 1983/ 1987/ 1991/ 1995
Different plans of sewer systems	1-5000	1996/97
Cross- sections for the Modau river	868	
Time series		
Precipitation (daily time steps)	Darmstadt Ober-Ramstadt Modautal	1949 – 1998 1941 – 1998 1971 – 1998
Precipitation (5 min time steps)	Darmstadt	1970 – 1986
Climate	Darmstadt	1949 – 1998
Discharge (daily timesteps)	Eberstadt	1955 – 1996
Discharge (continuous)	Eberstadt	1985 – 1996

2.2 Topographic Map

The topographic map was the base layer for all cartographic representations of the catchment such as urbanisation or flood plains. It was also used to digitise the community boundaries as well as the local and the global catchment boundaries. The Modau catchment is subdivided into 17 sub catchments with areas between 15,2 km^2 and 0,8 km^2 according to the inhomogeneity of basin characteristics.

2.3 Land Use and Related Changes

Spatial analysis of the catchment was based on multi-temporal air photographs and photographic maps from 1935, 1952, 1988 and 1997 which were digitised and categorised into three lands use types, forest, agricultural and urban areas, respectively.

The increase of urban areas is obvious, but more related to the communities in the northern part of the catchment. Also, Table 3 shows that the major increase of imperviousness occurred in the period between 1952 and 1988. However, further investigations indicated that the significant period was between 1953 and 1980.

Forested areas neither decreased nor increased significantly during the period analysed. The total difference is about 3.5 %, which is within the error range of digitalisation. Hence, land use changes are mainly related to increasing urbanisation and decreasing agricultural cultivation.

Table 3: Land use changes within the Modau catchment

Year	Forest [km^2 / %]	Urban area [km^2 / %]	Agricultural land [km^2 / %]
1935	29.2 / 32.3	2.6 / 2.9	58.5 / 64.8
1952	29.0 / 32.1	3.8 / 4.3	57.5 / 63.6
1988	32.3 / 35.8	12.0 / 13.3	46.0 / 50.9
1997	32.4 / 35.9	13.6 / 15.1	44.3 / 49.0

This coarse categorisation was insufficient, more detailed differentiation of cultivated agricultural areas was required. Agricultural statistical surveys from 1950, 1960 1970, 1983 and 1995 were used for this purpose, which were available from 6 communities. As statistical surveys are related to community instead of sub catchment boundaries, partial sub areas had to be assigned to communities. Finally, seven land uses and crops were extracted from statistical surveys:

- green cover (includes pastures, meadows etc.)
- special cultures (include fruit trees and nurseries)
- cereals
- root crops
- vegetables + garden crops
- fodder crops

- maize

The result of this analysis is data on the cultivated agricultural areas as part of each sub catchment for 1950, 1960, 1970, 1983 and 1995 respectively.

2.4 Soils

The large number of different soil types required careful data management for hydrological modelling to avoid infeasible computational efforts. Existing soil types in the region were reclassified according to the similarity of soil properties such as hydraulic conductivity and field capacity for each layer in a soil column.

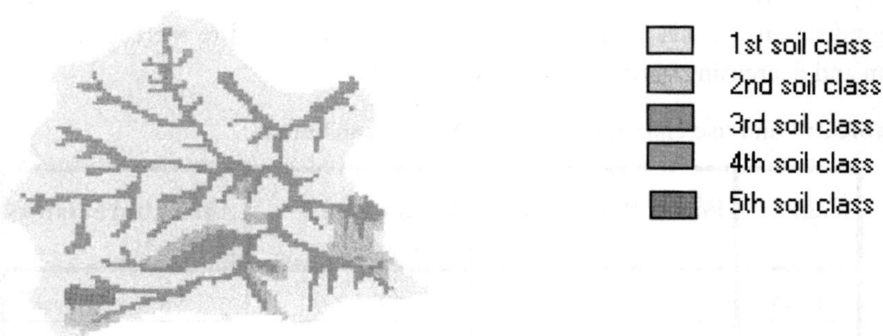

Fig. 3. Digital reclassified soil map of sub catchment Brandau

The reclassification led to a maximum of 5 different soil classes in a single sub catchment and 14 different soil classes within the whole watershed. The result of this analysis were new local soil maps for each sub catchment (Figure 3). They contain a significantly reduced number of data suitable for efficient data management and simulation, which is of high importance for parameter estimation.

2.5. Digital Elevation Model (DEM)

The digital elevation model was available with a resolution of 40mx40m. The DEM was used for the topographic analysis of the catchment. The program analysing the topography determines the complete drainage network including important parameters like slope, aspect and catchment area for each element according to Figure 4.

Based on the drainage network determined, an alternative approach was applied to derive sub catchment boundaries in order to compare the results with the sub catchments formerly digitised and to discover possible contradictions. It proved that the automatic delineation of watersheds is reliable.

Drainage network, slope and catchment area are essential input parameters for the raster based simulation of rural sub catchments.

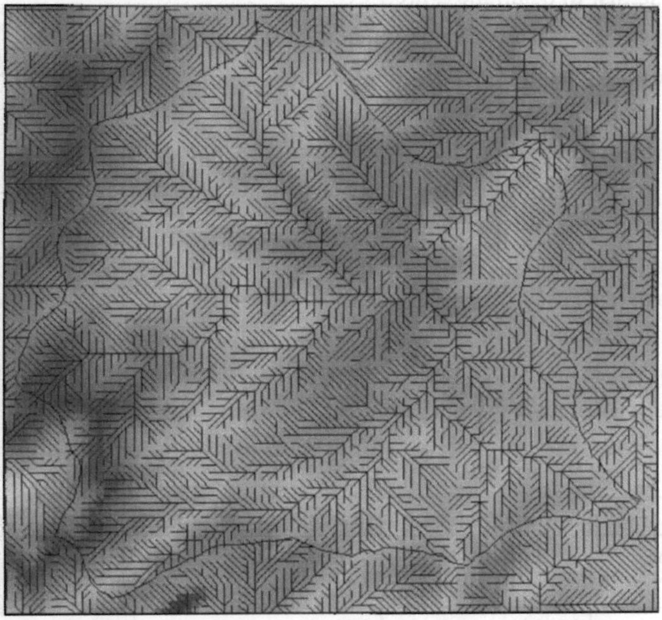

Fig. 4. Drainage network , subcatchment; Boundary and DEM of the Brandau subcatchment

2.6 The River Cross Sections

For the Modau River up to the gauging station at Eberstadt 868 river cross sections were available. They were partly derived through surveying the river channels including flood plains, partly through analysis of stereoscopic air photos. The latter also consider flood plains but are less accurate and do not consider the river itself. Based on this data a calibrated stationary hydraulic model for the computation of water surface levels became available. The Modau River was subdivided into river reaches with homogeneous longitudinal and cross sections. For each reach a representative stage discharge relationship was determined using a hydraulic model to provide input for the flood routing part of the hydrologic model. For the tributaries of the Modau river no cross sections were available. These had to be estimated during field visits.

2.7 Sewer System

Flow in the Modau River is significantly influenced by outflows from urban drainage systems. In detail these are:

- Outflows from waste water treatment plants
- Overflows from combined sewer systems
- Storm water outflows from separate systems
- Other outflows e.g. from industrial treatment plants of larger firms
- Central and de-central sewage treatment plants in rural areas

The part of the Modau River considered in this investigation receives discharges from 4 major community waste water treatment plants. Sewer systems are mostly combined systems, only very few and small areas are drained by separate systems. Some rural communities in the headwater catchment convey waste water treated in central or de-central septic tank systems to River Modau. Important information on the sewer systems are given in Table 4.

Table 4: Characteristics of major urban drainage systems

	Impervious area	Overflow structures	Storage volume
Brandau	13,2 ha	3 basins 2 overflows	936 m^3
Ernsthofen	13,5 ha	2 basins 1 overflow	530 m^3
Nieder-Ramstadt	211,7 ha	11 basins 30 overflows	5493 m^3
Eberstadt	181,5 ha	1 basin 5 overflows	1000 m^3

2.8 Hydraulic Structures

For the urban areas in the northern part of the catchment a flood control reservoir with a capacity of max. 400000 m^3 was built 300 m upstream of the community of Ober-Ramstadt. In addition multiple small water mills and weirs have been constructed which are of less importance for flood development, however.

2.9 Hydro-meteorological time series

Precipitation

Table 2 shows that daily precipitation data for the Modau catchment are available at 3 stations for varying periods. For model application it is required to extend the time series of the Modau station to the period of the other stations as it is representative for the headwater sub catchments. This was realised through correlation and regression analysis between the stations.

Discharge

The analysis of flows included an extreme value analysis of annual floods and a trend analysis. Several extreme value distributions were tested to determine expectation values of peak flows for chosen return intervals.

Table 5: Extreme value analysis flows

Flood frequency [a]	**2**	**5**	**10**	**25**	**50**	**100**
Time period	[m³/s]	[m³/s]	[m³/s]	[m³/s]	[m³/s]	[m³/s]
1956 - 1995	8,91	13,61	17,33	22,77	27,41	32,58
1956 - 1975	9,19	14,23	18,28	24,25	29,39	35,15
1976 - 1995	8,61	13,08	16,62	21,77	26,15	31,01

Table 5 shows the results of the Log-Gamma Distribution, which proved to fit best the sample at Eberstadt gage. The extreme value analysis proves that extreme flows in the period from 1956 to 1975 are significantly higher than those in the period from 1976 to 1995 as well as for the total observation period. This result is confirmed by a trend analysis. Over the whole observation period a slight non significant negative trend was determined. If further differentiated, the trend for 1975 to 1995 is negative, while it is clearly positive for 1956 to 1975. Based on the analyses described above the following conclusions can be drawn:

2.10 Interpretation of Statistical Analysis

Up to 1975 annual flood peaks increased in the Modau catchment. This can be assigned to accelerated urbanisation with maximum growth rates between 1952 and 1977. The rainfall pattern of single stations is inhomogeneous; thus their significance for the runoff pattern can not be evaluated. From 1980 peak flows as well as yearly runoff volumes decreased. At the same time, however, also yearly rainfall decreased which should be the main reason for smaller peaks and volumes. After 1980 growth rates of impervious areas have been very moderate, which in addition is compensated by retention basin volumes within urban drainage systems. It is also possible that peak flows from rural areas are reduced by accelerated runoff from urban areas now arriving earlier at the control gage. To a small extent increasing temperatures might have induced increasing evapotranspiration also reducing direct runoff.

3 Parameter Estimation and Verification

3.1 Parameter Estimation for Daily Time Step Simulation

A reliable model requires optimum parameters. The parameter estimation procedure very often is the most time consuming part of a hydrologic model based study. Also in this study the parameter estimation effort was considerable finally leading to reliable model parameters. It concentrated on the optimum estimation of soil parameters and retention constants of the linear and non linear storages representing the lateral transport processes.

The basin includes only one runoff gauge positioned on the outlet of the basin, producing discharge data useful for the calibration. Discharge data was available from 1951 to 1996. For calibration process hydrographs for the years 1994 and 1995 were computed and compared with the measured discharge. The goodness of fit of simulation and measurement was basically determined by means of three criteria: The total-volume-error, the coefficient of correlation and determination and the efficiency according to Nash / Sutcliffe.

Former experience with the WBM - model indicated that parameters describing unsaturated flow in the soil layers are most sensitive for the formation of direct runoff and for the correct computation of longterm water balances. For modelling purposes the soil column is divided into three layers with different depths depending on the vegetation and ground water level. Water flow between layers is base on non-linear relationships between infiltration, percolation, interflow and actual soil moisture. The major objective was to determine an optimum combination of these three functions.

Other important parameters were retention constants of the linear reservoirs representing lateral flows, which are surface and sub-surface flow components from pervious and impervious areas including interflow and baseflow.

First, constants were derived from measured hydrographs giving to receive good initial values. Based on these initial estimates optimal constants were iteratively found using the correlation of the hydrographs as a measure of goodness of fit. With these measures the results shown in Figure 5 for 1994 were obtained, Table 6 contains goodness of fit parameters.

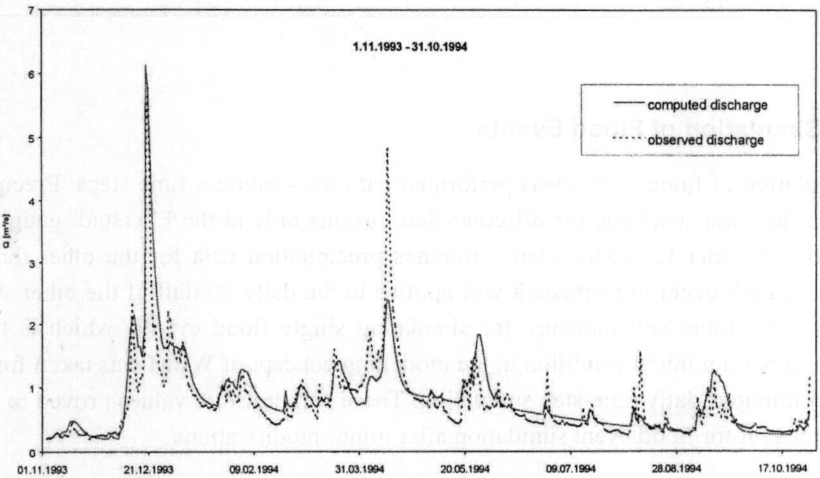

Fig. 5. Simulation of hydrological year 1994

Table 6: Goodness-of-fit parameters of calibrated hydrographs

	Total volume error [%]	Coefficient of determination [-]	Efficiency acc. to Nash/Sutcliffe [-]
Simulation of hydrological years 1994/95	3,74	0,822	0,647

3.2 Validation of the Calibrated Model for the Daily Time Step Simulation

To validate the parameters found during the calibration process the run-off from 1972 to 1995 was computed. The results according to goodness-of-fit parameters given in Table 7.

Table 7: Goodness of fit parameters for validation

	Total volume error [%]	Coefficient of determination [-]	Efficiency acc. to Nash/Sutcliffe [-]
Simulation of the years 1972 to 1995	4,4	0,805	0,394

3.3 Simulation of Flood Events

Calculation of flood events was performed with ten - minutes time steps. Precipitation data was available for different flood events only at the Eberstadt gauging station. In order to produce ten – minutes precipitation data for the other three stations each event in Darmstadt was applied to the daily rainfall of the other stations. The initial soil moisture for simulating single flood events, which is the most important initial condition in the modelling concept of WBM was taken from the continuous daily time-step simulation. These soil moisture values proved to be most useful for flood event simulation after minor modifications.

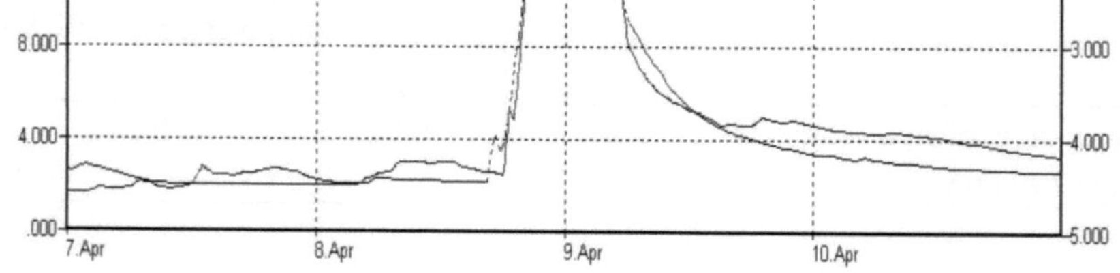

Fig. 6. Simulation of flood event in April 1983

Several events were analysed to determine optimum parameters being sensitive for flood simulation. While for the continuous daily time-step simulation soil parameters were most sensitive, for floods parameters describing conductivity and infiltration were more sensitive and had to be modified during calibration. Other flood sensitive parameters were retention constants and those being important for flood-routing such as roughness coefficients of the river bed and banks.

Results of two simulated flood events are presented in Figures 6 and 7. For all simulation runs the same set of parameters was used, only initial soil moisture was event dependent.

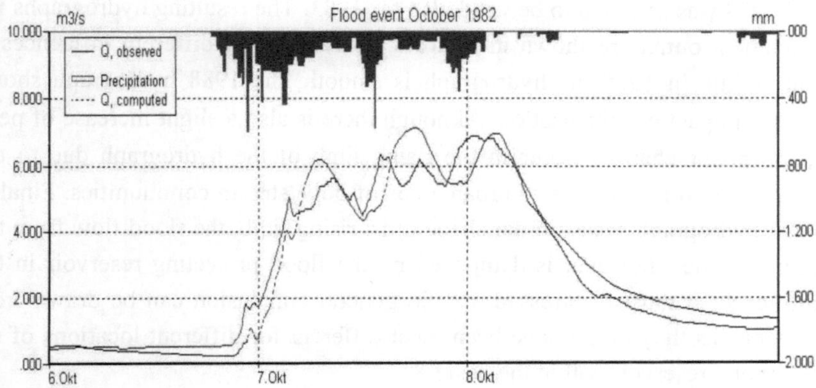

Fig. 7. Simulation of flood event in October 1982

Table 8: Goodness-of-fit parameters for flood events

	May 1978		October 1982		April 1983	
Mass balance error	3,3 %		2,9 %		-3,2 %	
Coefficient of correlation	0,881		0,971		0,980	
Nash-Sutcliffe coefficient	0,433	Good	0,760	Very good	0,932	Excellent
Schultz-criterion	11,9	sufficient	6,6	Good	4,1	Good

Results for goodness-of-fit parameters can be considered as good, most criteria are good or very results for the flood events in 1978, 1982 and 1983. Based on the calibrated and verified model the effect of anthropogenic impacts on floods could then be analysed.

4 Analysis of the Effects of Land Use Change

4.1 Analysis Based on the Historical 1983 Flood

In order to see the effect of different basin states in 1935, 88 and 97, the flood event in 1983 was also simulated for system states in 1935 and 1997, the system state of 1988 was assumed to be valid also for 1983. The resulting hydrographs for the catchment outlet are shown in figure 8. The impact of different influences is evident. While in 1935 the hydrograph is smooth, the 1988 hydrograph shows clearly the impact of urbanisation. Although there is also a slight increase of peak flow, the major changes occur in the rising limb of the hydrograph due to the faster reaction to rain storms of urban areas of downstream communities. Finally, the 1997 hydrograph has a similar shape in ist rising limb, the flood flow from the upstream reaches, however is dampened by the flood protecting reservoir in the middle reach. It must be stressed that no general conclusion can be drawn from these results as they might have been most different for different locations of urban areas and reservoir within the basin.

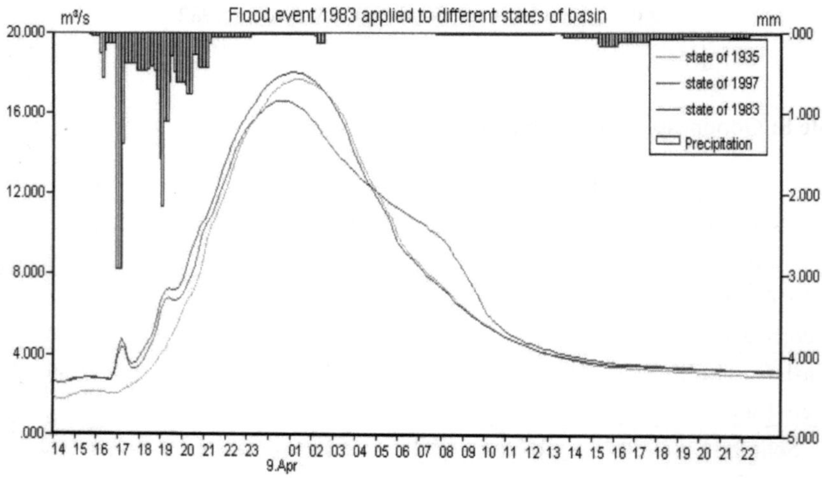

Fig. 8. Comparison of different states of the model by means of the flood 1983

4.2 Analysis of Land Use Effects Using Synthetic Rainfall Events

To get a more general base for the comparison of the different catchment states, a rainfall sample of different durations and depths was generated based on the information provided KOSTRA - Atlas, a comprehensive collection of local and regional tables of depth-duration of rainfall, provided by the German Weather Service DWD. For two different durations, 15 minutes and 2 hours, respectively, rainfall events were generated according a standardised temporal rainfall distribution as shown in Figure 9.

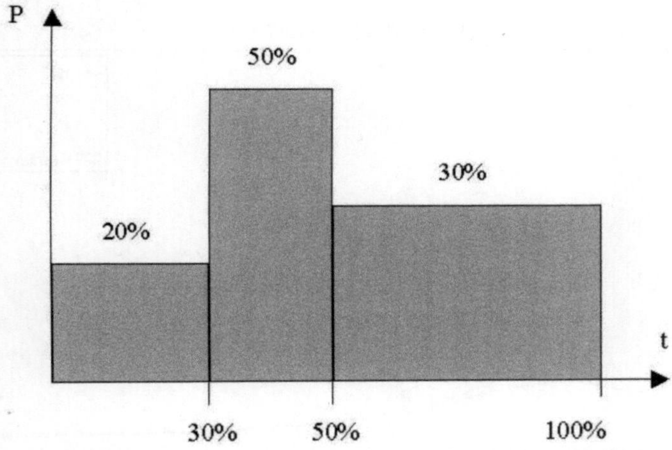

Fig. 9. Standard rainfall distribution

4.3 Comparison by Means of Synthetical Scenarios

The generated rainfall was applied to the model representing the different historic states 1935, 1988 and 1997. For all flood-event simulations initial soil moisture was chosen in such a manner, that the return period of the flood generated corresponded to the return period of the rainfall event.

4.3.1 Rainfall Duration 2hrs

Figure 10 and figure 11 show hydrographs generated with a 2h storm and return periods of 1 and 100, respectively. The effect of the sewer overflows is evident:

Floods with a return period of one year are generally produced by sewer overflows (see simulations of 1988 and 1997), the flood peak in 1997 is approximately five times higher than the peak in 1935 when no sewer system was constructed yet and hardly any flood flow from rural areas is produced. Floods produced for return periods of 100 years (figure 11) are also influenced by sewer overflows, producing a smaller early peak in the rising limb prior to the main peak. The main peak, occuring later, however originates from runoff from rural areas. One can see the dampening effect by 20 % of the peak due to the reservoir build in 1993. The flood frequency curve in figure 12 shows that flood peaks up to a return period of 10 years are produced by sewer systems.

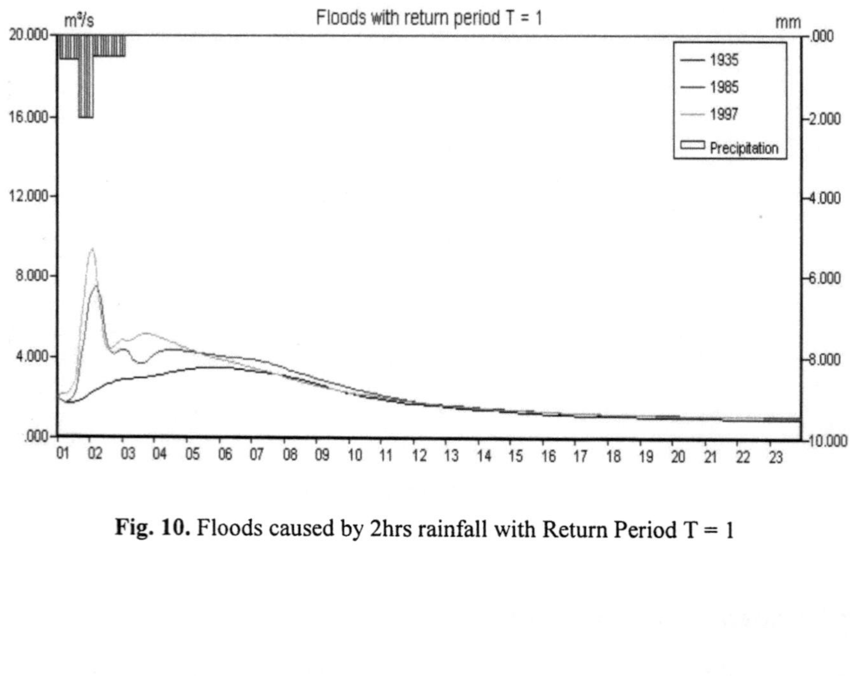

Fig. 10. Floods caused by 2hrs rainfall with Return Period T = 1

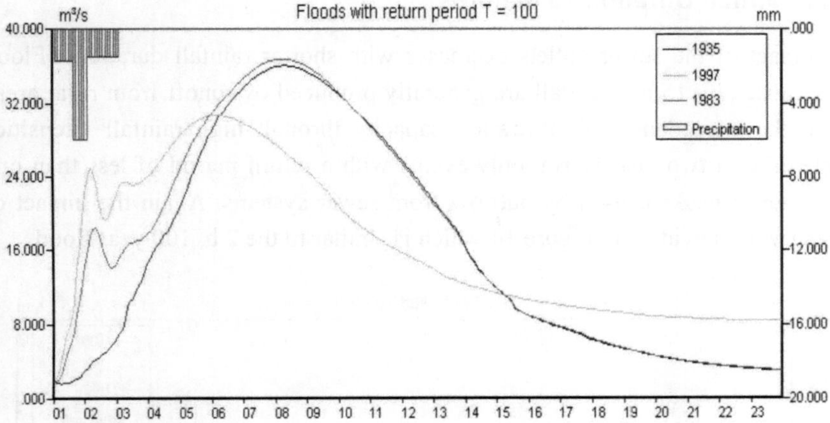

Fig. 11. Floods caused by 2hrs rainfall with Return Period T = 100

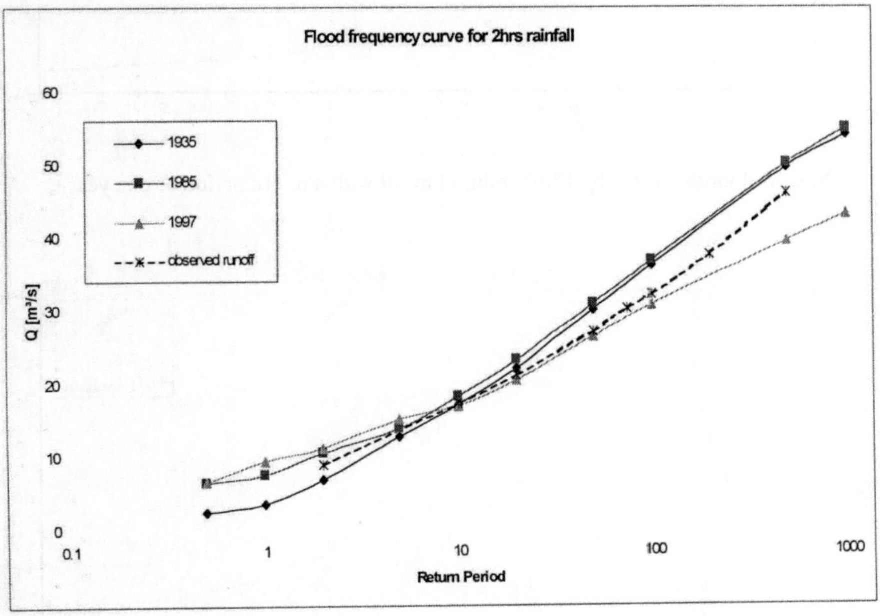

Fig. 12. Flood frequency curve for 2hrs rainfall

5.3.2. Rainfall Duration 15 Minutes

The impact of the sewer outlets decreases with shorter rainfall durations. Flood peaks caused by 15min rainfall are generally produced by runoff from rural areas due to the exceedance of infiltration capacity through high rainfall intensities compared to a two year storm, only events with a return period of less than one year produce peaks caused by outflow from sewer systems. Again the impact of the reservoir is evident in Figure 14 which is similar to the 2 h, 100 year flood.

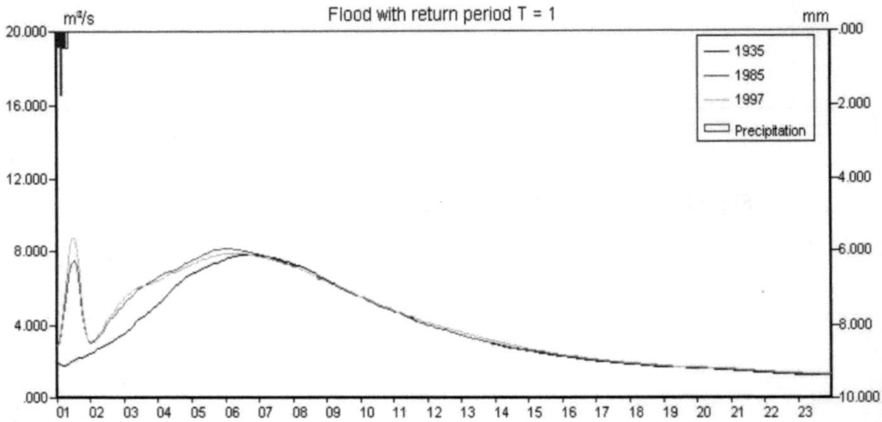

Fig. 13. Floods caused by 15min rainfall event with a return period of one year

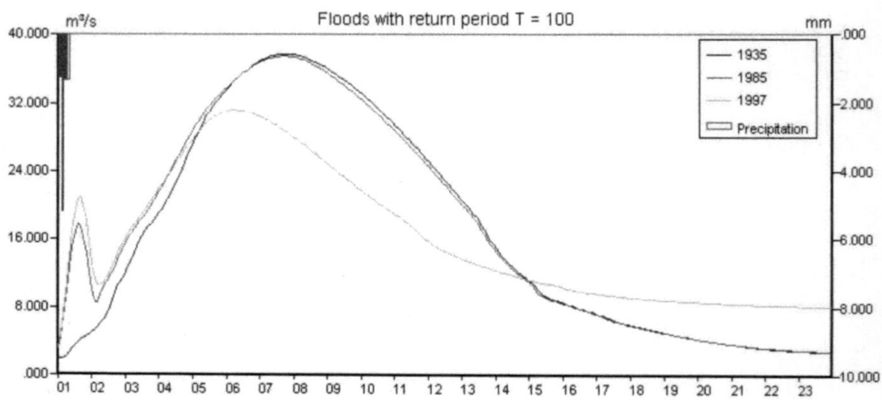

Fig. 14. Floods caused by 15min rainfall with Return Period T = 100

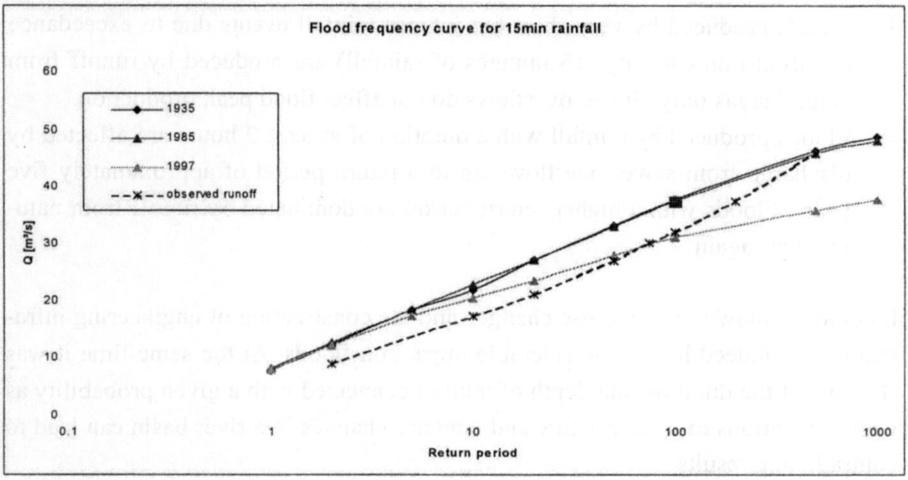

Fig. 15. Flood frequency curve for 15min rainfall

5 Conclusions

The Modau River basin was investigated to identify the impact of human activities on floods. For this purpose three system states in 1935, 1988 and 1997 were considered. Possible effects of the different states were analysed based on a historic flood event in 1983 and on synthetic storm events.

The following conclusions were drawn:

1. Increasing urbanisation has no influence on the flood peak of the 1983 event, which has approximately a return period of 50 years, but increases the discharge volume. Peaks of events with this return period are mainly produced by runoff from natural areas and the effects of changes in land use distribution are minor, one could even state that they are negligible. Increase of discharge volume is due to increasing impervious areas. A small peak prior to the flood peak is produced by sewer overflows, which are located near the outlet of the basin.
2. Implementation of the reservoir upstream of the highly urbanised areas would have reduced the flood peak. In 1983 the peak flow measured was 18 m^3/s, with the reservoir the peak would have been reduced to approximately 16 m^3/s.

3. Floods produced by very short but intense rainfall events due to exceedance of infiltartion capacity (15 minutes of rainfall) are produced by runoff from natural areas only. Sewer overflows do not affect flood peak production.
4. Floods produced by rainfall with a duration of at least 2 hours are affected by discharge from sewer overflows up to a return period of approximately five years. Floods with a higher return period are dominated by runoff from natural areas again.

It could be shown that land use changes and the construction of engineering infrastruture do indeed have a considerable impact on floods. At the same time it was shown that the duration and depth of rainfall connected with a given probability as well as locations of infrastructure and land use changes in a river basin can lead to contradicting results.

These results prove that general statements on the impact of human activities of floods can be misleading. Politicians and the public might draw wrong conclusions and take wrong actions. It must be kept in mind that each catchment has its individual structure and characteristics, which due to their interaction and superposition determine the formation and conveyance of floods. Only a reliable model is able to reproduce the important processes during a flood to predict their development in space and time qualitatively with high accuracy and quantitatively with limited accuracy.

6 Acknowledgment

This work was in part supported by the EC Environment and Climate Research Programme (Contract No. ENV-CT97-0529 Climate and Natural Hazards)".

7 References

Ostrowski MW, Lempert, M Erweiterte Anforderungen an deterministische mathematische Niederschlag-Abfluss-Modelle zur Berechnung des maximal moeglichen Abflusses, in press

De Michele C, Rosso R (1999) Self-affine and multi-affine concepts to model the space-time variability of extreme storm precipitation. Geoph. Res. Abstracts, Vol.1(4), 766 (abstract)

Salvadori, G, De Michele C (1999) Scaling laws in storm frequency analysis: the case study of the Bisagno basin. Geoph. Res. Abstracts, Vol.1(2), 345 (abstract)

Physical Modeling of a Glass Melter

Manfred Kelm

Introduction

Upscaling of a chemical process from a laboratory scale to a real plant scale is usually done stepwise in order to avoid misdirected investments. On the way to a full scale plant, pilot plants and mock-ups of the main components are operated to gain experience and to find suitable operating parameters. The linear scaling factor for the components is usually less than 5. Depending on the process, it can be difficult or even impossible to observe and measure directly interesting and important properties, even on a pilot scale. This is the case, e.g., for measurements at high temperature or in a corrosive environment, where sensors have short lifetimes. Furthermore, behavior such as flow or the kinetics of sedimentation processes can often not be studied by observation inside a full scale reactor but must be derived from other data.

The growing capability of computers and commercial software during the last 20 years allow nowadays the simulation of physico-chemical processes in a reactor using mathematical models. This method requires a detailed knowledge of all parameters which influence the process and a suitable code which describes it. Such modeling can be quite abstract. Most plant simulations are steady-state simulations. However, further computer and software development should allow dynamic simulations in the future.

A completely different but very instructive approach is to make processes visible and physical properties directly measurable by using a physical model of the process. Physical modeling of a chemical reactor requires that the model and the full scale apparatus (prototype system) are geometrically similar but not necessarily of the same size. The construction material of the model apparatus itself and the medium used can be chosen according to needs. They are usually different from that of the prototype, but they have to fulfill certain conditions. The chemical reactions occurring in the prototype reactor are not subject of such modeling. The aim of a physical model is to study physical properties like mass and heat flow or the electrical behavior.

In the following, the physical performance of a glass melter for the vitrification of radioactive waste will be derived from observing and measuring properties in a plexiglass melter model of reduced size.

1 Background

For the vitrification of high level radioactive waste, glass melters are developed which deviates significantly from furnaces for the glass industry in their design. The tank has a comparatively small volume and is lined with highly corrosion resistive ceramic. Under operation, heat is supplied solely by applying an electric current through the electrically conductive glass melt (Joule-heating principle). For this purpose, the furnace is equipped with a number of metallic electrode pairs and the maximum melter temperature is limited to about 1250°C. To prevent electrolysis at the electrodes supplying the electric energy, glass melters are generally operated with alternating current. Due to the positive temperature coefficient of the electrical conductivity of the glass melt, the electric current is kept constant for stable temperature conditions to develop.

Under operation, the glass former (solid glass frit) is fed into the melter, together with the aqueous waste solution, on top of the glass melt, thus forming a permanent cold cap. The product glass is drained through a bottom valve. Because of the cold cap, the melter content is never homogeneous. However, the drained product must not vary significantly in composition and must not contain unvitrified feed material. This requires good mixing within the glass melt. Unfortunately some noble metals in the waste solution to be vitrified are incompatible with the glass melt. They form separate phases with high electrical conductivity and are susceptible to sedimentation in the melt. A resulting reduced melting power due to an increased volume conductivity of the melt as well as modified temperature and flow profiles have to be considered.

A prototype melter for the vitrification of simulated high level radioactive waste was manufactured and operated according to the data given in Tab. 1. It is equipped with five electrode gaps on two levels. They are positioned such that mutual influence is minimized. Each gap is supplied by a separate transformer.

Table 1: Main data of prototype melter and melter model

Parameter	Prototype melter	Model
Tank length	1.4 m	280 mm
Tank width	1.0 m	200 mm
Tank depth	0.4 m	80 mm
Number of electrode gaps	5	5
Electrode surface (side)	254 cm^2	17.6 cm^2
Electrode surface (corner)	489 cm^2	31.1 cm^2
Max. voltage	200 V	150 V
Max. current density	1A cm^{-2}	
Max. power	150 kW	750 W
Surface cooling	Liquid feed, 80 lh^{-1}	Cooling plate
Nominal operation temperature	1150°C	79°C

2 Physical Model

2.1 Aim of Physical Modeling

To study the performance of the prototype melter by direct observation, a model is set up on the scale 1:5. The aim of the modeling is to visualize the glass flow in the melter under different operating conditions, e.g., different electrode circuitries, different power distributions between the electrode gaps and under the influence of precipitating electrically conductive particles. High velocities in the vortices indicate good mixing and thus good vitrification conditions while slow flow rates indicate areas with a tendency for sedimentation of unvitrified material. Furthermore, such a model allows direct measurement of the temperature distribution in the melt and detection of the electric field under operating conditions. The next chapters describe the construction and the operation of the model, how to derive prototype melter properties from observation of properties in the model and, as examples, flow pattern and appropriate temperature distributions under different operating conditions.

2.2 Mechanical Outline of the Model

The melter model is made of plexiglass, such that the entire tank volume can be observed visually (Fig. 1 and 2). Details of the model are listed in Tab. 1. All symbols used in the following are defined in the appendix. A transparent plexiglass

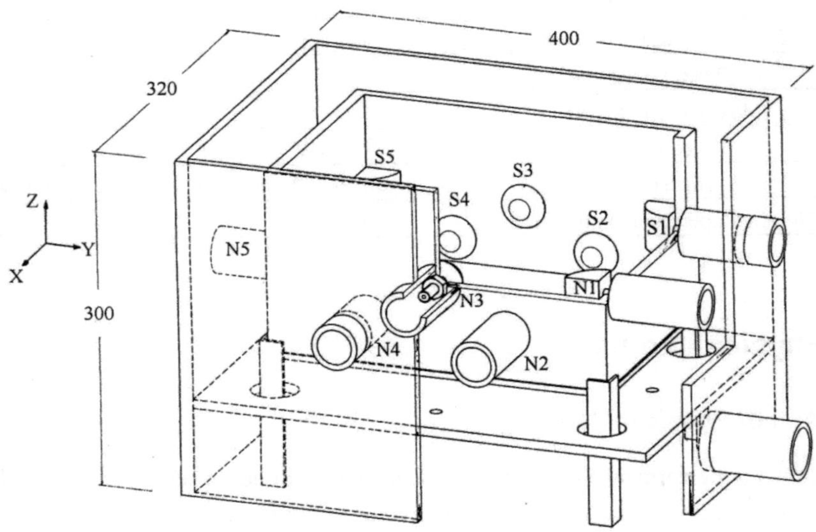

Fig. 1. Sketch of the melter model. Outer dimensions given in mm. Chamber dimensions: x= 200mm, y=280mm, z=80mm

jacket is installed around the model at a distance of about 50 mm. It allows to heat or cool the four sides and the bottom of the plexiglass chamber with air. Air temperature is controlled, such that heat losses of the prototype melter via the bottom and the walls can be simulated in the model. The cold cap in the prototype melter is modeled by a liquid cooled plate which is made of anodized (and thus electrically insulated) aluminum and which can be immersed up to a few mm into the bath to simulate heat losses occurring in the prototype melter due to the evaporation of feed solution. The cooling plate covers about 90% of the bath surface. Actual feeding and draining of the melter is not simulated. Temperature sensors can be inserted close to each other from the top to enable measuring the temperature distribution in the tank. Within the jacket, the melter model is based on a rack of L bars and flat bars, which is designed such that optical access is hardly disturbed. The 10 electrodes of the prototype furnace are also reduced to the scale of

1 : 5 in the model, made of copper and similarly positioned in the model (Fig.1). The electrodes are mounted liquid-tight to the model melter walls. For thermal insula-

Fig. 2. Photo of the melter model.

tion of the electrodes, wide plexiglass tubes separate them from the air flow in the jacket.

2.3 Electrical Set-up of the Model

The electrodes, which are all located on the same long side (all N or S electrodes, Fig.1) of the model melter, are switched in phase and each gap is fed from a separate AC supply with current control. The electrode current and voltage of all five gaps and six temperatures from sensors inserted into the tank at various positions are monitored simultaneously. The hot air jacket of the model is supplied from a temperature controlled hot air blower. To enable the calculation of heat loss from the bath surface via the cooling plate, the coolant flow rate and the temperature difference between inlet and outlet is measured.

2.4 Modeling Principle

For physical modeling, the prototype and the model melter have to be geometrically similar and for observation the model and the model liquid have to be transparent. Glycerol containing 7.5% LiCl is suitable as model liquid for modeling radioactive waste glasses. Glycerol has a proper viscosity and the LiCl makes it sufficiently conductive (see also Quigley and Kreid 1979, Noble et al. 1972, Routt 1983, Igarashi et al. 1993).

To infer the prototype melter behavior from the physical behavior of the model, the processes to be compared must be described by differential equations and boundary conditions of the same type. Differential equations describing flow and heat flux in a directly heated melter are based on the laws of conservation of mass, momentum, energy and charge. The task is to find a scaling factor (ratio between the value measured in the model and in the prototype) for each physical property of interest, that is constant over the temperature range under operation. For this purpose, dimensionless numbers are introduced. They appear in the differential equations as independent variables, if the equations are converted into a dimensionless form (Quigley and Kreid 1979). The dimensionless numbers can also be interpreted as ratios of energies, fluxes or forces etc.; if they are equal in different physical systems, they describe the same physical process.

In practice, those dimensionless numbers, which govern the physical behavior in the melter (but not all one can think of), are brought into agreement for the prototype and the model melter. This is often not easy and the difficulties associated with this process are outlined in detail by Buckingham (1937).

In a glass melter with low throughput (i.e., small forced flow), flow behavior is determined by the buoyancy forces resulting from the density differences in the melt. These conditions are described best by the Rayleigh number.

$$\mathrm{Ra} = \frac{\text{energy transport by bouancy}}{\text{energy transport by heat conduction}} \tag{1}$$

The Raleigh numbers of the prototype melter and the model should be made to agree. Further appropriate modeling criteria are the agreement of the Peclet numbers and power numbers.

$$Pe = \frac{\text{energy transport by convection}}{\text{energy transport by conduction}} \tag{2}$$

$$Po = \frac{\text{electric energy flux}}{\text{energy flux by conduction}} \tag{3}$$

To correctly reproduce the conditions prevailing at the inner wall of the bath, it is of crucial importance that the percentage of heat losses via the side walls, the bottom of the tank and the surface are in agreement for the model and the prototype melter. This is a weaker condition than the equality of Nusselt numbers.

$$Nu = \frac{\text{energy flux by convection}}{\text{energy flux by conduction}} \tag{4}$$

In spite of the few dimensionless numbers considered, correct modeling can only be achieved for a certain temperature, as all physical properties of the glass and the model liquid vary with temperature in a different manner. If the thermal conductivity (λ), electrical resistivity (σ) and viscosity (μ) for the glass melt and the model liquid vary in a similar extent and manner near the nominal temperatures, the error of modeling at different temperatures is small. This must be confirmed after the determination of the nominal model liquid temperature. Compared to the three parameters λ, σ and μ, the temperature dependence of other physical quantities relevant to modeling is rather small.

The requirements for modeling the melter can therefore be summarized as follows:

Equality of the Rayleigh number:

$$Ra = \frac{g \cdot \beta \cdot L^3 \cdot \Delta T}{\alpha \cdot \nu} \tag{5}$$

Equality of the Peclet number:

$$Pe = \frac{\rho \cdot C \cdot v \cdot L}{\lambda} = \frac{v \cdot L}{\alpha} \tag{6}$$

Equality of the power number:

$$P_o = \frac{\phi \cdot L^2}{\lambda \cdot \Delta T} \tag{7}$$

Equality of these three dimensionless numbers is the basis for calculating the scaling factors, as described in the following. After modeling the prototype melter according to that criteria, it must be demonstrated that the boundary condition for the heat loss (same percentage of heat loss over all six surfaces in the prototype melter and the model) is fulfilled.

2.5 Calculation of the Scaling Factors

The scaling factor is understood to be the ratio between the value of a physical property of the model and the value of the same property of the prototype melter.

$$\varepsilon_x = \frac{q_m}{q_p} \tag{8}$$

The length scaling factor is set arbitrarily to $\varepsilon_L = 0.2$. It gives the model a dimension which is easy to handle (Fig. 1 and 2). As both the property of the model liquid and the property of the glass melt depend on temperature, a nominal glass temperature has to be chosen, at which all scaling factors are to be exact and which is the normal operating temperature of the prototype melter. According to Tab. 1, this temperature is $T_{o,p}$=1150°C. For further calculations, several data for the glass melt at 1150°C and data sets for the model liquid at various temperatures are required. They are listed in table 2.

To determine the relation between the temperature scales for the model liquid and the glass melt, the dimensionless product $\beta \cdot \Delta T$ as part of the Rayleigh number is considered separately.

$$\Delta T_m \cdot \beta_m = \Delta T_p \cdot \beta_p \tag{9}$$

$$\varepsilon_{\Delta T} = \frac{\Delta T_m}{\Delta T_p} = \frac{\beta_p}{\beta_m} = \frac{8.51 \cdot 10^{-5}}{5.34 \cdot 10^{-4}} = 0.159 \tag{10}$$

A value of $\varepsilon_{\Delta T}$ = 0.1594 means that a temperature difference measured in the model corresponds to a $(0.1594)^{-1}$=6.3 times higher temperature difference in the glass melt. Using the equality of the second factor of the Raleigh number $g \cdot L^3(\alpha \cdot \nu)^{-1}$, an indirect temperature assignment can be made:

$$\frac{g \cdot L_m^3}{\alpha_m \cdot \nu_m} = \frac{g \cdot L_p^3}{\alpha_p \cdot \nu_p} \tag{11}$$

Table 2: Physical data of glass melt and model liquid

	Property	Symbol	Unit	Glasmelt at 1150°C
1	Thermal diffusivity	α	m^2s^{-1}	$3.47 . 10^{-7}$
2	Thermal conductivity	λ	$Wm^{-1}K^{-1}$	1.6
3	Density	ρ	kgm^{-3}	2450
4	Dynamic viscosity	μ	$kgm^{-1}s^{-1}$	5.6
5	Kinematic viscosity	ν	m^2s^{-1}	$2.29 . 10^{-3}$
6	Electrical resistivity	σ	Ωm	0.038
7	Specific heat	C	$Jkg^{-1}K^{-1}$	1880
8	Volume expansion coeff.	β	K^{-1}	$8.51 . 10^{-5}$

	Model liquid at different temperatures				
	50°C	60°C	65°C	70°C	80°C
1	$1.05 \cdot 10^{-7}$	$1.02 \cdot 10^{-7}$	$1.01 \cdot 10^{-7}$	$1.00 \cdot 10^{-7}$	$0.98 \cdot 10^{-7}$
2	0.324	0.322	0.321	0.320	0.319
3	1280	1274	1272	1269	1264
4	0.38	0.21	0.161	0.12	0.08
5	$2.97 \cdot 10^{-4}$	$1.65 \cdot 10^{-4}$	$1.29 \cdot 10^{-4}$	$0.95 \cdot 10^{-4}$	$0.63 \cdot 10^{-4}$
6	9.15	5.20	4.12	3.20	2.20
7	2407	2470	2487	2525	2575
8	$5.34 \cdot 10^{-4}$	$5.34 \cdot 10^{-4}$	$5.34 \cdot 10^{-4}$	$5.34 \cdot 10^{-4}$	$5.34 \cdot 10^{-4}$

Consequently, the length scaling factor is directly related to the scaling factors for α and ν:

$$\varepsilon_L = \frac{L_m}{L_p} = \frac{(\alpha_m \cdot \nu_m)^{\frac{1}{3}}}{(\alpha_p \cdot \nu_p)^{\frac{1}{3}}} \tag{12}$$

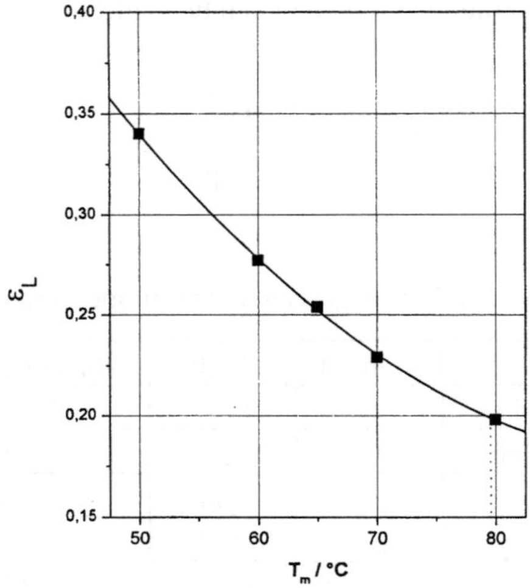

Fig. 3. Length scaling factor as function of model liquid temperature.

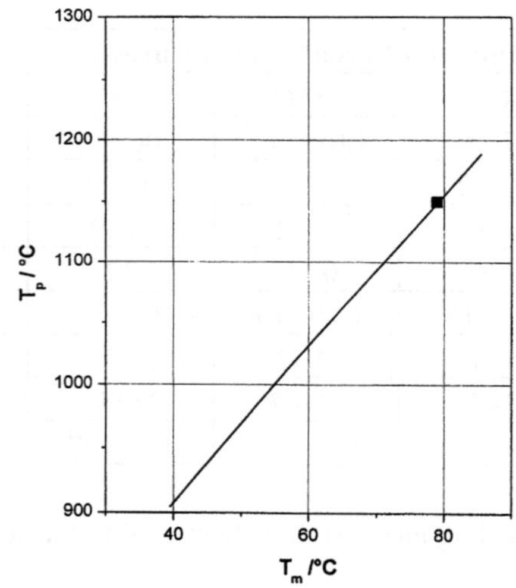

Fig. 4. Relation between glass melt and model liquid temperature.

With equation 12, ε_L-values are calculated using α_p and ν_p for the glass melt at 1150°C and products $\alpha_m \cdot \nu_m$ for the model liquid at various temperatures (Table 2).

From the appropriate graph (Fig. 3), one can determine that for model liquid data at 79°C an ε_L value of 0.2 is achieved. This indicates that a glass melt temperature of 1150°C is equivalent to 79 °C of the model liquid. Together with $\varepsilon_{\Delta T}$, the temperature scale of the model can now be converted into the temperature scale of the prototype melter (Fig. 4). The relationship between the two temperature scales enables us now to check to what extent the temperature dependence of the three physical quantities λ, σ and μ are similar for the glass melt and the model liquid. For this purpose, the normalized figures $q^*=q/q_0$, (data (q), relative to their value at nominal temperature (q_0)) are compared with each other as shown in Fig. 5. They are not equal for both liquids, but close to the nominal temperatures the deviations are small.

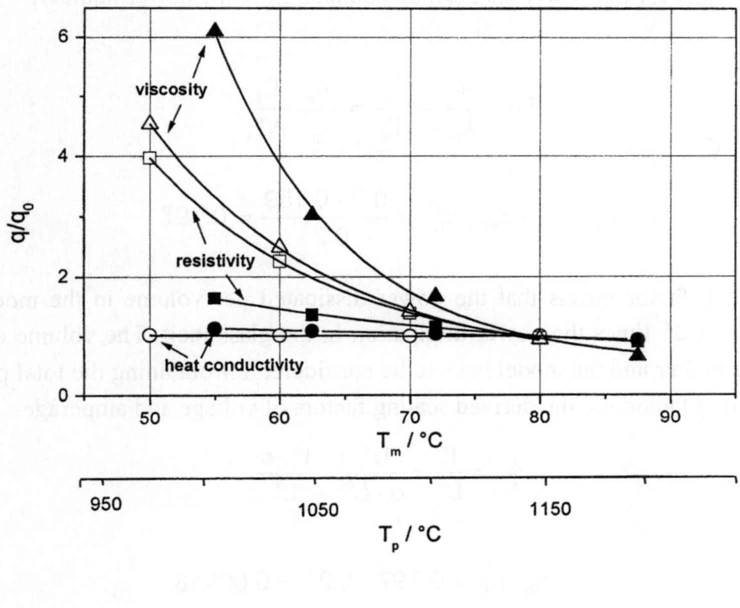

Fig.5. Normalized thermal conductivity λ, resistivity σ and viscosity μ of glass melt and model liquid. Filled symbols: glass melt (prototype), open symbols: model liquid

Other scaling factors can now be easily calculated for various properties using glass data at 1150°C and model liquid data at 79 °C. The equality of the Peclet numbers

$$\mathrm{Pe} = \frac{v_m L_m}{\alpha_m} = \frac{v_p L_p}{\alpha_p} \tag{13}$$

yields the scaling factor for velocity:

$$\varepsilon_v = \varepsilon_\alpha \cdot \varepsilon_L^{-1} = \frac{0.29}{0.2} = 1.48 \tag{14}$$

This means that observed velocities in the model correspond to a $(1.48)^{-1} = 0.66$ fold slower velocity in the glass melt.

Scaling factors for the power data can be obtained from the power number:

$$\mathrm{Po} = \frac{\phi_m \cdot L_m^3}{\lambda_m \cdot \Delta T_m} = \frac{\phi_p \cdot L_p^3}{\lambda_p \cdot \Delta T_p} \tag{15}$$

$$\varepsilon_\phi = \varepsilon_\lambda \cdot \varepsilon_{\Delta T} \cdot \varepsilon_L^{-2} = \frac{0.2 \cdot 0.159}{0.2^2} = 0.797 \tag{16}$$

This scaling factor means that the power dissipated per volume in the model is $(0.797)^{-1} = 1.25$ times the power dissipation in the glass melt. The volume of the prototype melter and the model have to be considered for obtaining the total power input scaling factor and the derived scaling factors of voltage and amperage.

$$\phi = \frac{P}{L^3} = \frac{U^2}{\sigma \cdot L^2} = \frac{I^2 \cdot \sigma}{L^4} \tag{17}$$

$$\varepsilon_p = \varepsilon_\phi \cdot \varepsilon_L^3 = 0.797 \cdot 0.2^3 = 0.00638 \tag{18}$$

$$\varepsilon_U = \varepsilon_\phi^{\frac{1}{2}} \cdot \varepsilon_\sigma^{\frac{1}{2}} \cdot \varepsilon_L = 0.797^{\frac{1}{2}} \cdot 60.5^{\frac{1}{2}} \cdot 0.2 = 1.39 \tag{19}$$

$$\varepsilon_I = \varepsilon_\phi^{\frac{1}{2}} \cdot \varepsilon_L^2 \cdot \varepsilon_\sigma^{-\frac{1}{2}} = 0.797^{\frac{1}{2}} \cdot 0.2^2 \cdot 60.5^{-\frac{1}{2}} = 0.00459 \tag{20}$$

These scaling factors mean that the electrical data measured for the model must be multiplied by 157 for total power, by 0.72 for voltage and by 218 for amperage to yield the corresponding data for the prototype melter. All scaling factors calculated are listed in Table 3.

Table 3: Scaling factors q_m/q_p to convert model data into prototype data

Property	Symbol	Scaling factor
Thermal diffusivity	α	0.282
Thermal conductivity	λ	0.200
Density	ρ	0.516
Kinematic viscosity	ν	0.0288
Electrical resistivity	σ	60.5
Specific heat	C	1.37
Volume extension coeff.	β	6.275
Length	L	0.2
Temperature difference	ΔT	0.159
Velocity	v	1.41
Time	τ	0.142
Power density	ϕ	0.797
Power	P	0.00638
Voltage	U	1.39
Current	I	0.00459

2.6 Steady-state Operation of the Model

For model experiments, about 4.5 l of the model liquid are filled into the melter model and heated with a hot air blower to about 70°C, which corresponds to about 1100 °C of the glass melt. When the temperature is reached, the hot air blower is switched off and direct heating over the five electrode gaps is started. The outer walls of the tank are then cooled using two additional blowers and the surface of the model liquid is cooled by immersion of the cooling plate. The mean coolant temperature is kept at about 0°C, which corresponds to about 670°C in the prototype melter.. After one to two hours, steady-state conditions are achieved with about 70°C in the center of the model and about 700 mA at 100 V at each electrode gap.

2.7 Thermal Balance under Steady-state Conditions

The heat supplied to the model under steady-state conditions can be calculated from the current and voltage data of the five electrode gaps

$$P_{tot,m} = 5 \cdot 0.7 \cdot 100 = 350 \text{ W} \tag{21}$$

This corresponds to the following data of the prototype melter:

$$P_p = \frac{350}{\varepsilon_p} = \frac{350}{0.00638} = 54.8 \text{ kW total power} \tag{22}$$

$$I_P = \frac{0.7}{\varepsilon_I} = \frac{0.7}{0.00459} = 153 \text{ A per electrode gap} \tag{23}$$

$$U_P = \frac{100}{\varepsilon_u} = \frac{100}{1.39} = 72 \text{ V} \tag{24}$$

The total power of 350 W dissipated in the model is carried off via the side walls and bottom and via the cooling plate. Keeping the outer walls of the tank at 20°C by the passing air flow and an inner wall temperature of 65°C, the heat loss via the bath walls and bottom is obtained from the thermal conductivity of plexiglass and the wall thickness according to equation 25.

$$\left(\frac{dQ}{dt}\right)_{W,B} = -\lambda_W \left(F_W \cdot \frac{\Delta T_W}{d_W} + F_B \cdot \frac{\Delta T_B}{d_B} \right) \tag{25}$$

The heat flux over the walls and the bottom is 94 W.

Heat is carried off from the model liquid surface via the cooling plate that is cooled by a flow of liquid coolant. The cooling power is obtained from the flux, the density and the specific heat of the coolant and from the temperature difference between coolant inlet and outlet as:

$$\left(\frac{dQ}{dt}\right)_K = \dot{V}_K \cdot \rho_K \cdot C_K \cdot \Delta T_K \tag{26}$$

This heat flux is 270 W. The total heat loss via the walls, bottom and the surface amounts to 364 W. This corresponds well to the electric power input of 350 W. The heat loss over the surface is about 70 – 75 % of the total heat input. In the prototype melter with liquid feeding more than 50% of the electric power input is spent for the water evaporation so that the total heat loss over the surface will be also close to 70%. This calculation confirms that the boundary conditions for the prototype melter and the model are fulfilled, i.e., the heat loss over the surface in percent is much higher than the heat loss over each side wall and the bottom. Hence, calculations of scaling factors are permissible, and temperature and flow profiles can be transferred from the model to the prototype melter.

2.8 Visualization of Flow

To visualize flow of the model liquid, anthracene particles, being insoluble in and having the same density as the model liquid, are used. Neither floating nor deposition of the anthracene crystals can be observed. For a single bath filling, several 10 mg are applied.

The bath is observed from one of the transverse sides, such that movement of the particles in the xz-plane (Fig. 1) can be seen. For resolution in y-direction, a section of the bath is illuminated from below in a dark room. The light source covers the bath on the entire xz-plane with the y-extension being about 1 cm.

The flow patterns are recorded successively with a y-spacing of 1 cm. A video camera is used to record film sequences from each position. These recordings allow direct observation of the flow patterns, and can be viewed in quick motion. For evaluation and documentation, photos are taken with a still camera with 22 s exposure time. In this case, individual particles blur to tracks with their track length indicating the velocity. To record flow direction and to ensure that the particles studied are located in the illuminated segment over the entire exposure time, exposure is interrupted once in the beginning and twice at the end of the exposure time. As a result, a core exposure time of 10 s is obtained. Usable particle tracks therefore are interrupted once in the beginning and twice at the end. All photos are taken with a telephoto lens to make perspective distortion negligible.

2.9 Simulation of Conductive Particles

To simulate the sedimentation of conductive particles in the prototype melter, a copper plate can be placed in the melter model. To allow the observation of flow under such conditions, the copper plate can be replaced by a copper mesh with a bottom coverage of 84%. Experiments using copper powder for the simulation of the conductive particles are found to be suitable for measuring the power and temperature variations only, as the model liquid becomes opaque by the metal.

3 Application of the Model

3.1 Steady-state Convection Pattern

The melter model is assumed to be in a steady state, when the macroscopic parameters U and I of the individual electrode gaps and the mean temperature no longer change with time. As an example, a steady state flow pattern is recorded in a position close to the corner electrodes (Fig. 6). For that experiment the electrodes, which are positioned opposite to each other, are fired from one supply each (electrode pairs S1/N1, S2/N2 etc., Fig.1) with about the same power of 70 W. The figure shows a number of smaller vortices, which covers only half the bath height. Their size and position vary with time. The character of flow is maintained over time with small vortices continuously forming and dissolving in the bulk of the bath. Only the zone close to the bottom shows invariably a horizontal flow towards the bath center. Using a different electrode circuitry (e.g., with the corner electrodes fired diagonally) makes no difference in the flow pattern. If the power input to the electrode gaps in the lower level (with the pairs N2/S2 and N4/S4) is reduced, the layer thickness in which horizontal flow prevails increases.

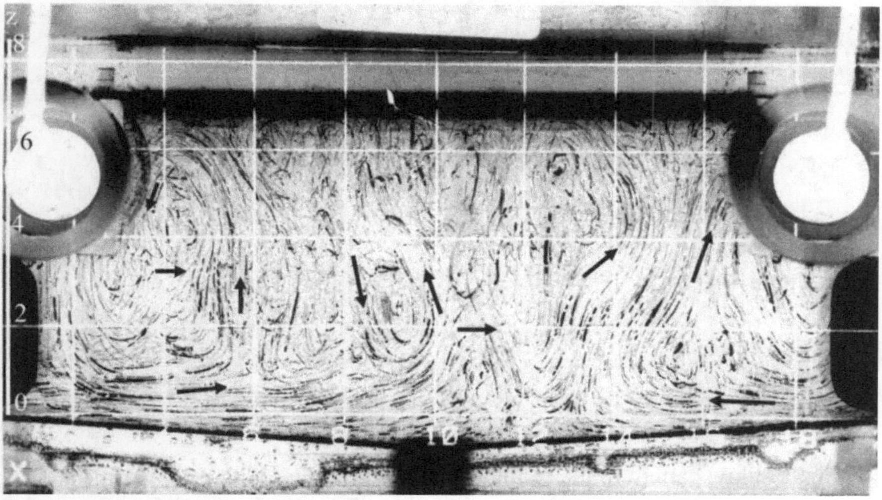

Fig.6. Steady state flow pattern in the model melter close to the corner electrodes. Arrows indicate flow direction.

The flow velocity in zones with horizontal flow is measured in the model melter to be about $3 \cdot 10^{-4}$ ms^{-1}. This corresponds to velocities in the glass melt of about $2 \cdot 10^{-4}$ ms^{-1}. Twice this velocity is observed in the vertical flows with peak values of $2.5 \cdot 10^{-3}$ ms^{-1} in the model ($1.5 \cdot 10^{-3}$ ms^{-1} in the prototype melter). The lowest flow velocity is observed in the center of the bath, near the bottom, where the horizontal flows meet. The results show that the kind of electrode circuitry is less important for good mixing in the course of the vitrification process as far as the electrodes in the lower level are supplied with sufficient power.

3.2 Effect of Conductive Particles on Total Power and Bath Convection

When the melter model is operated at steady state with all electrode gaps fired with 70 W and then disturbed by the insertion of a copper plate to simulate precipitated conductive particles, a dramatic decrease in power input is measured. The power input via the two electrode gaps on the lower level is reduced by about 25%, that of the electrode gaps on the upper level by about 16 %. The effect is even more dramatic if the model liquid is mixed with about 0.22 volume % of copper powder (which is about the same content of conducting particles as is found in radioactive waste glass) instead of inserting a copper plate. Especially the

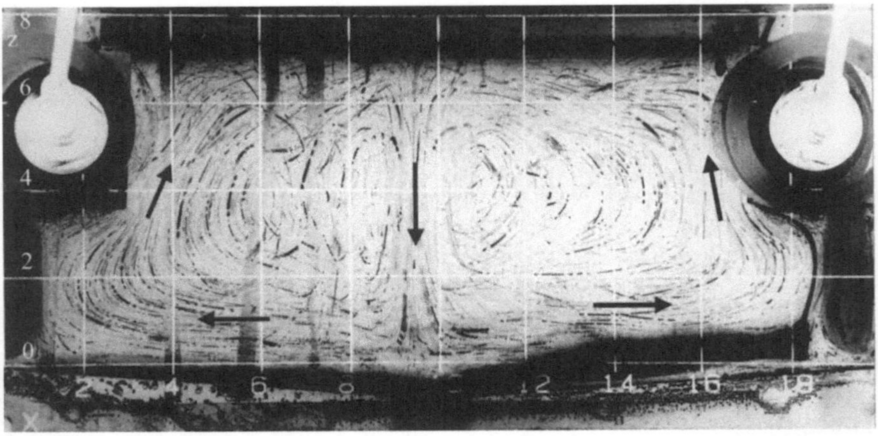

Fig. 7. Flow pattern in the center of the model melter with a copper mesh on the bottom, simulating precipitation of conductive particles. Arrows indicate flow direction

electrode gaps on the lower level show a power input reduction of 50 to 70 %, compared to the undisturbed case. The total power input decreases by about one third.

The effect on the flow pattern is also remarkable. With the copper mesh inserted to allow observation, the flow becomes very homogeneous. Two large flow rolls develop (Fig. 7) with a fast downward flow in the center ($2.2 \cdot 10^{-3}$ ms^{-1} corresponding to $1.5 \cdot 10^{-3}$ ms^{-1} for the glass melt) and a horizontal flow across the bottom towards the side walls. This flow direction is just opposite to that observed under normal conditions, i.e., without simulating conductive particles in the melter. From these observations, one can conclude that in the presence of conductive particles the homogenization of the melt takes more time due to poorer mixing and that the throughput is reduced because of reduced power input

3.3 Temperature Profiles

According to the flow conditions found in the model melter, the temperature in the bath volume cannot be expected to be constant with time at a certain location. This assumption is confirmed by temperature measurements. Fig. 8 shows the temperature behavior at different heights in the bath (z-levels, Fig. 1). Under normal operation condition, at a position near the corner electrodes, the temperature close to the bottom where the flow is horizontal and homogeneous is nearly constant at 65

°C. Up to a height in the model liquid of 50 % total height the temperature is about 70 °C, but there is a fluctuation of about 6K. Up to 75 % of bath height the temperature is also near 70°C, but the fluctuation increases to about 10K. Further up in the bath, close to the cooler, the temperature drops to about 40°C with nearly a 20K scatter. According to the scaling factor, the temperature variations in the glass melt are 6.3 times higher. With the copper plate on the bottom of the melter, creating a homogeneous flow in the bath (Fig. 7) there is a gradual temperature increase up to 75% of the bath height. There is much less temperature fluctuation.

This is especially pronounced in the melter center where a fast downstream flow is observed. Just below the cooler (94 % bath height) there is a sharp temperature drop to about 45 °C with a fluctuation of about 6K (Fig. 8).

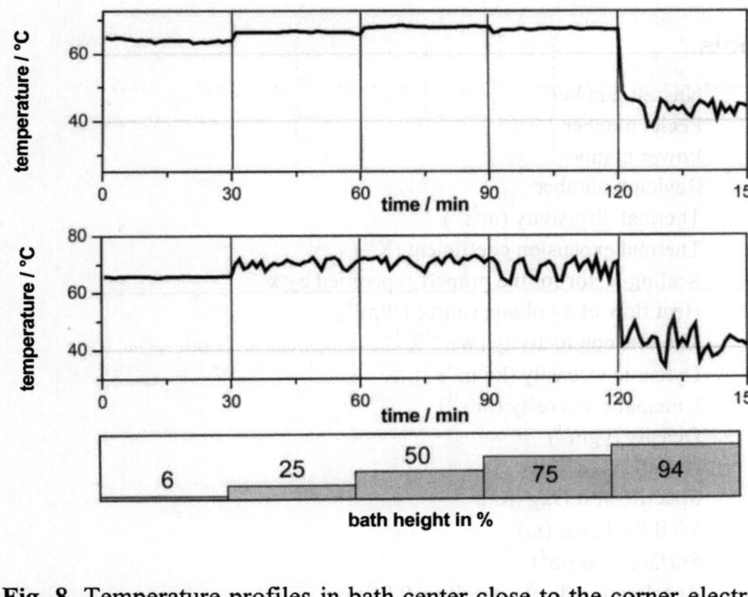

Fig. 8. Temperature profiles in bath center close to the corner electrodes (N1, S1). Upper: Copper plate on the bottom simulating precipitation of conductive particles. Middle: Normal operation. Lower: Bath heights in which temperature was measured over 30 minutes

4 Conclusions

The examples presented for a glass melter show that physical modeling is a very instructive method to study processes which usually cannot be directly observed.

The use of dimensionless numbers is a proved method for transferring properties of the model to the prototype. A general problem exists for complex systems like the glass melter in the selection of a few dimensionless numbers for which matching in the model and prototype system can be achieved. Experience is needed to select those numbers, which describe the 'governing' processes and neglect others which are less important. This methodology implicates always arbitrary components and can never be precise. On the other hand physical modeling is a method, which gives qualitative results within a short time span and requires no sophisticated instruments. Its advantage is to give fast and directly observable response to variations in the operating conditions.

5 Appendix

5.1 Symbols

Nu	=	Nusselt number
Pe	=	Peclet number
Po	=	Power number
Ra	=	Rayleigh number
α	=	Thermal diffusivity (m^2s^{-1})
β	=	Thermal expansion coefficient (K^{-1})
ε_x	=	Scaling factor for the property specified by x
ϕ	=	Heat flow of a volume source (Wm^{-3})
λ	=	Thermal conductivity ($Wm^{-1}K^{-1}$)
μ	=	Dynamic viscosity (kg $m^{-1}s^{-1}$)
ν	=	Kinematic viscosity (m^2s^{-1})
ρ	=	Density (kgm^{-3})
σ	=	Resistivity (Ωm)
C	=	Specific heat ($Jkg^{-1}K^{-1}$)
d	=	Wall thickness (m)
F	=	Surface area (m^2)
g	=	Acceleration due to gravity = 9.81 ms^{-2}
I	=	Electrode current (A)
L	=	Characteristic length (m)
P	=	Power (W)
q	=	Any physical quantity
Q	=	Amount of heat (Ws)
dQ/dt	=	Heat flux (W)
T	=	Temperature (°C)
ΔT	=	Temperature difference (K)
U	=	Electrode voltage (V)

v = Characteristic velocity (ms^{-1})
$\dot{V}$ = Volume flow (m^3s^{-1})
x, y, z = Coordinates in the bath of the melter model

5.2 Indexes

* = Normalized parameter
a = External
B = Bottom
K = Coolant
m = Melter model (model liquid)
o = At nominal temperature (1150°C for glass melt, 79 °C for model)
p = Prototype melter (glass melt)
tot. = Total
W = Wall (material)
x = Any physical property

References

Buckingham E (1937) The Use of Models for Studying the Circulation in Glass Tanks. J. Amer. Ceram. Soc. 20: 1-10

Igarashi H, Ayama Y, Ohuch J (1993) Thermal-hydraulic analysis of molten glass in a Joule-heated melter. 10. International High Level Radioactive Waste Management Conference Proceedings, Volume 2: 1741-1747

Noble JJ, Clomburg LA, Sarofim AF, Hottel HC (1972) Mathematical and Experimental Modeling of the Circulation Patterns in Glass Melts. J. Heat Transfer: 149-154

Quigley MS, Kreid DK (1979) Physical Modeling of Joule Heated Ceramic Glass Melters for High Level Waste Immobilization. PNL-2809

Routt KR (1983) Physical Modeling of a Glass Melter Designed for Vitrification of Defense Waste. In: Advances in Ceramics. 8, Nuclear Waste Management: 536 - 549, The American Ceramic Society Inc. Columbus Ohio

Modelling of Complexation Equilibria

Johannes Lützenkirchen and Bernhard Kienzler

Introduction

Aqueous and interfacial complexation equilibria are of interest for wide areas of science. Speciation of chemical elements, the distribution of an element among solute, interfacial and solid species, strongly affects their toxicity, availability and mobility. This has long been recognised and it is now becoming more and more standard to calculate an element's chemical speciation. Advances in aqueous and interfacial complexation models and their related databases open perspectives towards the application of multi-component reactive transport to estimate the potential hazard of environmental contamination or nuclear and chemotoxic waste repositories. Our main focus is on the latter with particular interest in deep geologic nuclear repositories, where the speciation of radionuclides is a major issue in controlling release of radiotoxic elements from the waste forms and subsequent interaction with engineered and geologic barriers and potential transport into the biosphere.

Up to now, no disposal for highly radioactive wastes is licensed world-wide. Deep underground disposal is considered for these wastes providing concentration and isolation of the hazardous radionuclides as well as prevention against intended and accidental access. Different types of host rock formations are investigated including rock salt (traditionally) in Germany, granite in Scandinavia and Canada, tuff in USA and argillaceous rocks in France, Belgium and Switzerland. For the assessment of the long-term behaviour of the wastes, the presence of water has to be taken into account. Due to its formation by cooling from magmatic melts, granite shows more or less pronounced fractures which might serve as groundwater flow paths. In argillaceous rocks with low permeability water is a constituent of clay minerals and in rock salt, water penetration is considered a possible accidental scenario. The German situation with respect to the storage of nuclear waste is presently characterised by a re-consideration of the available options. For the chemical speciation this means that a wide range of conditions need to be treated, ranging from granitic waters of low salt content to saturated salt solutions. Unfortunately, no comprehensive model is available to cover these conditions with respect to chemical speciation in solution and at interfaces.

The associated risk of a radioactive waste disposal system depends on the quantity of radionuclides mobilised in the case of groundwater (brine) intrusion to the waste. In the presence of water (brines), waste forms may be corroded and radioactive elements are leached. New solid reaction products will be formed, which are thermodynamically favoured under the given geochemical conditions. For performance assessment and long-term safety analyses, model calculations predicting the quantity of dissolved radionuclides in the solutions as a function of time are required. Within the time periods under consideration, the natural environment as well as engineered structures will change significantly and an unavoidable uncertainty regarding the future state or evolution of the repository system has to be taken into account. The mobility of radionuclides strongly depends on the chemical nature and stability of species in groundwater and the geochemical characteristics of various surfaces of minerals, rocks and engineered barrier materials (KIM and GRAMBOW 1999).

In this chapter the focus is on complexation reactions, which are most relevant to aqueous and surface speciation, but do not address interactions of solutes with organic matter. However, organic matter can in principle be either treated as a dissolved ligand or as an adsorbing solid phase within approaches and concepts discussed hereafter.

1 Importance of Chemical Speciation

The mobility of contaminants strongly depends on the speciation, i.e. the distribution of the contaminant among various species (such as in solution, at an interface or within a pure or composite mobile or non-mobile solid phase). Typical mobile species of radionuclides are complex ions or colloids. Key factors for their speciation are solution composition and the respective affinities towards accumulation in or at mobile or immobile solid phases. The tendency towards complex formation increases with pH and carbonate concentration due to the formation of hydroxo and carbonato complexes. Association with mobile colloids, which act as carriers, may increase with the concentration of e.g. organic material in the groundwater.

The oxidation state of many radiolements (so-called redox-sensitive elements) depends on the redox state of the geochemical environment. At the low redox potentials expected in the near field of a radioactive waste repository, the actinide elements (Np, U, Pu) and the fission product Tc may be reduced to the tetravalent state, becoming rather low in solubility but with a high tendency towards colloid formation. Systematic experimental and model investigation of the thermodynamic behaviour of actinides and long lived fission products in the natural environment is in progress.

Experimental and theoretical results relevant for quantifying the mobile species concentration (source term of the radionuclide concentrations) in the near-field of the waste forms are available. Important processes which control the radionuclide release from the waste forms are: (1) kinetics of waste matrix dissolution (leaching), (2) formation of secondary alteration products (controlling thermodynamic solubility), (3) sorption on surfaces in the engineered barrier system and (4) colloid formation. Quantification of these processes requires assessment of the energetics and dynamics of the various reversible and irreversible processes within an overall open non-equilibrium system. Environmental parameters affecting radionuclide release rates are the thermodynamic quantities temperature, pressure and solution and solid phase composition as well as radiation field, fluid volume and flow rates of groundwater and the degree of corrosion of container materials (iron etc.): *Kinetics*: The kinetics of waste matrix dissolution may be described with a general law of irreversible thermodynamics and transition state theory, as applied to mineral surfaces, according to AARGAARD and HELGESON (1982). *Secondary phases*: Geochemical modelling is based on the assumption of a "quasi closed" system including the waste forms and all other components present in the near-field. Parts of these components are the solutions, primary and secondary solid phases, and gaseous components. Due to an effective functioning of the geo-engineered near-field barriers, one can assume that exchange processes between this quasi closed system and the open natural system are limited and controlled by slow diffusive processes. The benefit from this approach is that thermodynamic reaction path models effectively describe the systems under consideration. The formation of new secondary phases, such as silicates, molybdates, uranates, carbonates etc. establishes a new geochemical barrier for re-immobilisation of radionuclides dissolved from the waste matrices. The phases formed are normally not pure but rather solid solutions of quite complicated composition. Solid solution formation is a beneficial effect for retention of radionuclides. *Sorption on surfaces of engineered barriers*: The presence of iron (corroding canister material) during waste form corrosion is expected to effectively reduce the solution concentration of radionuclides. Two effects might account for this finding: low solubilites of most long-lived nuclides at the low redox potential and the formation of corrosion products with high capacities to sorb radionuclides. *Colloid formation*: In the absence of colloid formation, formation of and adsorption onto secondary phases as well as sorption processes onto other available phases would control mobile radionuclide concentrations. However, the presence or generation of colloids, may strongly increase mobile radionuclide amounts in the aqueous phase contacting the waste forms. This is particularly the case for tri- and tetravalent actinides, which normally are sparingly soluble from a thermodynamic point of view.

All the speciation calculations discussed in this contribution are equilibrium calculations, i.e. no kinetics and effects of (partial) irreversibility are taken into account.

2 Speciation and Multi-component Transport

One major phenomenon responsible for the retention of contaminants within or outside a repository is sorption onto immobile sorbents. The term sorption is usually applied to processes responsible for contaminant removal from the aqueous phase and has traditionally been described by a partition coefficient (Kd). The Kd is defined via the following equation:

$$Kd = \{cont._{ads}\}/[cont._{aq}] \tag{2.1}$$

where $\{cont._{ads}\}$ is the amount of contaminant adsorbed (in mol contaminant per kg sorbent) and $[cont._{aq}]$ is the residual aqueous concentration of contaminant (in mol contaminant per liter aqueous phase). In this simple distribution concept no distinction is made between adsorption at interfaces (ion-exchange, surface complexation) and absorption in solids (solid solution formation). Not even true (bulk or surface) precipitation can be excluded since the batch experiments just yield removal of the added amount. Modern surface spectroscopy allows distinction between the various processes and even resolution of structure and bonding. Despite these advances, the Kd concept is widely used in performance assessment, on the one hand because this is a numerically simple approach, and on the other hand because speciation calculations are bound to fail whenever the thermodynamic databases are incomplete with respect to important species. The Kd concept has the huge disadvantages that (1) Kd-values are only valid for the conditions of measurement (2) usually the assumption of a linear sorption isotherm is not verified (it may be assumed to be valid for low contaminant concentration and excess sorbent for true adsorption) (3) a mixture of mechanisms (i.e. adsorption for low concentrations, i.e. at the onset of leaching of the waste versus absorption or precipitation for higher concentrations) cannot be described. In principle many experiments have to be carried out in order to account for variations in conditions and to assure validity of the assumptions inherent to the Kd concept. It would be much more favourable to have a conceptual and comprehensive approach that either allows for calculation of the Kd values (smart Kd values) as a function of the respective conditions or to have the conceptual model directly coupled to flow modules. For more conceptual models it is possible to safely interpolate between or - with the risk of failure - extrapolate beyond the calibration conditions. The latter is always done with the Kd approach when only one condition is studied as is frequently the case.

In the following we give a short overview on complexation in aqueous solution and surface complexation. Finally some applications are discussed for illustration. References to on-line databases and selected speciation tools are given after the conclusions.

3 Complexation in Aqueous Solution

3.1 General

The sum of the occurring or relevant equilibrium reactions describes a given chemical system at equilibrium. Here, we present a simple, classical way of considering aqueous complexation reactions. In that view, an equilibrium reaction yields the relationship between components A, B, C, ... and formed complexes $A_pB_qC_r...$:

$$pA + pB + rC \ldots \rightleftarrows A_pB_qC_r\ldots \quad (3.1)$$

where *p, q, r*,... are the stoichiometric coefficients in the reaction.

This formalism is directly extended to adsorption (surface complexation) or precipitation reactions. The formation (equilibrium) constant for the species formed is given in terms of concentrations by

$$K = [A_pB_qC_r...]\cdot[A]^{-p}\cdot[B]^{-q}\cdot[C]^{-r} \quad (3.2)$$

For new experimental data pqr... and K are fitted to the observations, and the resulting model can be corroborated by additional experiments (in particular via spectroscopy). Depending on the choice of the components the pqr-values may have positive or negative values, depending on whether the component in question is consumed or formed. As a rule, chemically well characterised, preferably mononuclear, species are chosen as components. The number of components is defined as the smallest number of species required to express all other species; i.e. the components are the independent variables used for solving mass balance equations. For example if H^+ is chosen as component, OH^- is considered as a complex (related to H^+ via the autoprotolysis reaction of water, K_w) and vice versa. The combination of the law of mass action for all chemical reactions considered and mass balances for all components allows an iterative solution of the chemical equilibrium problem.

In the systematic study of complexation equilibria it is necessary to characterise sub-systems before passing to multi-component systems. One important subsystem is concerned with the water dissociation. It is in principle a one-component system with usually H^+ as the component and OH^- as a dependent species. The importance of this sub-system stems from the potentiometric titration being one of the most important technique to study complexation reactions. It is generally necessary to account for OH^- ions in the interpretation of potentiometric data. Usually a background electrolyte is used to minimise activity coefficient variations for the solutes of interest. This introduces two more components. These are usually chosen such that no species are formed in the range of pH of interest with proton, hydroxyl and other solutes to be studied. pH is the negative logarithm of proton activity; note that various conventions may be used to calculate proton activity coefficients. System-build-up ideally involves the addition of no more than one component at a time. It also requires the study or knowledge of all sub-systems before studying a comprehensive system with all components. Species formed can be determined from pqr searches, but are best corroborated or even imposed via independent, other methods such as spectroscopy, solvent extraction or solubility measurements.

The concentration ratio in Eq. 3.2 defines a conditional equilibrium constant. Its value is valid only for the ionic medium in which it was determined and strictly speaking only for the conditions (i.e. concentration ranges of A, B, C...) studied. Such a conditional equilibrium constant can in principle be defined and determined for any medium. The nature/composition of the respective species formed in terms of p,q,r... may be inferred from certain (in particular spectroscopic) results and subsequently imposed on or fitted directly to the chemical data such as solvent extraction, solubility measurements or titration data, for which the concentrations and the ratios of A, B, C... have been systematically varied. For titration data alone, the values or p, q, r etc. would be varied until experimental data are best fit. Simultaneous fitting of data from different sources (titration, solvent extraction, solubility and spectroscopic or other data) is possible (INGRI et al. 1996).

The advantage of conditional constants is that no model for activity corrections is required in the calculations. The conditions must be such that variations of activity coefficients over the range of experimental conditions are small and can be achieved by using an "inert" background electrolyte of sufficiently high concentration. This approach is known as the constant ionic medium approach. The disadvantage of these constants is their conditional character. To be useful, thermodynamic constants should be extrapolated to the hypothetical condition of zero ionic strength, where species activities equal concentrations (or mole fractions). In the domain of aquatic chemistry it is common to use concentrations instead of mole fractions.

Extrapolations to zero ionic strength or vice versa application of tabulated zero ionic strength values to desired conditions are complicated by various facts: Many models exists that do not allow the calculation of activity coefficients for any condition (i.e. from very low to high ionic strength). Speciation models and equilibrium constants at zero ionic strength may differ depending on the ion-interaction model databases with respect to the various activity models are hardly ever complete in particular for the models involving specific ion interactions; as an example, in the case of U(VI) one may find e.g. $UO_2Cl_n^{(2-n)+}$ species, whereas the impact of these complexes in the Pitzer formalism is considered via UO_2^{2+}/Cl^- interaction terms.

Many books deal with these issues (e.g. PYTKOVICZ 1979) and therefore a detailed discussion would be beyond this chapter. As the extremes of the available models the classical Debye-Hückel for dilute electrolyte solution and the Pitzer model for brines can be seen. The Debye-Hückel theory relies entirely on the charge of an ion, whereas the Pitzer approach takes into account specific interactions between species. In between these extremes there are among others the extended Debye-Hückel equation, the Davies equation, the B-dot method or the specific interaction theory (SIT), all having their ranges of applicability.

The SIT is currently favoured in applications to nuclear repositories and is described in much detail elsewhere (GRENTHE and PUIGDOMENECH 1997).

We mostly rely on the Pitzer approach (PITZER 1973) for the modelling studies on saturated salt solutions, because it is the only formalism that can be extended to these harsh conditions (although the calibration range is exceeded, i.e. some unverified extrapolation is necessary) and on the Davies equation for adsorption modelling, where ionic strength rarely exceeds 0.5 M.
The Pitzer approach describes the thermodynamics of electrolyte mixtures in such a fashion that ion pair and complex formation among components are considered in case these species are evident. Less important species or species which are not determined explicitly are covered by interaction parameters. This explains differences in databases as compared to the diluted systems.

3.2 Thermodynamic Databases

There are various databases for aqueous solution speciation, ranging from the early one of SILLÉN and MARTELL (1964) over BAES and MESMER for the hydrolysis of cations (1976), to the six volumes of SMITH and MARTELL (1989) and the more current electronic databases of IUPAC (www.acadsoft.co.uk/scdbase/scdbase.htm) and NIST (www.nist.gov/srd/nist46.htm). In most of them all data are compiled without attempting a critical evaluation, whereas in others recommended values are published. Still other databases are built entirely on a theoretical basis with respect to gaps, extrapolating from existing data based on analogies and chemical properties (SHOCK and KORETSKY 1993, SHOCK et al. 1997). Most of these databases are restricted to standard conditions with respect to temperature and pressure, so that environmental conditions are not necessarily included. In all databases, there are substantial gaps, considering knowledge of certain systems as e.g. the aqueous complexes of metal ions with dissolved silica.

Special databases for a series of radionuclides have been started within the nuclear repository research community. The first database was published by GRENTHE et al. (1992) for uranium. Other databases are now available for americium (SILVA et al. 1995), technecium (RARD et al. 1999), neptunium and plutonium (LEMIRE et al. 2001). Previous work is constantly being updated (GUILLAUMONT et al. 2003) and work on other radioelements is in progress or planned within NEA/OECD programs. There are also critical reviews and database fragments available for pentavalent (NECK et al. 1995, FANGHÄNEL et al. 1995), tetravalent (NECK and KIM 2001) and trivalent (NECK and KIM 1998) actinides. Calculations for dilute solutions are either carried out using codes with incorporated databases, with the options to change equilibrium constants (such as MIN-

TEQA2) and to add or to remove species. In such codes the user defines components and the speciation code scans the database and sets up the equilibrium problem in terms of species and associated equilibrium constants. Some difficulties may arise with the first option and have been documented in the literature. Thus for the previous version of MINTEQA2, errors in the database itself were reported (SERKIZ et al. 1996), and worse erroneous readings from otherwise correct database entries have been observed (ONYSKO and MCNEARNY 1997). In other codes no data base is included and the user needs to set up the speciation scheme by himself (e.g. FITEQL).

Compared to the dilute solution, the situation for Pitzer calculations is more complex, since in principle interaction parameters between all entities defined as existing species are required. So adding only one species requires many measurements to estimate these parameters. Gaps in the Pitzer databases are expected to be much more common as compared to dilute solutions.

4 Surface Complexation

4.1 General

The starting point for quantitative surface complexation was work by Schindler (SCHINDLER and KAMBER 1968, SCHINDLER and GAMSJÄGER 1972) and Stumm (STUMM et al. 1972). Since then numerous surface complexation papers have been published. The modelling has in particular advanced since the incorporation of the surface complexation formalism into common speciation codes (MOREL et al. 1982). The whole spectrum of applicability of the theory is discussed elsewhere (STUMM 1992).

Surface complexation is considered as an adsorption process comparable to complexation reactions in aqueous solution. In the case of a complexation reaction of a metal ion a potential ligand in solution would correspond to a surface functional group at the interface. The representation of such surface functional groups is one fundamental problem of surface complexation theory. Determination of type and reactivity of the surface ligands is difficult, not only because in reality different types of sites will exist but also because the surface will include perturbations such as surface or bulk defects creating different reactivities. A major difference compared to the aqueous solution is the electric field that emerges at mineral electrolyte solution interfaces, which affects the adsorption reactions. The complexity of these issues is handled by more or less sophisticated approaches which will be discussed in the following.

4.2 Constituents of Surface Complexation Models

Surface complexation models (SCMs) consist of various components representing some distinct features. Here, we discuss the protonation mechanism of surface functional groups, the surface site heterogeneity, i.e. presence of various distinct surface functional groups with their distinct or continuously distributed reactivity, and interfacial electrostatics, i.e. the structure of the electric double layer, which is usually treated in a mean-field approach. The mean-field approach implies a smeared out potential which disregards phenomena such as discrete ion effects.

The above constituents are inherent to all SCMs except the non-electrostatic models. Non-electrostatic models are inherently incapable of describing mineral surface charge and are therefore not a realistic option. This statement can be based on classical solution chemistry, where the acid-base properties of a ligand have traditionally been considered as the basis for any metal-ligand study. An accurate model for the metal-ligand data requires an accurate model for the ligand acid-base properties. In surface complexation studies the surface functional groups take the part of the ligand and thus an accurate model for the surface acid-base properties should be a requirement. From this point of view mineral acid-base properties have a central importance in surface complexation modelling.

The components of SCMs are best discussed on models for the description of mineral surface charge. This surface charge arises from the (de)protonation reactions of surface functional groups. Protonation of and desorption of hydroxyls from surface functional groups, i.e. the reactions of the charge determining ions at the interface, cannot be distinguished in the macroscopic measurements - typically via potentiometric titrations - since their bulk solution concentrations are related via the ionic product of water.

The basic surface charge model is usually developed for the simplest experimental system involving - in known quantities - an inert electrolyte, the mineral and acid/base. From the measurement of pH or the logarithm of the proton concentration ($pH_c = pc_H = -\log[H^+] = ph$) it is possible via a simple mass balance to calculate the amount of protons ad- or desorbed. For the surface charge model different degrees of simplification are possible. For the available experimental data to be described the simplest model should account for generally accepted tendencies. This includes effects of variation of the experimental parameters, such as ionic strength or background electrolyte composition. Typically sufficiently simple models can be found for most systems and one requirement of such models that can be presently fulfilled is internal consistency, i.e. proper sub-system build-up, c.f. LÜTZENKIRCHEN (2001a, 2001b).

There are numerous reviews of surface complexation modelling by authors interested in colloid chemistry, environmental applications or catalysis (HEALY and WHITE 1978, JAMES and PARKS 1982, WESTALL 1986, SCHINDLER and STUMM 1987, DZOMBAK and MOREL 1990, DAVIS and KENT 1990, STUMM 1992, GOLDBERG 1995, ARDIZZONE and TRASATTI 1996, KOOPAL 1996, BLESA et al. 1997, LÜTZENKIRCHEN 2002a, LÜTZENKIRCHEN 2002b). In the following sections the different components of SCMs are discussed in some detail.

Protonation Mechanism

The protonation mechanism is discussed for "generic" surface sites (SOH^n, n being the charge of the site), which are currently the most frequently used representations of surface functional groups. Opposed to generic sites are "mechanistic" surface sites, which can be justified based on knowledge of surface structure. For generic sites two major approaches exist to describe the acid base behaviour of minerals in aqueous media by a protonation mechanism:

The *2-pK approach* involves two consecutive protonation steps on one generic surface group:

$$SOH_2^+ \leftrightarrow SOH^\circ + H^+ \leftrightarrow SOH^- + 2H^+ \quad (4.1)$$

SOH° is the "generic" surface site (for a one-site model), which can yield or take up a proton. This approach, which has been originally proposed by PARKS (1965), is the conventional protonation mechanism, which is usually mentioned in the major text-books.

The *1-pK approach* is characterised by the $SOH_2^{0.5}$ "generic" surface site, to which one de-protonation step is applied:

$$SOH_2^{+0.5} \leftrightarrow SOH^{-0.5} + H^+ \quad (4.2)$$

The 1-pK formalism was introduced by van Riemsdijk and co-workers (BOLT and van RIEMSDIJK 1982, van RIEMSDIJK et al. 1986, van RIEMSDIJK et al. 1987, HIEMSTRA et al. 1987) and can be interpreted as a simplification of a very complex reality without substantial loss of accuracy compared to the equivalent 2-pK model, but with a significant reduction in adjustable parameters. Thus, for the mineral goethite the 1-pK model can also be interpreted as a special case of a more comprehensive model using mechanistic sites (BOILY et al. 2001).

When related to the comparable reactions in solution the second de-protonation for the $SOH_2^{+0.5}$ species (i.e. the formation of the $SO^{-1.5}$ species) is the analogue of an aquo-group such as H_3O^+. Based on such a comparison, consecutive de-protonation of one functional aquo-unit in solution is typically separated by more than 10 pH units, so that the 1-pK mechanism would be a more reasonable analogue to aqueous solutions than the 2-pK mechanism. In models based on the latter consecutive protonation typically occurs within a too narrow range of pH.

Another major advantage of the 1-pK model is that the logarithm of the equilibrium constant corresponding to the surface chemical reaction is directly related to the measurable point of zero charge (PZC). For the 2-pK model two equilibrium constants must be determined by fitting experimental data. These two pK-values are often chosen to be symmetrical around the PZC, so that a reduction to one adjustable is possible. The PZC can be measured as a function of temperature or predicted by theory[1] (SCHOONEN 1994, LÜTZENKIRCHEN 2003) so that thermodynamic functions are much easier to obtain for the 1-pK approach.

Activity corrections are typically only applied to the aqueous species using some variant of the Davies equation (DAVIES 1962). For the activity corrections for surface species no sound approach exists, which can be verified experimentally. Often, the assumption is made that the activity coefficients of surface species cancel in mass law equations.

Structure of the Double Layer

A second distinction among SCMs is due to the postulated structure of the electrical double layer. The models have been extensively discussed by LÜTZENKIRCHEN (2002a, 2002b), where references to related work can be found.

Various electrostatic models are implemented in standard speciation codes allowing for surface complexation such as FITEQL (1982), MINTEQA2 (1991), ECOSAT (1999) or SOLGASWATER (1979). The structure of the electric double layer defines model inherent charge potential relationships (see figure 5.5). These allow calculation of the respective potentials from a balance over the surface species, which contribute to the charge in the respective planes. Thus the electrostatic effects on the surface chemical equilibria are accounted for in some model dependent fashion. The constant capacitance model (CCM) is a good approximation

---

1 Problems with respect to the determination of PZCs are in all models crucial to obtain reliable model parameters.

at high ionic strength and furthermore compares to the constant ionic medium approach often practised in comparable solution studies.

The other extreme of sufficiently low ionic strength makes the diffuse layer model (DLM) a good approximation. Formation of ion-pairs between charged surface groups and the ions of the electrolyte is usually taken into account in electrostatic models involving more layers such as the Basic Stern model (BSM) or the triple layer model (TLM). The interactions of charged surface groups with the "inert" electrolyte ions is expected to be weaker, more of an electrostatic nature, compared to those with protons. The ion-pairs contribute to the surface charge and are involved in the electroneutrality condition for the particles. This condition involves that the sum of the charges on the particle surface and within the electrical double layer are net zero.

The more elaborated the structure of the interface becomes in a model, the more realistic the model is expected to be within the respective assumptions such as the mean-field approximation. The mean-field approximation considers a smeared-out potential in the respective adsorption planes and disregards such phenomena as discrete ion effects. A multi-layer model would be judged to be more realistic because it includes various phenomenologically expected aspects:

Anion and cation of the electrolyte have different sizes and thus should be placed in different planes; consequently models with more planes of adsorption are required compared to the triple layer model, where anio and cation are placed in one plane of adsorption .

Asorbed ions might have substantial size compared to the proton and charges may be placed in different positions in the surface complex; this is taken into account in the charge distribution (CD) principle. CD becomes more flexible the more planes of adsorption are available.

For high ionic strength the Gouy-Chapman equation is no longer valid. Another drawback is that in most speciation codes the Gouy-Chapman equation is only valid for symmetrical electrolytes (these are z:z electrolytes, like the typical 1:1 background electrolytes such as NaCl). An improvement is possible by including the correct equations and an additional ionic strength loop in the computations, whereas the former aspect is more difficult. Theories exist to account for the ion-ion correlations in the diffuse layer such as the hypernetted chain (hnc) approximation (LARSON and ATTARD 2000). The hnc approach has recently been applied with SCMs (LÜTZENKIRCHEN 2002b). More experimental and theoretical work in this area is required to obtain a sound basis and to come to a more comprehensive parameterisation of the hnc.

Surface Heterogeneity

The above considerations for simple acid-base reactions at an interface invoked only one type of "generic" surface group. In reality, it is expected that several distinct surface groups exist.

This issue may be further diversified by considering distributed properties with multiple sites or/and patches of surfaces, where electrostatics may create even more complications (LÜTZENKIRCHEN and BEHRA 1996). In the following sections discrete and continuous heterogeneity are discussed.

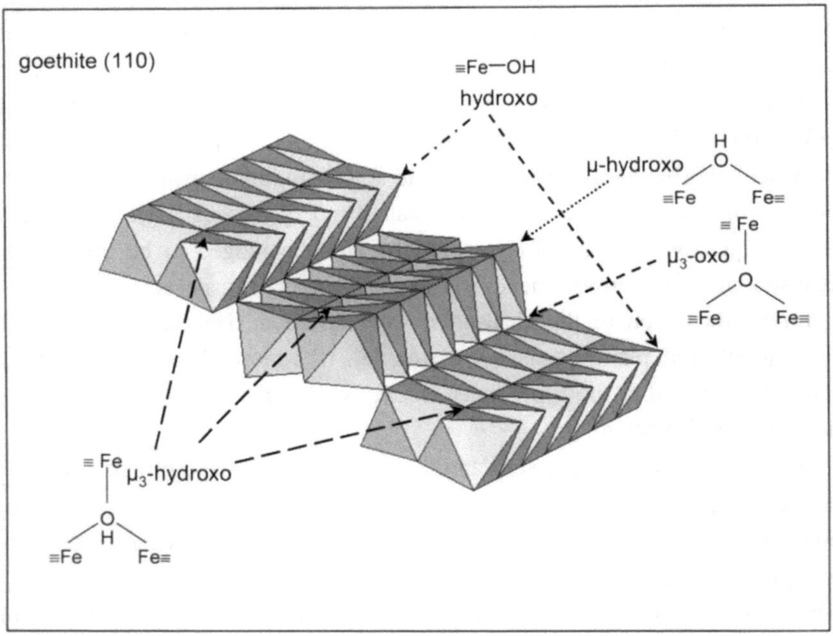

Fig. 4.1. Surface functional groups on the (110) crystal plane of goethite.

Discrete heterogeneity is best discussed with an "ideal" crystal plane of some solid. Such an approach assumes the solid to be free of defects, which might influence the reactivity of the different sites.

Figure 4.1 shows the predominant goethite crystal plane for most synthetic goethite preparations. The individual proton affinity constants for the different sites must be known for each crystal plane of importance. At present two major theoretical approaches have been applied to calculate these affinity constants. van Riemsdijk and co-workers (VENEMA 1997) consider the structure of the sorbent and estimate the proton affinities of the various sites from the bond valence principle. Rustad and co-workers (RUSTAD et al. 1996a, 1996b, FELMY and RUSTAD 1998) apply molecular modelling techniques to obtain proton affinities.

Figure 4.2 shows the surface charge density of the goethite crystal plane calculated with both options. The overall fits to the surface charge curve are equally good, but the contribution of the different groups to the overall surface charge, i.e. the surface speciation, is very different:

With the model parameters from van Riemsdijk and co-workers (VENEMA 1997) only the singly co-ordinated surface (hydroxo) group shows variable pH behaviour (Figure 4.2a). The other mechanistic sites only adjust the overall charge level such that calculated and measured PZC coincide. Not all triply co-ordinated sites behave equally. Thus nominally identical surface groups may have variable affinities to some solute (e.g. to a proton) and it is better to distinguish between them (as in figure 4.1). This may be due to differences in hydrogen bonding properties of e.g. triply co-ordinated hydroxyls on goethite (VENEMA 1997) or caused by defects in the structure of a sorbent close to the surface. Therefore, heterogeneity of surface sites should be considered in terms of "discrete" differences in properties (i.e. different surface sites) and/or "distributed" (i.e. continuous) properties of otherwise identical surface sites (affinity distributions). Two kinds of triply co-ordinated hydroxyls are present. The complexity of this mechanistic approach can be simplified: only one "semi-generic" surface site is used with a proton affinity which corresponds to the overall PZC of this crystal plane. This site is interpreted as being equivalent to the singly co-ordinated group and one third of the triply co-ordinated groups. To what extent this approximation can actually be used in further going adsorption studies has to be decided by the modeller: whenever surface spectroscopy shows that several kinds of surface functional groups are involved in the formation of an important surface complex, the full model would ideally be required in a mechanistic SCM interpretation of such data.

Based on Rustad and co-workers (RUSTAD et al. 1996a, 1996b, FELMY and RUSTAD 1998) a completely different surface speciation is predicted (Figure 4.2b). All surface groups show pH-dependent behaviour. Three types of triply co-ordinated sites are distinguished. One of the surface hydroxyls violates the 1-pK dogma, i.e. two protonation steps of one oxo-group occur within a small pH range. Due to this latter feature , the present authors favour the bond-valence approach inherent to the MUlti SIte Complexation (MUSIC) model.

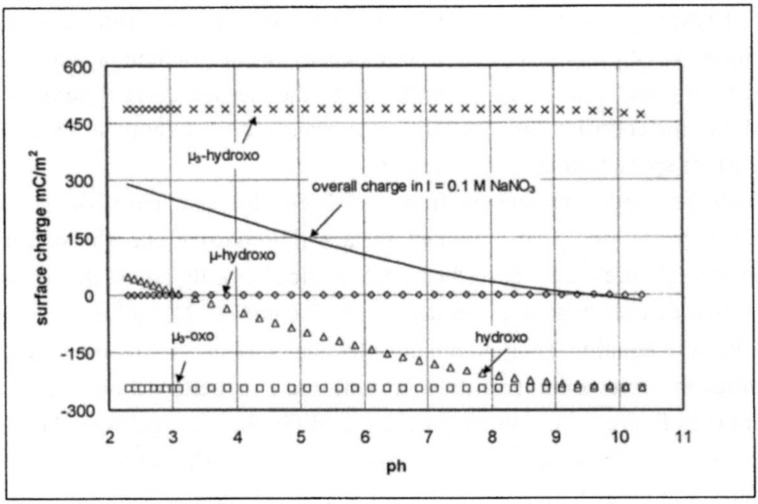

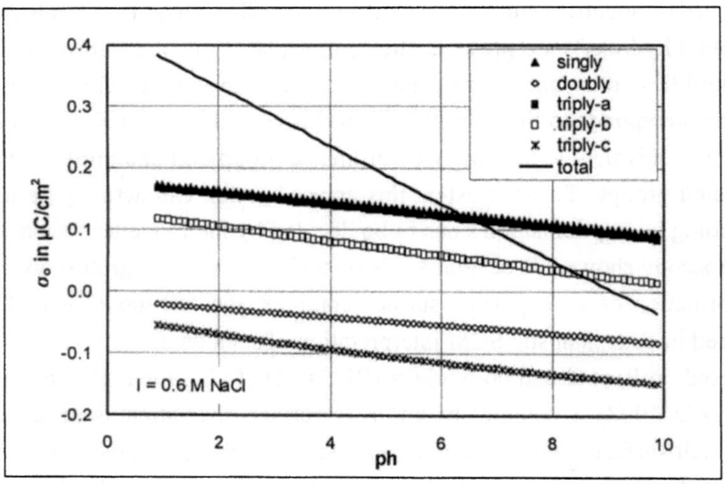

Fig. 4.2. Surface speciation of the (110) crystal plane of goethite as predicted by a) the MUSIC model (bond valence approach, HIEMSTRA et al. 1996) and b) molecular dynamics (RUSTAD et al. 1996a, 1996b, FELMY and RUSTAD 1998). Symbols are the contributions from the individual sites and line is the overall charge.

Some crucial aspects inherent to the two approaches should be addressed: The MUSIC model proposes the interface to be a consequence of the crystallographic structure. For the predictions of proton affinities for sorbents with hydrogen bonds in the structure of the sorbent it is crucial whether or not a hydrogen bond persists at the interface (VENEMA 1997). Furthermore, distances between atoms are affected by surface relaxation causing different distances between atoms at the surface as compared to the respective atoms in the structure. This will affect discrete proton affinities (BOILY et al. 2001). Molecular modelling calculations are typically carried out on clusters which are considered to be representative of the sorbent. In such calculations the methods, the basis sets, the size of the cluster and the residual charge at peripheral atoms may affect the results. Results may be verified by changing the methods, the basis sets and the cluster size. The role of the solvent (water molecules) has been largely disregarded in the evaluation of the reactivity of surface functional groups. In the absence of water molecules gas phase acidities have been obtained and are applied to the acidities in an aqueous medium containing electrolyte ions. Despite such shortcomings, more extensive basis sets and the ever increasing computational power will make these approaches more important. Ultimately they will also allow the consideration of the solvent and electrolyte ions with the advent of ab initio molecular dynamics.

Results from both approaches stress the presence of different functional groups. These are typically considered as distinct entities, which can be described by:

$$\theta_{t,\,solute} = \Sigma_j(f_j \cdot \theta_{j,\,solute}\{K_{j,\,solute},\ [solute]\}) \tag{4.3}$$

Here, [solute] is the molar solute concentration in solution, θ is the surface coverage (adsorbed amount of solute divided by total amount of surface sites), f is the fraction of surface sites of class j, $K_{j,solute}$ is the affinity constant for the local sorption isotherm describing the interaction between the solute and the surface site j; the index t stands for the overall (total) adsorption.

The usual approach for continuous heterogeneity starts from a classical SCM and a distribution of site properties, usually the affinity constant, i.e. surface sites are not considered energetically homogeneous as in the previous equation. Applications have typically been restricted to "generic" surface sites in combination with different electrostatic models or to non-electrostatic models with multiple sites. The following equation describes continuous heterogeneity for one site:

$$\theta_{t,\,solute} = \int_{range}(\theta_{solute}\{K_{solute},\ [solute]\}\,\Xi\{\log K_{solute}\})\,d\log K_{solute} \tag{4.4}$$

The equation involves $\Xi\{\log K_{solute}\}$, a continuously distributed affinity constant of the solute for that site. Integration over the range of affinity constants requires a distribution function for any affinity constant. Combining the previous equations allows for several distinct sites with affinity distributions:

$$\theta_{t,\,solute} = \Sigma_j(f_j \int_{range}(\theta_{j,solute}\{K_{j,solute}, [solute]\} \times \Xi_{j,solute}\{\log K_{j,solute}\})d \log K_{solute}) \quad (4.5)$$

There are some major problems in treating this equation. Information on the presence of different surface groups is often restricted to well defined solids. For powders, which are more frequently used in adsorption studies than well-defined surfaces, the situation is more complicated. The determination of crystal planes becomes a formidable task, and this is even more the case for environmentally relevant samples, such as coatings of reactive minerals on sand. For those, not only the characterisation will be problematic, but also properties may be spatially distributed (e.g. along flow paths). Different affinity distributions can be assumed. At present the determination of affinity distributions from experimental surface charge data involves such questionable assumptions as the absence of electrostatics.

Besides the discrete sites, which may be obvious from the surface structure of the sorbent, defects probably exist on most surfaces to some extent and will result in very reactive sites, which might be of particular importance for the interaction with trace components as long as other significantly competing adsorbing solutes are absent.

Framework of Surface Complexation Theory

The combination of a (de)protonation mechanism of one (or more) surface site(s) with distinct or distributed properties and a model for the electric double layer results in a formalism to calculate intrinsic equilibrium constants for surface species. The term intrinsic refers to a state for which there is no gradient of electrostatic potential between the bulk solution and surface plane(s) relevant for the surface species.

For the simple example of proton adsorption to some generic surface functional group, $\equiv SG^n$, with charge n, the interaction of this group with a proton "close to the surface", H^+_s, can be written as a surface chemical reaction:

$$\equiv SG^n + H^+_s = \equiv SGH^{1+n} \quad (4.6)$$

A simple splitting of the process involves the transfer of the proton from the bulk of the solution to some plane "close to the surface" and the actual reaction of the surface functional group with H^+_s to form a protonated surface group. Since for both the particles and the bulk solution an electroneutrality condition must hold, co-adsorption of one negative charge unit is required. This is considered to result in adherence of an anion from the background electrolyte to the particle. The overall free energy of adsorption for this reaction would be

$$\Delta G_{ads} = \Delta G_{bond} + \Delta G_{elec} \tag{4.7}$$

This separation of intrinsic and electrostatic contributions depends on conditions (i.e. pH, ionic strength for acid-base problems) and on the electrostatic model. A mass law equation can be written as

$$K_+^{int} = [\equiv SGH^{1+n}] \cdot [\equiv SG^n]^{-1} \cdot [H^+_s]^{-1} \tag{4.8}$$

K_+^{int} is the intrinsic equilibrium constant. It is in principle experimentally accessible for $[H^+_s] = [H^+]$ and accounts for ΔG_{bond}. The overall equilibrium constant is given by the following equation and is valid for charged interfaces

$$K_+ = [\equiv SGH^{1+n+}] \cdot [\equiv SG^n]^{-1} \cdot [H^+]^{-1} \tag{4.9}$$

K_+ can be measured if surface site concentration is known. Via the equation

$$[H^+_s] = [H^+] \cdot exp\{-F \cdot \Psi/(R \cdot T)\} \tag{4.10}$$

the proton concentration "close to the surface" and the electrostatic contribution ΔG_{elec} are defined, once the surface potential, Ψ, is known. Ψ is usually calculated from charge potential relationships, which are defined throught the structure of the electrical double layer. The surface charge is calculated from a balance over all surface sites. The Boltzmann-factor represents the energy required for the transfer of one (mol/liter) proton(s), here equivalent to one (mol/liter) charge unit(s), from the bulk solution to the interface. R,T,F are gas constant, absolute temperature and Faraday's constant, respectively.

The above mass law equations are written in terms of concentrations, which is appropriate for the aqueous species in a constant ionic medium. For the proton in solution and at the interface and for the surface functional groups in principle activity corrections are required for extrapolation of K_+^{int} to zero ionic strength. Different authors consider some or all (even solution) corrections to be included in the Boltzmann-factor. A sound thermodynamic treatment suffers from the lack of experimental data on activity coefficients for solutes "close to the surface" and for the surface functional groups. All may vary with pH and ionic strength. The small distances between reacting entities at surfaces as compared to sufficiently dilute bulk solutions result in discrete ion effects and give rise to further deviations from ideal behaviour.

Combining the various options from the three individual constituents of SCMs yields a wide range of model variations (LÜTZENKIRCHEN 2002a, 2002b). Since there are no criteria for selecting a unique electrostatic model and due to the variety of possible electrostatic models database development is hampered. Furthermore comparison of the quantitative interpretation of otherwise comparable systems is difficult.

4.3 Databases

The issue of databases is already quite complex for aqueous solutions because no consistent model and respective database cover the whole range of possible conditions with respect to electrolyte concentration for a wide range of chemical components. The situation is even more complicated for SCMs. For the DLM consistent databases exist for HFO, hydrous ferric oxide, (DZOMBAK and MOREL 1990) or manganese oxide (PRETORIUS and LINDNER 2001). However, it is known that acid-base properties of the important minerals are not well described by the DLM (VENEMA 1997, LÜTZENKIRCHEN 1998). It was also shown that predictions based on linear free energy relationships provided by DZOMBAK and MOREL (1990) completely failed for the adsorption of dissolved silica onto HFO (LÜTZENKIRCHEN 1996). Other approaches can be cited for various models with focus on the TLM such as SVERJENSKY (1993) for cation adsorption or Sverjensky and co-workers (SAHAI and SVERJENSKY 1997, SVERJENSKY and SAHAI 1998) for acid base properties.

All these efforts are at present restricted to 2-pK approaches, although 1-pK approaches are not only simpler but also facilitate the prediction of some temperature dependencies.

5 Application Examples

In the following examples we illustrate complexation equilibria in solution, at interfaces and in dynamic systems. In all of them the proton concentration occurs as one important variable via the pH value. It is important to note that pH may have been calculated based on different conventions. For the dilute solutions the proton activity coefficient is calculated from the Davies equation (DAVIES 1962). For electrolyte concentrations in excess of 0.1 M tabulated values from DZOM-

BAK and MOREL (1990) are used in the adsorption systems. The Pitzer convention in EQ3/6 (1990-1993) was applied for the concentrated salt solutions. Some groups prefer the use of log $[H^+]$ = ph, i.e. a concentration scale, instead of the activity scale (cf. Figure 4.2), most probably best for a constant ionic medium approach.

5.1 Americium in Saturated Salt Solutions

This example is intended to show the importance of chemical speciation and the effects of ionic strength on Am solubility. Figure 5.1 shows the solubility of Americium as a function of the NaCl concentration as calculated with EQ3/6 (1990-1993) using ion interaction parameters and equilibrium constants given by NECK et al. (1998) and $Am(OH)_3$(cr) as the dominating solid. For a constant total Am concentration in the system the ionic strength effects on the solubility product and the ionic product of water cause higher aqueous Am concentrations by two orders of magnitude and thus enhanced Am mobility.

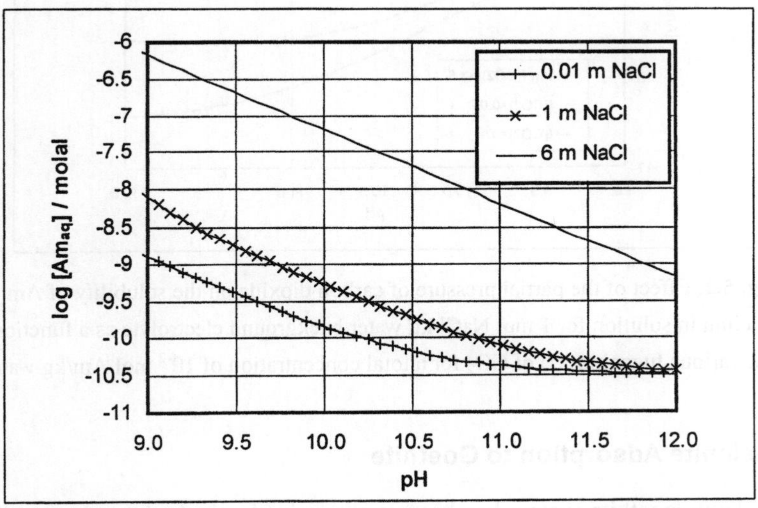

Fig. 5.1. Effect of ionic strength (i.e. NaCl concentration) on the solubility of Am(III) for a total concentration of 10^{-2} mol Am/kg water.

5.2 Influence of Carbon Dioxide on Americium Speciation and Solubility

Figure 5.2 is similar to figure 5.1, but a plot at constant NaCl concetration for various partial pressures of carbon dioxide. The calculations were done using EQ3/6 (1990-1993) and parameters from NECK et al. (1998). Carbonate forms aqueous complexes with Americium and thereby increases its solubility as compared to a carbonate free system. Again the complexation equilibria with carbonate enhance Am mobility and thus the potential release from a nuclear waste repository.

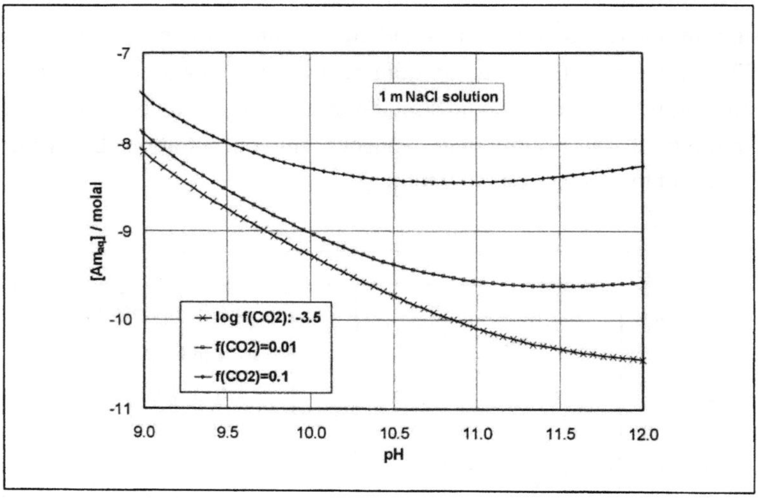

Fig. 5.2. Effect of the partial pressure of carbon dioxide on the solubility of Am(III). Americium in solution for 1 mol NaCl/kg water background electrolyte as a function of pH at various fugacities (f) of CO_2 for a total concentration of 10^{-2} mol Am/kg water.

5.3 Selenite Adsorption to Goethite

The selenite/goethite system has been studied via classical adsorption studies and by surface spectroscopy. The adsorption studies by HAYES (1990) are used to show an ideal modelling approach for macroscopic data. The spectroscopic (EXAFS) results (HAYES et al. 1987) suggest a bidentate surface complex and may have been corroborated by infrared results (DRIFT) on dry samples (SU and SUAREZ 2000).

Subsystem Analysis

Prior to the selenite adsorption model, the acid-base properties of goethite and the aqueous speciation of selenite must be considered. Figure 5.3 shows a model for the surface charge data of goethite (HAYES 1990). A 1-pK BSM was used with two adjustable parameters. Aqueous speciation suggests that SeO_3^{2-} is present in the pH range of the adsorption edges observed where the change from 100 % to 0 % adsorption occurs. Below pH 8 the $HSeO_3^-$ and below pH 4 the H_2SeO_3 species dominate, respectively.

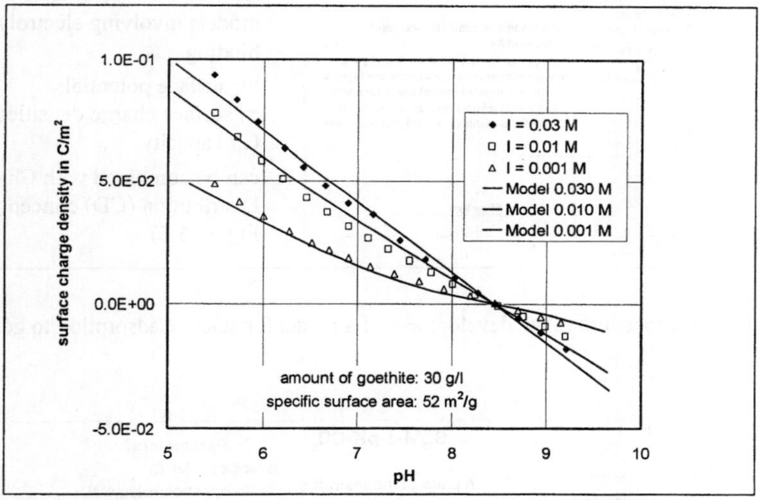

Fig. 5.3. Acid/base model for the goethite titration data published by HAYES (1990).

System Analysis

The present system allows a combination of spectroscopic and macroscopic data to leave only one degree of freedom in the description of the adsorption data, namely the equilibrium constant for the selenite-goethite surface complex. The procedure is summarised in Figure 5.4. Additionally to the bidentate nature, we impose an ideal Charge Distribution (CD), which is explained in Figure 5.5. This is in agreement with suggestions by RIETRA et al. (1999). The model results are shown in figure 5.6. A good fit to the data is obtained with a total of three adjustable. One important feature of this example is that Kd values may for one system easily vary over orders of magnitude for a small variation in pH so that a Kd concept is out of question for variable pH conditions.

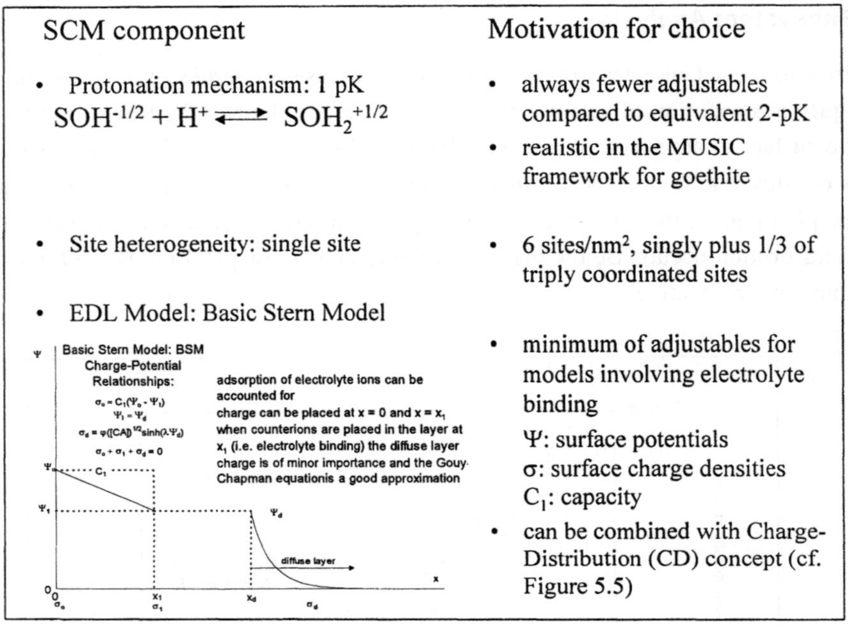

Fig. 5.4. Procedure in the development of a model for selenite adsorption to goethite.

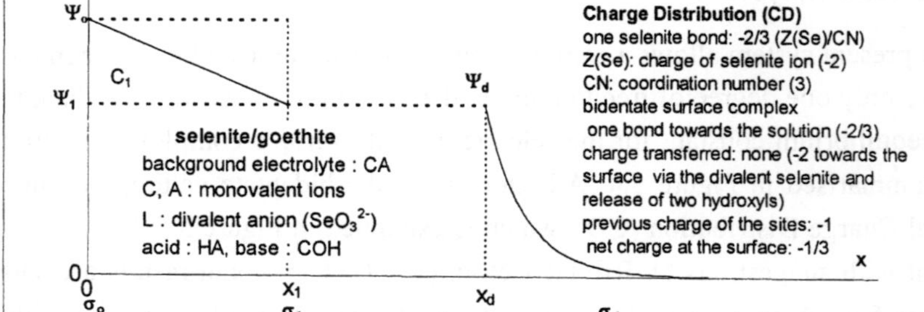

Fig. 5.5. Explanation of charge distribution. The elctrostatic model shows features discussed in the text.

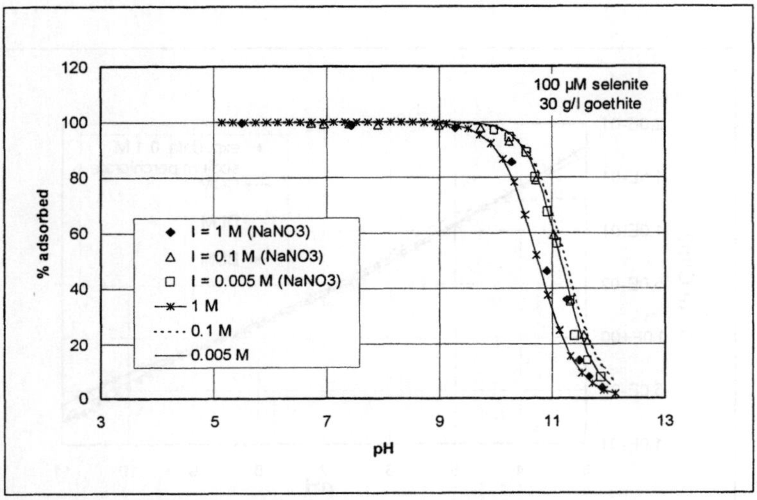

Fig. 5.6. Output of a quantitative model for selenite adsorption onto goethite. Symbols are experimental data (HAYES 1990). Lines are model calculations.

5.4 Coupling of Speciation with Water Flow

This example on the europium/alumina (γ-Al2O3) system shows that description of reactive transport is possible with SCMs. The importance of accurate subsystem analysis will also become obvious. As explained in the previous example Kd may vary strongly for only slight variations of pH. In transport problems, where a pH front runs through the system, the transport of protons is a crucial component. The scenario for the transport calculations is such that the column is equilibrated with Eu solution at pH 6.5 and at t=0 a solution of pH 2.6 is injected. Because of this strong acid pulse, the subsystem (e.g. the underlying acid-base properties of the system) must be accurately described. Only this allows a calculation of the correct pH at any time and place. A small variation in pH may make the difference between, say 20 or 80 % adsorption or release of Eu. For such dynamic problems the description of Eu adsorption alone is not sufficient.

Figure 5.7 shows the application of different generic 2-pK SCMs to acid-base properties of alumina. The CCM parameters were fitted to experimental titration data (RABUNG et al. (1999). Model data were then used to obtain DLM and BSM surface charge parameters. CCM and BSM perform similar, although the BMSThe

DLM clearly does not perform well, which is now widely acknowledged (VENEMA 1997, LÜTZENKIRCHEN 1998).

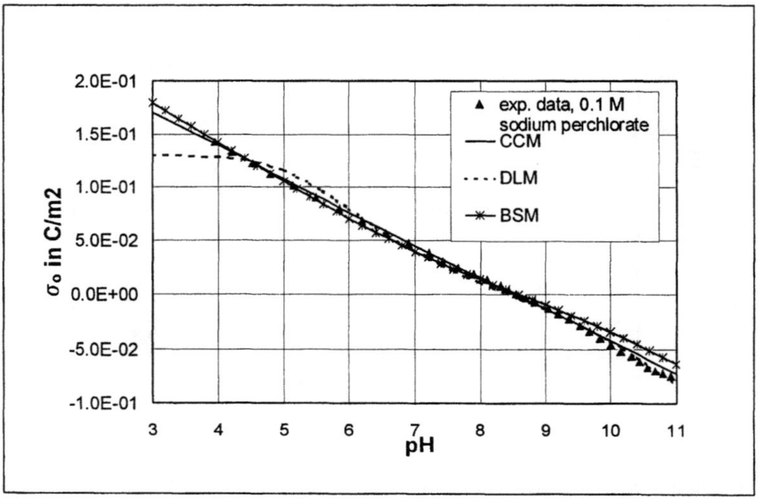

Fig. 5.7. Models for the acid/base properties of γ-Al_2O_3. Experimental data from RABUNG et al. 1999.

The acid-base models are then used to describe Eu adsorption data. The reference model is shown in figure 5.8. The CCM reference model was fitted to experimental data of RABUNG et al. (1999). The calculated CCM data are in all cases taken as the reference. They have been used to determine the parameters for the DLM and the BSM. The other two model variants yield the same goodness of fit for the adsorption data (not shown). The simplest surface species required for a good fit to the experimental or CCM data was chosen for the individual models. Thus in the overall system only the underlying Al_2O_3 acid-base models differ with respect to goodness of fit. The range of Eu concentration is such that the Kd concept would hold for pH = const. But with increasing pH, Kd strongly increases for constant aqueous Eu. This tendency is opposite to that of anions (c.f. figure 5.6).

The SCMs were then applied to a dynamic system. The elution of Eu from a hypothetical column containing alumina as the effective sorbent was predicted for all three model variations.

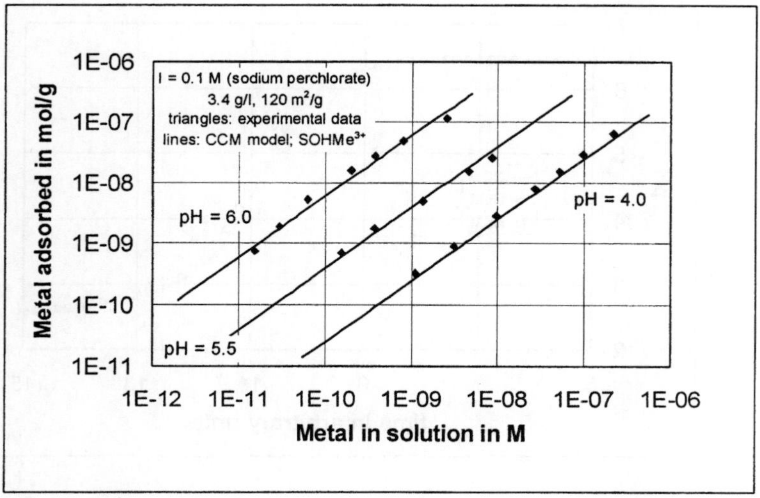

Fig. 5.8. CCM description for the adsorption of Eu onto γ-Al_2O_3. Experimental data from RABUNG et al. 1999.

Figure 5.9 shows the breakghtrough of the pH pulse (figure 5.9a) and the corresponding elution curves for Eu (figure 5.9b). The CCM is again taken as the reference. For all calculations identical initial and boundary conditions and hydrodynamic parameters were used. As stated above, the only difference is the performance of the respective acid-base model. The submodel performance causes a difference in Eu elution. With the bad DLM acid-base description of the sorbent, a retardation of the pH pulse is obtained. This keeps the pH in the column higher and favours Eu retention. Thus, the DLM would predict a contaminant release later than the reference model. As a consequence, not only for database development but also for any individual modelling study the goodness of fit in the underlying subsystems for adsorption data should be stressed. Acceptance of such a requirement would exclude any non-electrostatic model as well as the DLM from modelling systems with potential pH variation, i.e. transport problems with pH fronts.

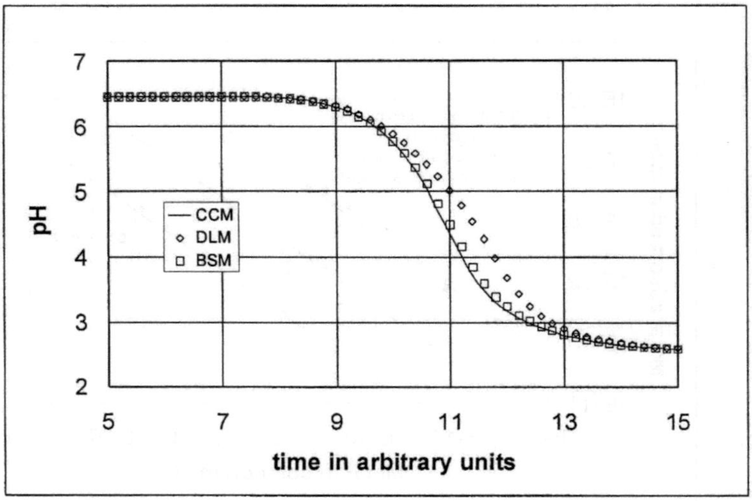

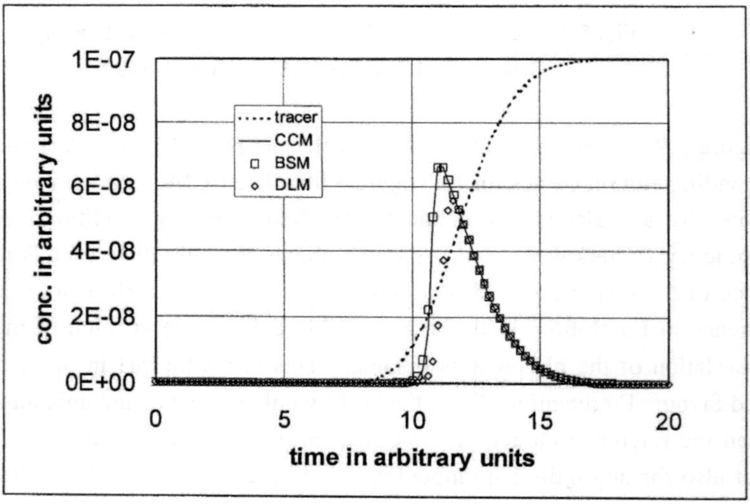

Fig. 5.9. Reactive transport calculations for the H/Eu/γ-Al_2O_3 – system.
a) pH breakthrough front
b) Eu leaching in response to the pH front.
The curve tracer indicates the breakthrough of a non-adsorbing substance.

6 Conclusions

We have addressed formalisms to calculate chemical speciation in aqueous solutions and at interfaces. For most situations a unique speciation formalism is possible with the notable exception of highly concentrated salt solutions. At present no comprehensive formalism exists to cover the extreme situations of very dilute and highly concentrated solutions. With the distinct models available for the individual situations the calculations are restricted by gaps in the thermodynamic data, i.e. comprehensive databases do not exist. SCMs, i.e. (semi) mechanistic adsorption models, can be placed within the same formalism as that for dilute aqueous solutions. Different approaches exist with respect to the treatment of functional group protonation behaviour, heterogeneity and electrical double layer. This makes database development difficult. Existing databases are largely built with a 2-pK, single (and double) site DLM. Unfortunately, this electrostatic model yields the worst performance in the description of the acid-base properties (the mineral subsystem). Some attempts to include higher electrolyte concentrations have focussed on the hypernetted chain approximation, for which more parameters should be determined numerically thus allowing a broader range of applicability.

The application examples have shown how important speciation may be for the estimation of radionuclide release from waste repositories. They also stressed the importance of subsystem understanding and modelling. Without accurate subsystem models, it is very difficult to rely on the overall model. The combination of modern spectroscopic methods with adsorption methods permits modelling close to the mechanistic level. The closer a surface complexation reaction reflects the underlying adsorption mechanism, the more confidence is built in the thermodynamic parameters.

With respect to safety analysis or performance assessment, speciation models are still not in use. On the one hand, the Kd concept is usually much less time consuming in the tedious calculations, while on the other hand, the use of SCMs would require extensive characterisation of potential sorbents as well as the respective initial conditions such as occupancy of surface sites with omnipresent adsorbing components such as carbonate, silica etc. Also the compositions of the waters is often complex and issues such as complexation of radionuclides with omnipresent dissolved silica are not settled. Use of smart Kds might be a compromise allowing for e.g. calculating Kd for respective conditions, and associated uncertainties via the uncertainties given in the thermodynamic databases or in the transient composition of the water. It is certainly a necessary step to leave behind the classical Kd concept, since speciation is one key factor for the mobility of radionuclides and other contaminants. Modelling concepts and tools are available in principle and with only few exceptions the phenomenology can be described. The difficulty to predict situations and conditions as well as nature's diversity and heterogeneity will constrain the application of those concepts and tools.

Codes and Databases

ECOSAT: Keizer MG, van Riemsdijk WH ECOSAT, vers. 4.7, Subdepartment of soil science and plant nutrition, Wageningen University, 1999.

EQ3/6: EQ3/6, A Software Package for Geochemical Modelling Version 7.2a. University of California, Lawrence Livermore National Laboratory. Software produced under the Designated Unclassified Subject Area (DUSA) "Yucca", 1990-1993.

FITEQL: Westall JC, Herbelin A FITEQL3.2, Department of Chemistry, Oregon State University, Corvallis, Oregon, 1995. Westall JC, Herbelin A FITEQL3.1, Department of Chemistry, Oregon State University, Corvallis, Oregon, 1994. Westall JC Fiteql 2.1, Department of Chemistry, Oregon State University, Corvallis, Oregon, 1982.

IUPAC: The IUPAC Stability Constants Database, SC-Database and Mini-SCDatabase, http://www.acadsoft.co.uk/scdbase/scdbase.htm, stability constants database, including 77903 experiments and 17227 references covering literature up to 1993.

MINTEQA2: Allison JD, Brown DS, Novo-Gradac KJ, "Minteqa2/Prodefa2, A Geochemical Assessment Model for Environmental Systems: Version 3.0" US EPA, Athens, GA, 1991. A more recent and verified version exists.

NIST Standard Reference Database 46: NIST Critically Selected Stability Constants of Metal Complexes: Version 6.0, http://www.nist.gov/srd/nist46.htm

References

Aagaard P, Helgeson HC (1982) Thermodynamic and kinetic constraints on reaction rates among minerals and aqueous solutions. 1. Theoretical considerations. Am. J. Sci 281: 237-285.

Ardizzone S, Trasatti S (1996) Interfacial properties of oxides with technological impact in electrochemistry. Adv. Colloid Interface Sci. 64: 173-251.

Baes CF jr, Mesmer RE (1976) The Hydrolysis of Cations, John Wiley, New York.

Blesa MA, Magaz G, Salfity JA, Weisz AD (1997) Structure and reactivity of colloidal metal oxide particles immersed in water. Solid State Ionics, 101-103: 1235-1241.

Boily JF, Lutzenkirchen J, Balmes O, Beattie J, Sjöberg S (2001) Modelling proton binding at the goethite (α-FeOOH)-water interface. Colloids and Surfaces A 179: 11-27.

Bolt GH, van Riemsdijk WH (1982) Ion adsorption on inorganic variable charge constituents. In. Bolt and van Riemsdijk (Ed.) Soil Chemistry. B. Physicochemical models, 2nd ed. Elsevier, Amsterdam.

Davies CW (1962) Ion association, Butterworths, London.

Davis JA, Kent DB (1990) Surface complexation modelling in aqueous geochemistry. Reviews in Mineralogy, 23: 177-279.

Dzombak DA, Morel FMM (1990) Surface complexation modelling – Hydrous Ferric Oxide. Wiley, New York.

Fanghänel T, Neck V, Kim JI (1995), Thermodynamics of Neptunium(V) in Concentrated Salt Solutions. II. Ion Interaction (Pitzer) Parameters for Np(V) Hydrolysis Species and Carbonate Complexes, Radiochim. Acta 69: 169-176.

Felmy AR, Rustad JR (1998) Molecular statics calculations of proton binding to goethite surfaces: thermodynamic modelling of the surface charging and protonation of goethite in aqueous solution. Geochim. Cosmochim. Acta 62: 25-31.

Goldberg S (1995) Adsorption models incorporated into chemical equilibrium models. In Chemical equilibrium and reaction models. SSSA Special Publication 42. Soil Science Society of America, American Society of Agronomy: Madison, 75-95.

Grenthe I, Fuger J, Konings RJM, Lemire RJ, Muller AB, Nguyen-Trung C, Wanner H, Chemical Thermodynamics of Uranium, Elsevier, Amsterdam, The Netherlands, 1992.

Grenthe I, Puigdomenech I (eds.) (1997) Modelling in aquatic chemistry, OECD-NEA, Paris.

Guillaumont R, Fanghänel T, Fuger J, Grenthe I, Neck V, Palmer DA, Rand MH (2003). Update of the Chemical Thermodynamics of Uranium, Neptunium, Plutonium, Americium and Technetium, in preparation.

Hayes KF, Roe AL, Brown GE jr, Hodgson KO, Leckie JO, Parks GA (1987) Insitu X-ray absorption study of surface complexes – selenium oxyanions on α-FeOOH, Science, 238: 783-786.

Hayes KF (1990) Ph.D. thesis, Stanford University.

Healy TH, White LR (1978) Ionizable surface group models of aqueous interfaces. Adv. Colloid Interface Sci. 9: 303-345.

Hiemstra T, van Riemsdijk WH, Bruggenwert MGM (1987) Proton adsorption mechanism at the gibbsite and aluminium oxide solution interface. Netherlands Journal of Agricultural Sciences 35: 281-293.

Hiemstra T, van Riemsdijk WH, Bolt GH (1989) Multisite proton adsorption modelling at the solid/solution interface of (hydr)oxides: a new approach. I. Model description and evaluation of intrinsic reaction constants. J. Colloid Interface Sci. 133: 91-104.

Ingri N, Andersson I, Pettersson L, Yagasaki A, Andersson L, Holmström K (1996). LAKE-a program system for equilibrium analytical treatment of multimethod data, especially combined potentiometric and nuclear magnetic resonance data. Acta Chemica Scand. 50: 717-734.

James RO, Parks GA (1982) Characterization of aqueous colloids by their electrical double layer and intrinsic surface chemical properties. In Surface and Colloid Science; Matijevic, E., Ed; Plenum: New York, pp. 119-216.

Kim JI, Grambow B (1999) Geochemical assessment of actinide isolation in a German salt repository environment. Engineering Geology 52: 221-230.

Koopal LK (1996) Mineral hydroxides: from homogeneous to heterogeneous modelling. Electrochim. Acta 41: 2293-2306.

Larson I, Attard P (2000) Surface charge of silver iodide and several metal oxides. Are all surface Nernstian. J. Colloid Interface Sci. 227: 152-163.

Lemire RJ, Fuger J, Nitsche H, Potter P, Rand MH, Rydberg J, Spahiu K, Sullivan JC, Ullman WJ, Vitorge P, Wanner H (2001). Chemical Thermodynamics Vol. 4. Chemical Thermodynamics of Neptunium and Plutonium. Elsevier, North-Holland, Amsterdam.

Lützenkirchen J (1996) Ph.D. thesis, University of Strasbourg.

Lützenkirchen J, Behra P (1997) A new approach for modelling potential effects in cation adsorption onto binary (hydr)oxides. J. contam. Hydrol. 26: 257-268.

Lützenkirchen (1998) Comparison of 1-pK and 2-pK versions of surface complexation theory by the goodness of fit in describing surface charge data of (hydr)oxides, Environmental Science and Technology 32: 3149-3154.

Lützenkirchen J (2001a) A discussion of the surface complexation modelling in the paper by Sarkar et al. (1999). Soil Sci. Soc. Am. J. 65: 1348-1349.

Lützenkirchen J (2001b) Oxide surface behavior in aqueous electrolytes: An example of inconsistent modelling. J. Colloid Interface Sci. 237: 297-298.

Lützenkirchen J (2002a) Surface complexation models of adsorption, in Encyclopedia of Surface and Colloid Science, Ed. Arthur Hubbard, Marcel Dekker, New York.

Lützenkirchen J (2002b) Surface Complexation Models of Adsorption: A Critical Survey in the Context of Experimental Data, in Adsorption Theory, Modelling, and Analysis, Ed. J Toth, Marcel Dekker, New York.

Lützenkirchen J (2003) Determination of points of zero charge of minerals, experimental and computational approaches, Research trends, in preparation.

Morel FMM, Westall JC, Yeasted JG (1981) Adsorption models: a mathematical analysis in the framework of general equilibrium calculations. In Adsorption of Inorganics at solid-liquid interfaces; Anderson, M.A.; Rubin, A.J., Eds.; Ann Arbor Science: Ann Arbor, pp. 223-245.

Neck V, Fanghänel T, Kim JI (1998) Aquatische Chemie und thermodynamische Modellierung von trivalenten Actiniden. Report FZKA 6110.

Neck V, Fanghänel T, Rudolph G, Kim, JI (1995). Thermodynamics of Neptunium(V) in Concentrated Salt Solutions: Chloride Complexation and Ion Interaction (Pitzer) Parameters for the NpO_2^+ Ion, Radiochim. Acta 69: 39-47.

Neck V, Kim JI (2001). Solubility and hydrolysis of tetravalent actinides. Radiochim. Acta 89: 1-13.

Onysko SJ, McNearny RL (1997) GIBBTEQ: A MINTEQA2 thermodynamic data error detection program, GROUND WATER 35: 912-914.

Parks GA (1965) The isoelectric points of solid oxides, solid hydroxides, and aqueous hydroxo complexes. Chem. Rev. 65: 177-198.

Pitzer KS (1973) Thermodynamics of Electrolytes. I. Theoretical basis and general equations, Journal of Physical Chemistry 77: 268-277.

Pretorius PJ, Lindner PW (2001) The adsorption characteristics of manganese dioxide: a collection of diffuse double layer constants. Applied Geochemistry 16: 1067-1082.

Pytkowicz RM (ed.) (1978) Activity coefficients in electrolyte solutions, 2 volumes CRC Press.

Rabung T, Stumpf T, Geckeis H, Klenze R, Kim JI (2000) Sorption of Am(III) and Eu(III) onto alumina: experiment and modelling. Radiochim. Acta 88: 711-716.

Rard JA, Rand MH, Anderegg G, Wanner H (1999). Chemical Thermodynamics Vol. 3., Chemical Thermodynamics of Technetium. Elsevier, North-Holland, Amsterdam.

Rietra RPJJ, Hiemstra T, van Riemsdijk WH (1999) The relationship between molecular structure and ion adsorption on variable charge minerals. Geochim. Cosmochim. Acta 63: 3009-3015.

Rustad JR, Felmy AR, Hay BP (1996a) Molecular statics calculations of iron oxide and oxyhydroxide minerals: toward a flexible model of the reactive mineral-water interface. Geochim. Cosmochim. Acta 60: 1553-1562.

Rustad JR, Felmy AR, Hay BP (1996b) Molecular statics calculations of proton binding to goethite surfaces: a new approach to estimation of stability constants for multisite surface complexation models. Geochim. Cosmochim. Acta 60: 1563-1576.

Sahai N, Sverjensky DA (1997) Solvation and electrostatic model for specific electrolyte adsorption. Geochim. Cosmochim. Acta 61: 2827-2848.

Schindler PW, Gamsjäger H (1972) Acid-base reactions of the TiO_2 (Anatase)-water interface and the point of zero Charge of TiO2 suspensions. Kolloid-Z. u. Z. Polymere 350, 759-763.

Schindler PW, Kamber HR (1968) Die Acidität von Silanolgruppen. Helvetica Chimica Acta 51: 1781-1786.

Schindler PW, Stumm W (1987) The surface chemistry of oxides, hydroxides and oxide minerals. In Aquatic Surface Chemistry: Chemical processes at the particle-water interface; Stumm, W., Ed.; Wiley: New York.

Schoonen MAA (1994) Calculation of the point of zero charge of metal oxides between 0 and 350 °C, Geochim. Cosmochim. Acta, 58: 2485-2493.

Serkiz SM, Allison JD, Perdue EM, Allen HE, Brown DS (1996) Correcting errors in the thermodynamic database for the equilibrium speciation model MINTEQA2, Wat. Res. 30: 1930-1933

Sillén LG, Martell AE (1964) Stability Constants, Special Publication No. 17, The Chemical Society, London.

Shock EL, Koretsky CN (1993) Metal-organic complexes in geochemical processes: calculation of standard partial molal thermodynamic properties of aqueous acetate complexes at high pressures and temperatures, Geochim. Cosmochim. Acta, 57: 4899-4920.

Shock EL, Sassani DS, Willis M, Sverjensky DA (1997) Inorganic species in geologic fluids: correlations among standard molal thermodynamic properties of aqueous ions and hydroxide complexes, Geochim. Cosmochim. Acta, 61: 907-950.

Silva RJ, Bidoglio G, Rand MH, Robouch P, Wanner H, Puigdomenech I (1995). Chemical Thermodynamics Vol. 2., Chemical Thermodynamics of Americium. Elsevier, North-Holland, Amsterdam.

Smith RM, Martell AE (1989) Critical Stability Constants, vol. 6, Plenum, New York.

Stumm W (1992) Chemistry of the solid-water interface. Wiley: New York.

Stumm W, Huang CP, Jenkins SR (1970) Specific Chemical Interactions Affecting the Stability of Dispersed Systems. Croatica Chemica Acta 42: 223-245.

Su CM, Suarez DL (2000) Selenate and selenite sorption on iron oxides: an infrared and electrophoretic study. Soil Sci. Soc. Am. J. 64: 101-111.

Sverjensky DA (1993) Physical surface-complexation models for sorption at the mineral-water interface. Nature 364: 776-780.

Sverjensky DA, Sahai N (1998) Theoretical prediction of single-site enthalpies of surface protonation for oxides and silicates in water, Geochim. Cosmochim. Acta, 62: 3703-1715.

van Riemsdijk WH, Bolt GH, Koopal LK, Blaakmeer J (1986) Electrolyte adsorption on heterogeneous surfaces: adsorption models. J. Colloid Interface Sci. 109: 219-228.

van Riemsdijk WH, de Wit JCM, Koopal LK, Bolt GH (1987) Metal adsorption on heterogeneous surfaces: adsorption models. J. Colloid Interface Sci. 116: 511-522.

Venema P, Hiemstra T, van Riemsdijk WH (1996) Comparison of different site binding models for cation sorption: description of pH dependency, salt dependency and cation-proton exchange. J. Colloid Interface Sci. 181: 45-59.

Venema P (1997) Ph.D. thesis, Wageningen University, The Netherlands.
Westall JC (1986) Reactions at the oxide-solution interface: Chemical and electrostatic models. Am. Chem. Soc. Symp. Ser. 323: 54-78.

Weather Prediction by Numerical Modelling of Atmospheric Processes

Gerhard Adrian and Detlev Majewski

Introduction

The prediction of the future state of the atmosphere made considerable progress during the recent decades: A five-day forecast of the flow pattern over Europe now has the same accuracy as a one-day forecast 40 years ago. This improvement in the quality of weather forecasts is mainly due to the introduction of numerical weather prediction models at many meteorological services worldwide. Nowadays these models form the basis of all forecast products like general prediction maps for a whole country or detailed information for a small region. The accurate prediction of high-impact weather, for example heavy precipitation, freezing rain or gale-force winds, is the main task of national weather services, and numerical models are indispensable tools. Thus numerical weather prediction is a main activity of all modern weather services.

Moreover, numerical weather prediction models provide the meteorological input data for a wide range of application models: The wind field is used in sea wave prediction models, the transport of pollutants is computed by dispersion models, a UV-B index calculation procedure provides the public with information about the impact of solar radiation on the skin, and other models predict plant diseases or different agrometeorological parameters.

All numerical models of the atmosphere are based upon the same set of governing equations which can be derived from fundamental physical principles like the conservation of mass, momentum and energy. These Euler Equations describe the temporal evolution of principal prognostic variables like temperature, pressure, density and wind components, which characterise a fluid (or gas) subjected to some external forces. The same set of equations which have been described by the Swiss mathematician Leonhard Euler in the mid 18th century form the basis of all computational fluid dynamics (CFD) models, too. For atmospheric applications, certain extensions and modifications of this general set of equations have to be introduced. Firstly, the equations are written in a rotating coordinate frame to take into account the rotation of the earth. Secondly, water in all its phases, namely vapour, liquid and frozen, plays an important role in many atmospheric processes and is an important forecast parameter (e.g. clouds and precipitation), too. Therefore conservation laws for these water phases have to be added to the "dry" Euler equations.

Numerical weather prediction (NWP) models differ in the approximations and assumptions made in the applications of these original equations, the numerical algorithms employed to solve the equations, and in the way physical processes like radiation or turbulence are handled. Most NWP models assume that the atmosphere is a rather thin layer of air of about 40 to 60 km depth around the earth which has a mean radius of some 6370 km. Modelling the atmosphere with the aim of weather forecasting is rather different from a typical CFD problem like the simulation of the airflow around the wing of an aircraft, mainly due to four reasons:

- Atmospheric processes cover spatial and temporal scales of several orders of magnitude ranging from planetary waves with a wave length of 10,000 km or more down to micro-scale turbulence of 1 mm scale around a leaf. The temporal scales range from several weeks to less than a second.
- Many meteorological phenomena like deep convective clouds are strongly linked to water in all its phases, and the latent heat released during condensation of water vapour forms an important source of energy for atmospheric systems.
- External forcing like the radiation from the sun and the interaction with the underlying surface, play a significant role, and
- The accurate determination of the initial state has a crucial impact on the quality of the forecast.

Thus a numerical model of the atmosphere for NWP purposes consists of several components (Fig. 0.1.) which must all interact with another during the forecasting process. Physical parameterization is the numerical representation of a physical process like radiation or turbulence in an atmospheric model. Many of these processes occur in the atmosphere on a scale too small to be directly resolved by the model. The dynamics of an NWP model are the governing equations without the forcing terms due to the physical parameterization.

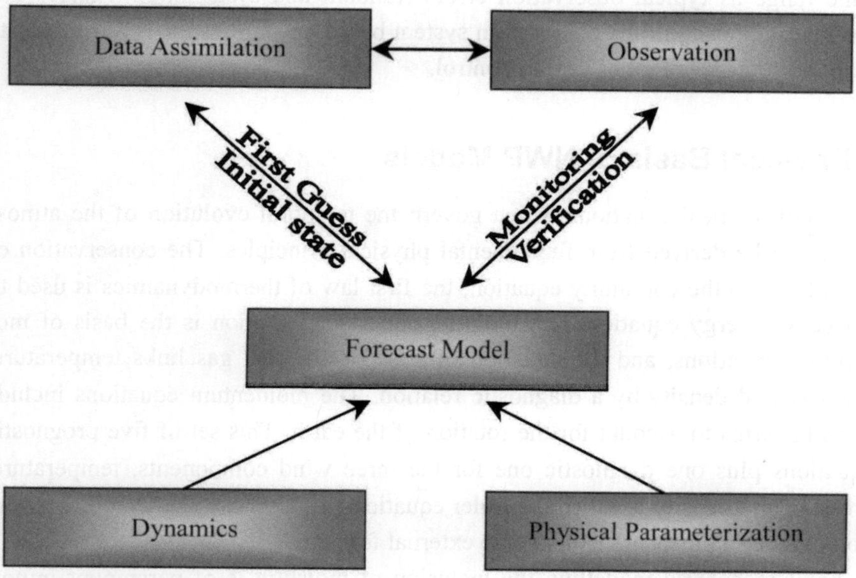

Fig. 0.1. Main components of an NWP model

Section 1 and 2 describe the dynamics, physical parameterizations and numerical methods employed in the regional NWP model LM developed at Deutscher Wetterdienst (DWD). At DWD two NWP models, a global one called GME, and a local one, called LM, are delivering operational forecasts several times the day. The role of the abundant observations of the atmosphere and the underlying surface which are taken all over the world and from space, is twofold. The initial state of each model forecast run is determined by a process called data assimilation (see section 3), which merges the information from the model first guess with the observations in some optimal way. Secondly, observations serve as the ultimate benchmark for the quality of the NWP model. Thus at first glance the verification of NWP models (see section 4) seems to be straightforward because there's an almost infinite number of observations against which the model results could be compared. But since the quality of an NWP model crucially depends on the initial state of the model as well as on the formulation of the NWP model itself, a clean separation of the forecast errors into two parts, one depending on the initial state, and the other determined by the NWP model alone, is hardly ever possible. Moreover, short-range forecast errors of a state-of-the-art NWP model are in the same range as typical observation errors (random and systematic). Therefore a careful monitoring of the observation system based on short-range NWP forecasts is an indispensable tool of quality control.

1 Physical Basis of NWP Models

The mathematical equations which govern the temporal evolution of the atmosphere can be derived from fundamental physical principles. The conservation of mass leads to the continuity equation, the first law of thermodynamics is used to derive an energy equation, Newton's second law of motion is the basis of momentum equations, and the equation of state of an ideal gas links temperature, pressure and density by a diagnostic relation. The momentum equations include Coriolis terms to account for the rotation of the earth. This set of five prognostic equations plus one diagnostic one for the three wind components, temperature, pressure and density is called the Euler equations. It describes the temporal evolution of a dry atmosphere under some external forcings.

For atmospheric modelling the inclusion of moisture is of paramount importance. Current operational NWP models differ significantly with respect to the level of complexity in the treatment of moisture. Some models take only water vapour into account whereas others include liquid (cloud water) and frozen (cloud ice) water phases, too. Sophisticated research models treat rain, snow and graupel particles prognostically. Even more complex atmospheric models which aim to simulate the temporal evolution of individual clouds, include additional equations for the size distribution of cloud droplets, ice, rain and snow particles. Recently, certain important chemical constituents of the atmosphere, like ozone, are treated in some models like the global forecast model GME as prognostic variables with additional budget equations as well.

The "Lokal-Modell" (local model LM) of DWD (Doms and Schättler, 1997) is based on the following set of coupled partial differential equations for three wind components (u, v, w), temperature (T), pressure (p), density (ρ), the specific contents of water vapour (q_v), cloud liquid water (q_c), and cloud ice (q_i):

$$\frac{\partial u}{\partial t} + v \cdot \nabla u - \frac{uv}{a} \tan\varphi - fv = -\frac{1}{\rho a \cos\varphi} \frac{\partial p}{\partial \lambda} + M_u \qquad (1.1)$$

$$\frac{\partial v}{\partial t}+\mathrm{v}\cdot\nabla v+\frac{u^2}{a}\tan\varphi+fu = -\frac{1}{\rho a}\frac{\partial p}{\partial \varphi}+M_v \tag{1.2}$$

$$\frac{\partial w}{\partial t}+\mathrm{v}\cdot\nabla w = -\frac{1}{\rho}\frac{\partial p}{\partial z}-g+M_w \tag{1.3}$$

$$\frac{\partial \mathrm{T}}{\partial t}+\mathrm{v}\cdot\nabla\mathrm{T} = \frac{1}{\rho c_{pd}}\left(\frac{\partial p}{\partial \mathrm{t}}+\mathrm{v}\cdot\nabla p\right)+Q_T \tag{1.4}$$

$$\frac{\partial p}{\partial t}+\mathrm{v}\cdot\nabla p = -\left(c_{pd}/c_{vd}\right)pD+\left(c_{pd}/c_{vd}-1\right)\rho c_{pd}Q_T \tag{1.5}$$

$$\rho = p\left\{R_d\left(1+\left(R_\upsilon/R_d-1\right)q_v-q_c-q_i\right)T\right\}^{-1} \tag{1.6}$$

$$\frac{\partial q_v}{\partial t}+\mathrm{v}\cdot\nabla q_v = -\left(S_c+S_i\right)+M_{q_v} \tag{1.7}$$

$$\frac{\partial q_c}{\partial t}+\mathrm{v}\cdot\nabla q_c-\frac{1}{\rho}\frac{\partial P_c}{\partial z} = S_c+M_{q_c} \tag{1.8}$$

$$\frac{\partial q_i}{\partial t}+\mathrm{v}\cdot\nabla q_i-\frac{1}{\rho}\frac{\partial P_i}{\partial z} = S_i+M_{q_i} \tag{1.9}$$

where λ is longitude (zonal direction), φ is latitude (meridional direction) of an orthogonal spherical horizontal coordinate system, z is the geometrical height above mean sea level (vertical direction), t is time, $\mathrm{v}\cdot\nabla$ is the three-dimensional advection operator, a is the radius of the earth (assumed to be a sphere of mean radius a = 6371229 m), f is the Coriolis parameter, g is the constant mean value of absolute gravity acceleration, c_{pd} is the specific heat of dry air at constant pressure, c_{vd} is the specific heat of dry air at constant volume, R_d is the gas constant for dry air, R_v is the gas constant for water vapour, M_x is the source term due to turbulent mixing of quantity x, D is the three-dimensional divergence of the wind field, Q_T is the diabatic heating term, and $S_{c,i}$ is a hydrological source term.

For atmospheric modelling an external forcing like solar radiation and the interaction of the flow with the underlying surface plays a dominant role for a realistic simulation of weather systems. Operational NWP models have to predict especially near-surface weather conditions properly because people want to have a forecast of maximum and minimum temperatures, wind, gusts, rain and snow near the ground for several days ahead. Therefore NWP models contain separate modules like soil and surface models which predict temperature and moisture in the ground as well as at the surface. The temporal evolution of snow cover on the surface is also an important forecast parameter. Finally, vegetation and its significant impact on the flow near the ground have to be taken into account.

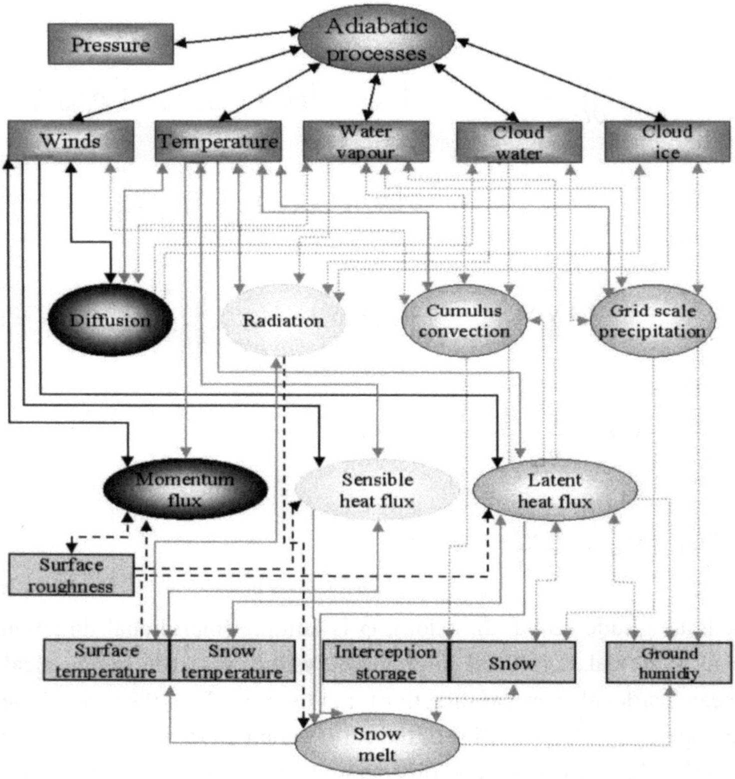

Fig. 1.2. Main prognostic variables and physical processes of the local model LM

Many of these physical processes occur on a (spatial) scale too small to be directly resolved by the NWP model. They are represented by so-called physical parameterizations. Fig. 1.2. gives an overview of the main prognostic variables (in rectangular boxes) and physical parameterizations (ellipses) and their interactions (arrows) included in LM. There are many positive and negative feedback loops visible which make it very hard to anticipate the result of a change in a certain component of the model.

The following physical phenomena are simulated by dedicated parametrization modules in LM:

- Radiative transfer of solar (wave length: 0,25 to 4,6 μm) and thermal (wave length: 4,6 to 104,5 μm) radiation in clear and cloudy atmospheres (Ritter and Geleyn 1992). The shortwave radiation from the sun is energy input in the earth-atmosphere system. At the top of the atmosphere, the net solar radiation input is balanced globally by thermal radiation of the earth; but the differential heating with an average gain in tropical regions and a loss in the extra-tropics is the source of atmospheric motion systems which aim to reduce the temperature difference between equator and poles which is caused by the radiation.
- Deep and shallow convection based on a mass flux approach (Tiedtke 1989). Since most of the solar heating occurs at the earth surface, the atmosphere is heated from below. Thus the stratification in the vertical direction may become unstable giving rise to thermal plumes and convective clouds of different horizontal and vertical extension (between 500 m and 15 km in both directions). Non-precipitating fair weather cumulus clouds, precipitating showers, thunderstorms, squall lines and tornados are examples of such atmospheric motion systems which transport heat and moisture from the surface to higher levels of the atmosphere.
- Vertical turbulent fluxes (Müller 1981), based on Louis (1979) in the Prandtl layer close to the earth surface, and a diagnostic level-two scheme based on Mellor and Yamada (1974) for the boundary layer and the free atmosphere. The vertical fluxes of heat, moisture and momentum at the surface of the earth occur due to turbulent eddies at very small horizontal and vertical scales. These eddies are responsible for the gustiness of the wind or visible as dust devils in hot deserts.
- Cloudiness is derived from specific cloud liquid water and cloud ice contents, relative humidity, convective activity and stability. Larger scale systems like frontal clouds or cloud clusters of several hundred kilometer extension may also give rise to precipitation which is simulated based on parameterized cloud microphysics (Doms and Schättler 1997).
- Finally, a surface and soil model (Jacobsen and Heise 1982) describes the conditions at the ground and in the earth. The vertical temperature and moisture fluxes in the ground control to a large extent the near-surface weather

conditions. The inclusion of vegetation is important due to evaporation of soil moisture from deep layers in the ground which helps to cool the air close to the surface and is a significant source of water vapour in the atmosphere.

Each parameterization module consists of a set of equations linking the prognostic variables of the NWP model with fluxes which will appear on the right-hand-side of the prognostic model equations as source terms. Many modules cannot be derived rigorously from fundamental physical principles but contain some "tuning" constants which have to be fixed by observational data and process studies.

2 Numerical Solution of Continuous Equations

Equations 1.1 to 1.9 form a closed set of coupled nonlinear partial differential equations which can be solved analytically only for highly simplified, "academic" flow situations. For real weather prediction purposes a numerical solution of these equations has to be derived by approximating the analytical form by an algebraic one.

2.1 Spatial Discretization

To transform the model equations into a set of algebraic relations a three-dimensional spatial grid is introduced. In the horizontal direction (Fig. 2.1.), the grid interval is $\Delta\lambda$ in zonal and $\Delta\varphi$ in meridional direction.

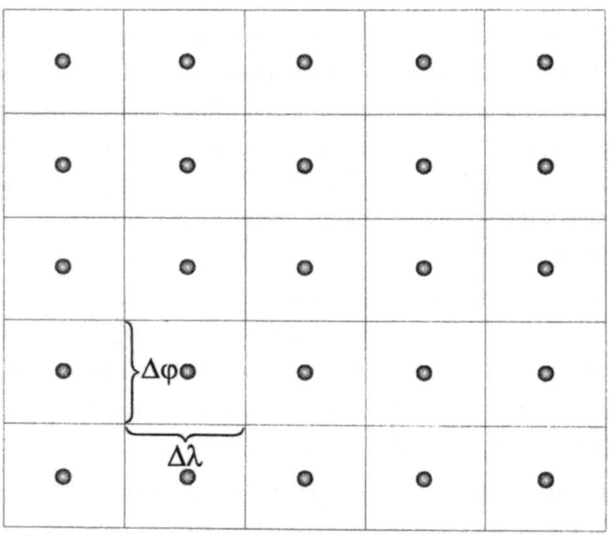

Fig. 2.1. Horizontal grid of LM, grid points are represented by dots

For $\Delta\lambda = \Delta\varphi$ the grid consists of square grid boxes where each box covers an area $F = a^2 \cos\varphi\, \Delta\lambda\, \Delta\varphi$. In the case of LM with $\Delta\lambda = \Delta\varphi = 0.0625°$ the average grid box area is 49 km^2, for GME which has a coarser grid the cell area is approximately 3100 km^2. In the vertical, a terrain following coordinate (ζ) system is introduced where the lowest surface of constant vertical coordinate becomes conformal to the orography (Fig. 2.2.).

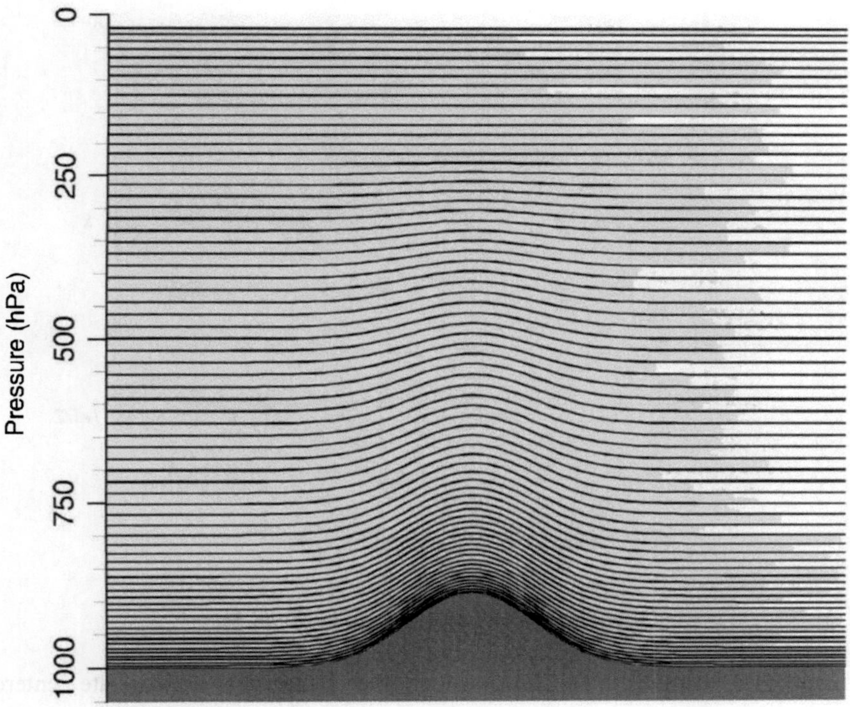

Fig. 2.2. Vertical grid of LM

The vertical grid is not equidistant in physical space but there are more layers close to the ground than higher up, because a detailed simulation of physical processes close to the earth surface is important for accurate weather prediction. The transformation of Eq. 1.1 to 1.9 to the generalised vertical coordinate ζ is performed such that the irregular grid associated with the terrain-following system in physical space is mapped onto a rectangular as well as regular computational grid with constant increments $\Delta\lambda$, $\Delta\varphi$, $\Delta\zeta$. The computational (λ, φ, ζ)-space is then represented by a finite number of grid points (i, j, k), where i corresponds to the λ-direction, j to the φ-direction, and k to the ζ-direction. Every grid point (i, j, k) represents the centre of an elementary rectangular grid volume with side length Δλ, Δφ and Δζ. The model variables of LM are staggered on an Arakawa-C/Lorenz grid with scalars like temperature (T) or pressure (p) at the centre of a grid box and the normal velocity components (u, v, w) defined on the corresponding box faces (Fig. 2.3.).

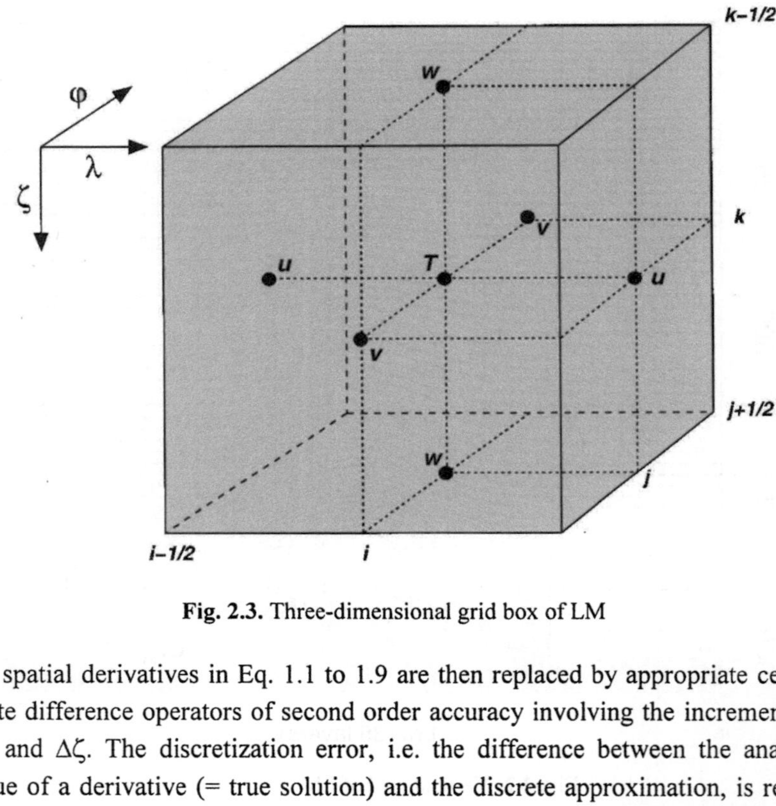

Fig. 2.3. Three-dimensional grid box of LM

All spatial derivatives in Eq. 1.1 to 1.9 are then replaced by appropriate centered finite difference operators of second order accuracy involving the increments Δλ, Δφ and Δζ. The discretization error, i.e. the difference between the analytical value of a derivative (= true solution) and the discrete approximation, is reduced by a factor of four if the mesh size (Δλ, Δφ and Δζ) of the model is reduced by a factor of two.

2.2 Temporal Discretization

All temporal derivatives in Eq. 1.1 to 1.9 are also replaced by a centered finite difference operator involving a time step Δt. Because the equations describe a compressible model atmosphere, meteorologically unimportant sound waves are also part of the solution. As such waves propagate very fast they severely limit the stable time step of an explicit time integration scheme. To improve the numerical efficiency, the mode-splitting time integration method proposed by Klemp and Wilhelmson (1978) is employed. This technique is based on a separation of the prognostic equations into terms which are directly related to sound wave modes and into terms which refer to comparatively slowly varying modes of motion like horizontal advection. The time step for stable explicit integration of the slow modes is then subdivided into a number of small time steps, and the acoustically active terms are updated every small time step while all other terms related to the slow modes are computed only once every big time step. Consequently, only the small time step size is limited by the stability criterion for sound wave propagation. As not the complete but only a reduced set of equations is evaluated on the small time steps, the mode-splitting technique makes the explicit time integration more efficient.

Another time integration method is the semi-implicit scheme. Here the forecast values after a big time step which is stable for the explicit integration of the slow modes is corrected by a modification which is due to an implicit integration of the fast modes. This involves the costly solution of a full three-dimensional Helmholtz equation, usually by an iterative method.

2.3 Computational Cost of NWP Models

At each grid point of a state-of-the-art NWP model like LM about 4000 floating point operations (flop) have to be performed to advance the forecast by one time step. Thus the total computational cost C_{tot} (flop) of a 24-h forecast is given by

$$C_{tot} = N_i \cdot N_j \cdot N_k \cdot 24 \cdot 3600\, s/\Delta t \cdot 4000\, (flop) \tag{2.1}$$

where N_i is the number of grid points in λ-direction, N_j is the number of grid points in φ-direction, N_k is the number of layers (vertical direction), and Δt is the time step in seconds.

Usually mesh size and time step of an NWP model cannot be chosen independently but have to fulfil the Courant-Friedrich-Levy criterion which limits the stable time step for a given mesh size. For example, LM with a horizontal mesh size of 7 km has to use a big time step of 40 s, i.e. a 24-h forecast consists of 2160 time steps. Thus with $N_i = 325$, $N_j = 325$, and $N_k = 35$ a 24-h forecast of LM is equivalent to $C_{tot} = 32 \cdot 10^{12}$ flop.

For operational NWP, a 24-h forecast must be completed in a short time period, e.g. at DWD about 30 minutes at most, to be useful for forecasting purposes. The sustained speed of the computer running the LM operationally must thus exceed about $17 \cdot 10^9$ flop/s = 17 Gflop/s. Therefore high-speed supercomputer resources are required for operational NWP. At the DWD, an IBM RS/6000 SP has been installed at the end of 2001. This number cruncher with 80 nodes, 1280 Power III processors has a sustained performance of more than 150 Gflop/s. At the current operational resolution, i. e. 60-km mesh, 31 layers for GME and 7-km, 35 layers for LM, GME is using 6 nodes and LM 11 nodes.

In the second half of 2003, the mesh size of GME will probably be decreased from 60 to 40 km and the number of layers will be increased from 31 to 42, and LM will cover a larger portion of Europe with $N_i = 440$, $N_j = 488$ and also 42 layers. These upgraded model versions of GME and LM will need about 30 nodes each for operational forecast.

3 Determination of the Initial State

For many CFD problems like the airflow around the wing of a plane, the exact initial state is not very crucial because a steady state solution is sought. Similarly, in climate modelling the simulated long-term climate statistics should not depend on the initial state of the climate run but only on the external forcing and the internal feedback of the model. For NWP however, the situation is totally different: Here the quality of the initial state, i.e. the correct analysis of the state of the atmosphere at the start of the forecast run, determines the accuracy of the prediction to a large extent, especially for the first few days ahead. The solution of an NWP model is so strongly dependent on the initial state of the forecast because many atmospheric structures like cyclones or convective cells develop due to the sudden release of (baroclinic or convective) instabilities in the initial flow field. Very small changes in the initial state may give rise to a completely different temporal evolution of the simulation. Therefore the suitable determination of the initial state of the forecast forms an integral part of any NWP system, and is as important as the correctness of the prediction model.

3.1 Types of Observing Systems

The national weather services world wide operate a complex and quite expensive observing system comprised i. a. of

- About 7.000 land stations (SYNOP) where measurements of pressure, wind, temperature and humidity close to the ground are taken at hourly intervals. Manned stations additionally report cloud type and cover as well as the current weather situation.
- Commercial ships (SHIP) and drifting and moored buoys (BUOYS) provide similar observations over the oceans.

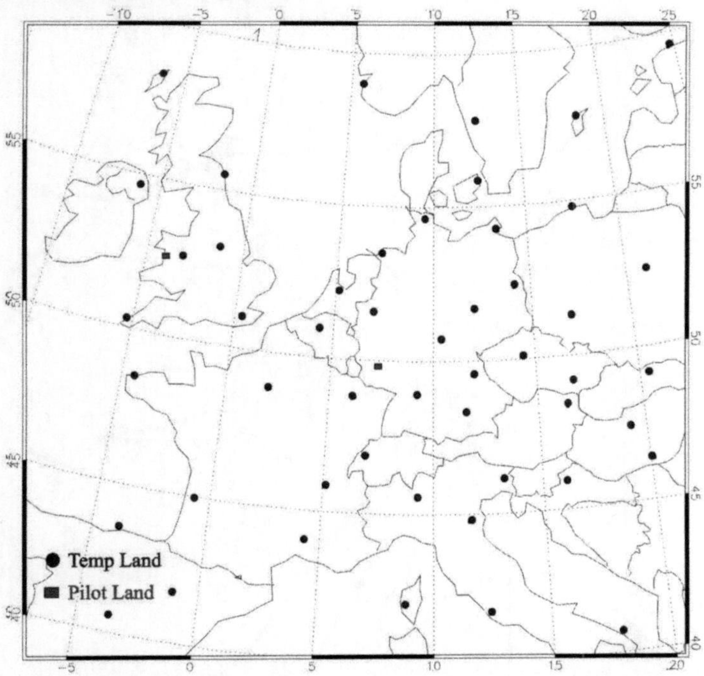

Fig. 3.1. Network of upper air stations in the LM-domain

- At about 700 stations world wide a balloon carries an instrument (RADIOSONDE) from the surface up to about 30 km height at least twice daily (at the main synoptic times 00 and 12 UTC). The resulting profiles of temperature, humidity and wind (Temp) provide detailed information about the vertical structure of the atmosphere. At some stations, only the vertical profile of the horizontal wind component (Pilot) is measured. Fig. 3.1. shows the network of 50 RADIOSONDE stations in the LM-domain; the average distance between the stations exceeds 300 km.

- The large data void regions in between these stations can be partly filled by aircraft measurements (Fig. 3.2.), at least during daytime, because more and more commercial aircrafts are equipped with systems (AMDAR) taking measurements of pressure, wind and temperature during flight at cruising level, i.e. at 10 to 12 km above mean sea level, as well as during take off and landing.

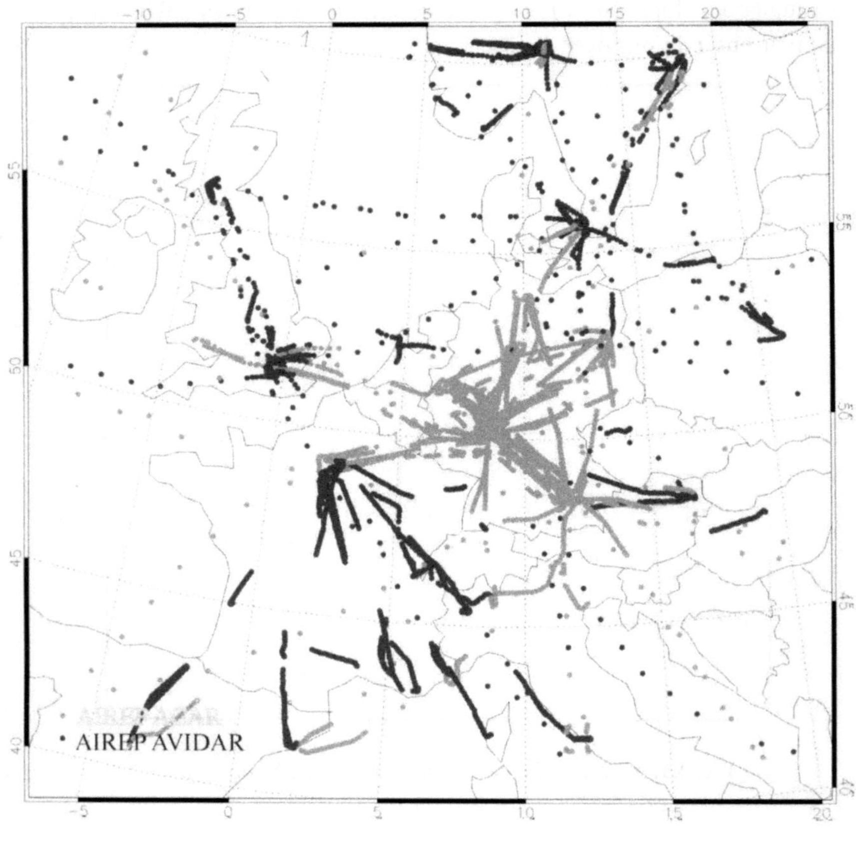

Fig. 3.2. Upper air measurements by commercial aircrafts on 17-08-2001 in the LM-domain

- Finally, two types of meteorological satellites, polar-orbiting and geostationary ones, fill the large gaps in the conventional network of in-situ measurements. Polar-orbiting satellites spin around the earth at a height of 800 km and pass over a certain point of the earth surface twice daily. Geostationary satellites at a height of 36.000 km above the equator rotate around the earth axis at the same rate as the earth. Thus they seem to be fixed relative to the rotating earth surface. Both types of satellites provide remote sensing observations, namely measurements of radiation at different wave length, e.g. for the visible, infrared or water vapour channels, either reflected or emitted by the earth-atmosphere system. Different products are derived from these satellite measurements: Atmospheric motion (SATWIND) can be estimated from successive geostationary satellite images by tracking the translation speed of individual clouds. Radiance measurements from polar-orbiting satellites may be converted to vertical profiles of temperature and humidity (SATEM). These profiles are valuable information in data sparse regions like the oceans.

All observations are exchanged globally between the national weather services connected to the Global Telecommunication System.

3.2 Outline of a Data Assimilation Scheme

To analyse the initial state of a high-resolution NWP model like LM requires the specification of the eight prognostic variables (u, v, w, T, p, q_v , q_c , q_i) at each grid point; thus N_i x N_j x N_k x 8 = 29.6 Million variables (degrees of freedom) have to be set. However, there are insufficient observations at any time to determine the state of the atmosphere. For example, at 12 UTC there are about 3000 SYNOP, 200 SHIP, 50 RADIOSONDE, 2500 AMDAR, 50 SATOB and 50 SATEM reports in the LM domain. To analyse the initial state of the atmosphere one has therefore to solve a severely underdetermined mathematical problem. Additional information must be incorporated into the solution process, namely the temporal evolution of the atmosphere embodied in the forecast model. Data assimilation is then the process of finding a model state which is most consistent with the observations available over a certain period of time. The forecast model organises and propagates forward this information.

LM uses a continuous nudging scheme (Fig. 3.3.) to modify the model state by incorporating new observations. The aim here is to adapt the model to a new state which is most consistent with the latest observations as well as the previous ones carried forward in the forecast model.

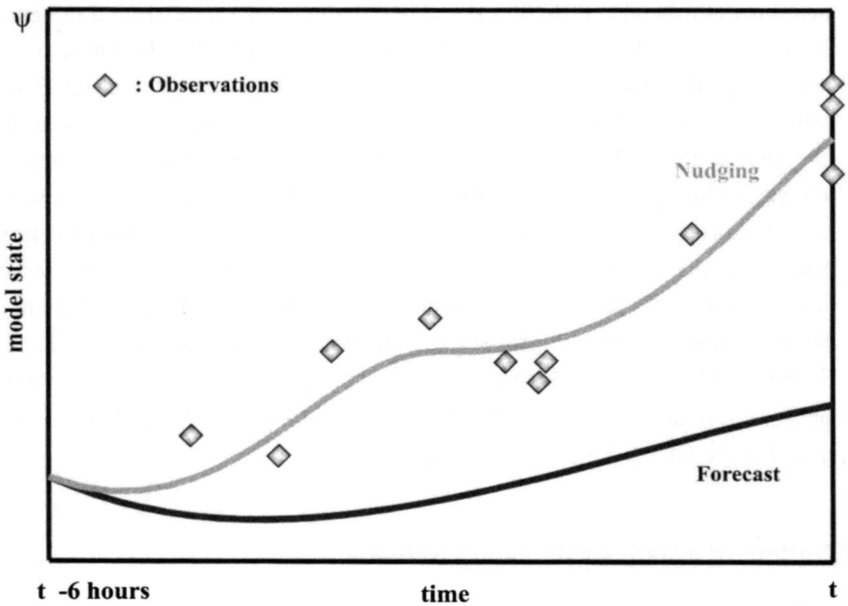

Fig. 3.3. Analysis of a model variable ψ by the continuous nudging scheme of the LM

Mathematically the nudging scheme leads to an additional (artificial) forcing term on the right-hand-side of the prognostic equations for any variable ψ (e. g. temperature) with the general form

$$\frac{\partial}{\partial t}\psi(\lambda,\varphi,z,t)=F\left(\underline{\psi},\lambda,\varphi,z,t\right)+G_{\psi}\cdot\sum_{k=1}^{N}W_k\cdot\left[\psi_k-\psi\left(\lambda_k,\varphi_k,z_k,t\right)\right] \quad (3.1)$$

$$W_k=\frac{w_k}{\sum_{j=1}^{N}w_j}\cdot w_k \quad (3.2)$$

$$w_k=w_t\cdot w_{\lambda,\varphi}\cdot w_z\cdot\varepsilon_k \quad (3.3)$$

where

$\underline{\psi}$ is the vector of all variables of the model, W_k is the weight of a particular observation ψ_k, and N is the number of observations influencing the model variable $\psi(\lambda, \varphi, z, t)$. G_ψ determines the characteristic time scale for the relaxation.

The weight W_k for the model grid point (λ, φ, z) at time t depends on

- time difference to the observation (w_t)
- space distance to the observation ($w_{\lambda,\varphi}$, w_z)
- observation and model errors (ε_k).

This forcing term nudges the prognostic variables of LM towards the observed values which have been interpolated to the model grid points.

Most observations contain contributions from different atmospheric pattern, some of which are on spatial or temporal scales too small to be resolved by the NWP model. For example, LM with a 7-km mesh cannot resolve horizontal structures which are smaller than 14 km. Any smaller scale structure which is sampled by an observation, but LM is unable to represent, is called representativeness error and should not be assimilated into the model. This error is treated in the data assimilation scheme of LM by incorporating it into the overall observation error which is instrument error plus representativeness error. The impact of a specific observation on the state of LM is proportional to the inverse of the overall observation error (see Eq. 3.1). This error is based on instrument calibration measurements and long-term statistical evaluations; but it can only provide an estimate of an average observation error, not the actual one of an observation at a specific place and time.

Therefore an automatic quality control system which is incorporated in the data assimilation system has to perform a careful statistical analysis of probable observation errors as well as the model (background) error. Each observation which will be assimilated into the LM is first compared to the model forecast which has been interpolated to the observation site. If the difference between model value and observation exceeds certain thresholds which depend on observation type, parameter and height, the observation will be flagged dubious. Another important check consists of comparing one observation to neighbouring ones. If the difference exceeds certain thresholds again, this observation will not be used during the data assimilation process at all. To eliminate observations which are grossly in error from the data assimilation process is necessary in order to avoid bad forecasts.

4 Verification of NWP Models

Observations of the atmosphere form the empirical basis of any verification scheme for NWP models. The comparison between the observed and predicted weather parameters seems to offer a simple way of estimating the quality of an NWP model. But in practise there exist several intrinsic shortcomings of this procedure which are hard to overcome:

- NWP models predict volume average quantities; the grid cell size of GME for example is $3100km^2$ x 60m close to the ground and $3100km^2$ x 800m at a height of 10km above mean sea level. Contrary, observations, especially in situ ones, represent spot values with rather limited spatial averaging.
- Most observing systems (see section 3) sample a wide range of atmospheric scales, some of which are not resolved by the NWP model under consideration.
- On the other hand, the spatial resolution of many observing systems (like RADIOSONDES, see section 3) is too coarse to verify small-scale weather pattern predicted by the NWP model.
- Finally, the quality of the numerical prediction does not only depend on the correctness of the NWP model but to a large extent on the quality of the initial state of the forecast. Because the data assimilation scheme tries to merge information from the NWP model and observations in a consistent way to determine the initial state, a clean verification of the NWP model alone is not possible (see section 3).

Despite these shortcomings verification of NWP forecasts against observations is the basis of all improvements of the NWP system, and provides important information to the users of NWP products. At operational NWP centres most routine verification is performed on a monthly or seasonal basis to reduce the impact of random errors and to highlight any systematic bias. To estimate the quality of an NWP system for particular weather situations like developing cyclones or convective cloud clusters the verification results could be also grouped with respect to general weather pattern. This case-dependent verification is of great value to model developers who try to improve the simulation of particular weather systems.

There exists a multitude of indices which are useful to quantify the verification results. Each index has certain merits and disadvantages; thus only a combination of different indices will usually give a true picture of the quality of an NWP system. For most users the "hit rate" of a forecast is the most important verification index, i.e. the number of cases that the forecast matches the observation within a predefined interval, relative to the total number of forecasts. For near-surface temperature forecasts this interval is usually set to 2K; thus for an observed temperature of 20°C, forecasted temperatures in the range of 18°C to 22°C are considered "correct". Fig. 4.1. shows the hit rate of maximum and minimum near-surface temperature forecasts of LM at the observing station Frankfurt/Main during each months of 2001 and forecast ranges between 18 and 30 hours. Between 39 and 85% of the LM temperature forecasts at this station are correct according to this

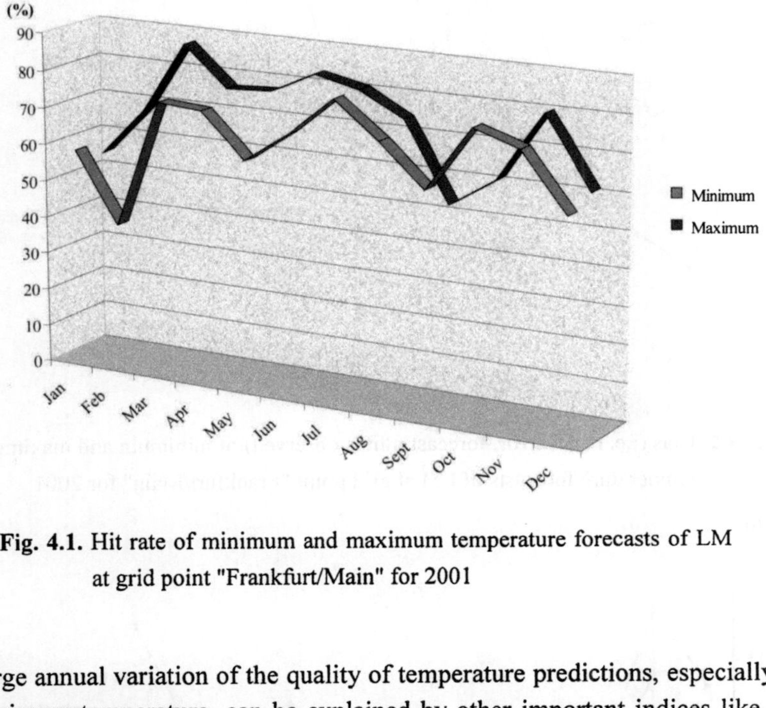

Fig. 4.1. Hit rate of minimum and maximum temperature forecasts of LM at grid point "Frankfurt/Main" for 2001

index.

The large annual variation of the quality of temperature predictions, especially for the minimum temperature, can be explained by other important indices like bias and root-mean-square-errors (RMSE), which are defined as:

$$\text{Bias} = \frac{1}{N}\sum_{i=1}^{N}\left(\psi_{\text{fct}}^{i} - \psi_{\text{obs}}^{i}\right) \quad \text{and} \quad \text{RMSE} = \sqrt{\frac{1}{N}\sum_{i=1}^{N}\left(\psi_{\text{fct}}^{i} - \psi_{\text{obs}}^{i}\right)^{2}} \tag{4.1}$$

where

ψ_{fct} are the predicted values, ψ_{obs} are the observed ones, and N is the number of cases.

Fig. 4.2. and 4.3. show Bias and RMSE of maximum and minimum near-surface temperature forecasts of LM at the observing station Frankfurt/Main for January to December 2001.

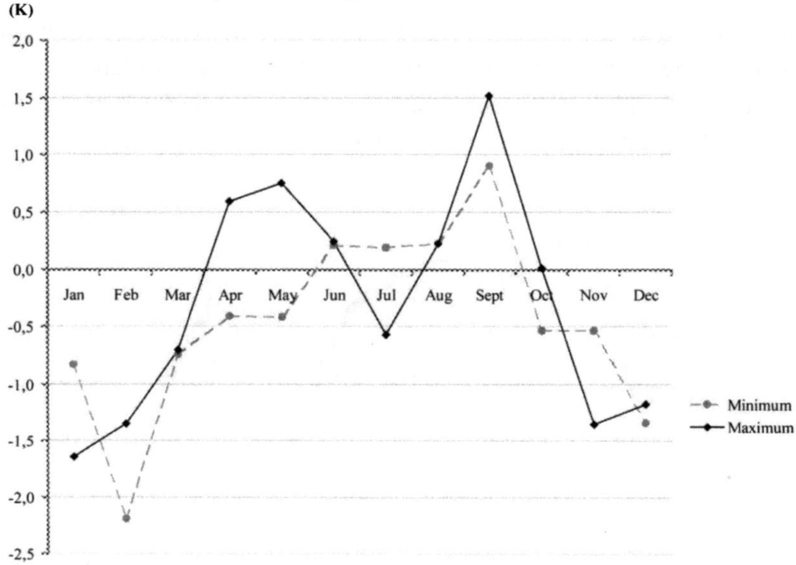

Fig. 4.2. Bias (i.e. mean error, forecast minus observed) of minimum and maximum temperature forecasts of LM at grid point "Frankfurt/Main" for 2001

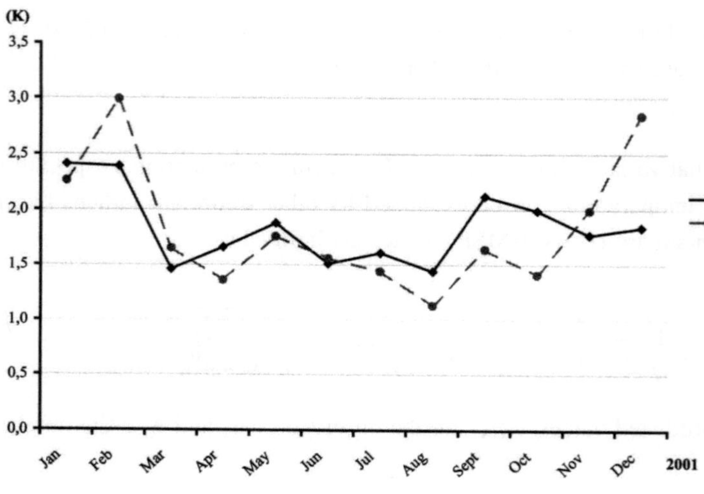

Fig. 4.3. Same as Fig. 4.2., but root-mean-square error (RMSE)

During late autumn and winter a pronounced negative bias of the minimum temperature of up to –2.2K develops in LM. This systematic error is mainly due to the inability of the model to properly simulate the stable boundary layer close to the ground which is usually characterised by very weak turbulent fluxes. For the spring and summer months, temperature forecast errors are much smaller because of the positive impact of a special analysis of the soil moisture of LM. The diurnal variation of the near-surface temperature during these months is controlled by the available soil moisture to a large extent. Fortunately, the systematic part of the temperature forecast error can easily be corrected by statistical methods. Thus the quality of the final forecast product for the public is much higher with hit rates exceeding 85 % throughout the year.

Outlook

As in the past, the further improvement of the overall quality of weather forecasts will be closely linked to the operational usage of better observing systems, sophisticated data assimilation schemes and more realistic models which as closely as possible resemble the real atmosphere. This deterministic modelling approach will be more and more complemented by ensemble prediction systems (EPS) which produce a set of forecasts based on slightly different initial states and model versions. The EPS aims at providing the users of NWP products not only with a single deterministic forecast but an estimate of its reliability based on the spread of the individual EPS members. These probabilistic forecast systems form a natural extension of usual deterministic modelling, and take the "chaotic" behaviour of atmospheric flows fully into account.

References

Doms G and Schättler U (1997) The nonhydrostatic limited-area model LM (Lokal-Modell) of DWD. In: Part I: Scientific Documentation. Deutscher Wetterdienst (DWD), Offenbach, March 1997.

Jacobsen I and Heise E (1982) A new economic method for the computation of the surface temperature in numerical models. In: Beitr. Phys. Atm., 55, No. 2, 128-141.

Klemp J B and Wilhelmson R B (1978) The simulation of three-dimensional convective storm dynamics. In: 10.conf.on severe local storms, oct.18-21,1977, omaha,ne. - boston/ma: amer.meteorol.soc. pp 283-290.

Louis J-F (1979) A parametric model of vertical eddy fluxes in the atmosphere. In: Boundary-Layer Meteor., 17, 187-202.

Mellor G L and Yamada T (1974) A hierarchy of turbulence closure models for planetary boundary layers. In: J. Atmos. Sci., 31, 1791-1806.

Müller E (1981) Turbulent flux parameterization in a regional-scale model. In: ECMWF Workshop on planetary boundary layer parameterization, 193-220.

Ritter B and Geleyn J-F (1992) A comprehensive radiation scheme for numerical weather prediction models with potential applications in climate simulations. In: Mon. Wea. Rev., 120, 303-325.

Tiedtke M (1989) A comprehensive mass flux scheme for cumulus parameterization in large-scale models. In: Mon. Wea. Rev. - 117. 8. - pp 1779 – 1800

Simulation of Hydrogen Behaviour During a Nuclear Power Plant Accident

Ulrich Bielert, Wolfgang Breitung and Alexei Kotchourko

Introduction

Severe accidents in nuclear power plants are very unlikely events, but when they occur, they may be associated with high damages. Because of the combination of very low probability and very high damage potential commonly used risk assessment strategies are not adequate for the analysis of severe accidents in nuclear power plants.

Accident management in nuclear power plants is based on the 'defence in depth' principle: several physically independent safety barriers are established against catastrophic failure of the total system. Even when one safety system should fail, diverse or redundant backup systems cope with this situation. The last line of defence in reactor accidents is the containment building. The purpose of this building is to ensure that under all circumstances radioactive material is not released into the environment. During the licensing procedure of a nuclear power plant the containment integrity must be proven mechanistically for a certain set of accident scenarios, the so-called design basis accidents. There is a general consensus in the nuclear community that further development of nuclear power into a safe, economically competitive, clean, sustainable and accepted energy source, must include control of severe beyond-design accidents, even if their statistical probability of occurrence is extremely low. This new safety goal requires that accident consequences remain essentially restricted to the plant site itself, which means that containment integrity must be proven for severe accidents.

A containment building size, which can reach 100,000 m^3, makes it prohibitive to demonstrate the containment integrity experimentally. Modelling of a severe accident is the only feasible way. A major load mechanism, which poses a possible threat to the containment during a severe accident, is hydrogen explosion. A typical accident leading to large amounts of hydrogen in the containment building is a 'loss of coolant accident'. A major water supply line rupture and the unavailability of all safety injection systems would cause the reactor core to overheat and dry-out. During this phase or if in a later stage the reactor core is flooded, the hot metallic materials in the core can react with steam. The metals (mainly zircaloy and stainless steel) are oxidized by steam producing large amounts of hydrogen. This hydrogen is subsequently released through the break into the containment where it mixes with air. The burning of this hydrogen-air mixture will cause a transient pressure load. The question that needs to be answered is then: can the containment withstand the loads associated with the hydrogen explosion?

The whole accident scenario covers a wide range of time and length scales and involves many different physical processes. Presently it is not feasible to cover the whole problem by a single integrated numerical simulation based entirely on first principles. Instead, different stages of the accident scenario are modelled by specialized codes, which focus on the essential features of that stage. The whole suite of codes eventually describes the complete sequence of physical events.

In section 1 we will describe the various modes of combustion and will present criteria to identify the different combustion regimes and describe how they fit into the analysis procedure of a severe accident scenario. For the remainder of this paper we will then focus on the turbulent combustion code COM3D [9, 21]. In section 2 we summarize the main features of this code. Section 3 describes the verification and the validation procedure of the code. We will close with an example of a turbulent combustion process simulation in a nuclear power plant containment.

1 Analysis Procedure for Severe Accident Scenarios

1.1 Combustion Regimes of Hydrogen Air Mixtures

In hydrogen-air mixtures, combustion can proceed in distinctly different combustion regimes. If the hydrogen is ignited immediately after begin of the release a diffusion flame will develop. Normally the flame will attach to the release location and will continue as long as new hydrogen is released. Since the process is relatively slow and the combustion product (steam) can condensate in colder parts of the containment, the pressure rise in the containment is small. However, the diffusion flame is burning for an extended period resulting in high local thermal loads to the interior structures in the vicinity of the release location. The thermal power of a standing diffusion flame in a severe accident can reach 100 MW.

If the ignition is delayed the hydrogen will first mix with air. After ignition of the hydrogen-air mixture an initially laminar flame can develop. Due to intrinsic instabilities of the laminar flame it can get into so-called 'self-turbulizing' regime. In practical situations this happens relatively rarely, since the flame usually meets upstream already turbulent flow. This flow can interact with the flame causing it to wrinkle and thereby increasing its area and thus its reaction rate. The increasing reaction rate in turn creates a stronger flow field. This positive feedback can cause the flame to accelerate and become highly turbulent. Turbulent premixed combustion is the predominant combustion mode in almost all practical applications.

As long as the flame propagates with velocities well below the speed of sound in the burned gas, the pressure waves created by the combustion process are relatively small. This regime is usually referred to as slow deflagration. Propagation velocities in this regime are of the order of 100 to 200 m/s and overpressures are about 1 bar for atmospheric initial conditions. A fast turbulent deflagration is observed when the flame accelerates almost to the speed of sound in the burned gases. Overpressures in this case can reach up to 10 bar for initially atmospheric conditions. Typical flame velocities in hydrogen-air mixtures for this regime are about 800 m/s. During detonation, which is a supersonic regime of combustion, of atmospheric hydrogen-air mixtures the flame velocities can achieve 2000 m/s producing overpressures of about 20 bar.

1.2 Selection Criteria for Different Combustion Regimes

Clearly the mode of combustion determines the maximum load on the containment. A first approach would be to select always the most severe combustion mode possible. This would require a completely new, 'revolutionary' containment design, which is in principle possible [7]. An alternative, more 'evolutionary' approach would be to reduce the loads by countermeasures to a level, which can be safely sustained with contemporary containment technology. This way would allow also back-fitting of existing containments.

For realistic and not too conservative load estimates it is essential to predict the appropriate combustion mode for a given situation. For the purpose of our containment analysis we use three criteria, which generalize experimental observations of combustion phenomena, which not fully understood by theory or could not be resolved by numerical simulations on containment scale.

The first process after an ignition in hydrogen-air-steam mixture, e.g., by a weak spark, is the growth and stability of the initial flame kernel. Depending on the local flow conditions and mixture properties, the kernel can survive and evolve to a macroscopically stable flame or it can be quenched. For large scale containment calculations simple criteria are needed, which allow to predict if an ignition would result in stable flame propagation or not [1]. The second process is the transition from initially laminar flame to the fully developed turbulent regime, because it is influenced by details of the turbulence field and small sub-grid flow obstacles. An empirical criterion is required to decide on the probability of flame acceleration for a given mixture and scale. The third complex and not yet completely understood process is the transition from a fast turbulent deflagration to a fully developed detonation. Local hot spots either in the flame brush or in shock focusing details of the geometry can trigger the onset of a detonation, if the conditions in the surrounding of the local explosion support its further amplification. Again a criterion is needed to decide if such a process could be possible or not.

The modelling situation can be summarized as follows: the regimes of slow laminar deflagration, fully developed turbulent deflagration and detonation can be described numerically with reasonable accuracy. The transitions between them cannot, therefore we developed empirical criteria, which allow to judge transition probability. If the probability is high, the faster combustion mode is chosen in the numerical simulation.

Ignition Limit

An ignition of a fuel-air mixture is only possible when the fuel concentration is within certain limits. For hydrogen-air mixtures under atmospheric conditions combustion is only possible for hydrogen concentrations between approximately 4% and 76% of hydrogen [12]. The exact limits depend on the propagation direction of the flame. A flame propagating upwards with buoyancy support needs a somewhat lower hydrogen concentration than a flame propagating downwards against buoyancy. However, the values given above are conservative limits. Whenever the hydrogen concentration at an ignition location is within the ignition limits of the hydrogen-air-steam mixture a combustion simulation is performed.

Flame Acceleration Criteria

The acceleration of a premixed flame is caused by a positive feedback mechanism between flow field, turbulence and chemical reaction. The flow field also depends strongly on the boundary conditions imposed by the geometry of the system. Therefore flame acceleration is also influenced by the geometry.

In order to find a criterion for flame acceleration, which is generally applicable, experimental data from a wide variety of test facilities with different length scales and geometric configurations were analysed. This analysis showed that flame acceleration was only observed for sufficiently sensitive mixtures. The controlling parameter was identified as the expansion ratio σ [5, 2], which is the ratio between the specific volumes of the burned and unburned gases at constant pressure. The combustion induced increase in specific volume is responsible for the displacement flow, which causes turbulence and flame acceleration. From the experimental observations a critical value σ_{crit} was found, below which a fast turbulent deflagration was never observed. Note that an expansion ratio above σ_{crit} is only a necessary criterion for flame acceleration, but not sufficient. This means in a given geometrical configuration a sensitive mixture can still burn in a slower combustion mode. In this sense the use of the σ-criterion is a conservative approximation.

Detonation Criteria

On a large scale a detonation front appears as a one-dimensional discontinuity in the direction normal to the propagation path. However, closer inspection of a detonation front reveals a complex three-dimensional structure with transverse waves. These transverse waves form a characteristic structure called a detonation cell. The detonation cell size λ depends on the properties of the gas mixture, e.g., λ decreases with increasing sensitivity of the mixture. Experimental evidence shows that a detonation can only develop if the combustible mixture can accommodate at least 7 of these detonation cells. Thus the λ-criterion states that a minimum size of the combustible cloud is a necessary condition for the onset of a detonation [4]. As a rule the detonation develops by transition from a fast turbulent deflagration, therefore the σ-criterion should be fulfilled also.

1.3 Analysis Procedure for Hydrogen Release Scenario

Fig. 1 shows a schematic representation of the analysis procedure for a hydrogen related accident scenario [2]. This procedure was initially developed for studies of severe accidents in nuclear power plants but it is also generally applicable to release scenarios in non-nuclear applications.

The first left block of Figure 1 shows the steps, which are necessary to analyse the combustible mixture generation. First, a plant design must be chosen. Next, possible mitigation systems must be selected. In a nuclear power plant such mitigation systems could be igniters, which ignite the mixture as it reaches combustion limits, or recombiners, which reduce the hydrogen amount using a catalytic surface oxidation. In addition, an accident scenario must be selected. An example for an accident scenario would be the 'loss of coolant accident' mentioned above.

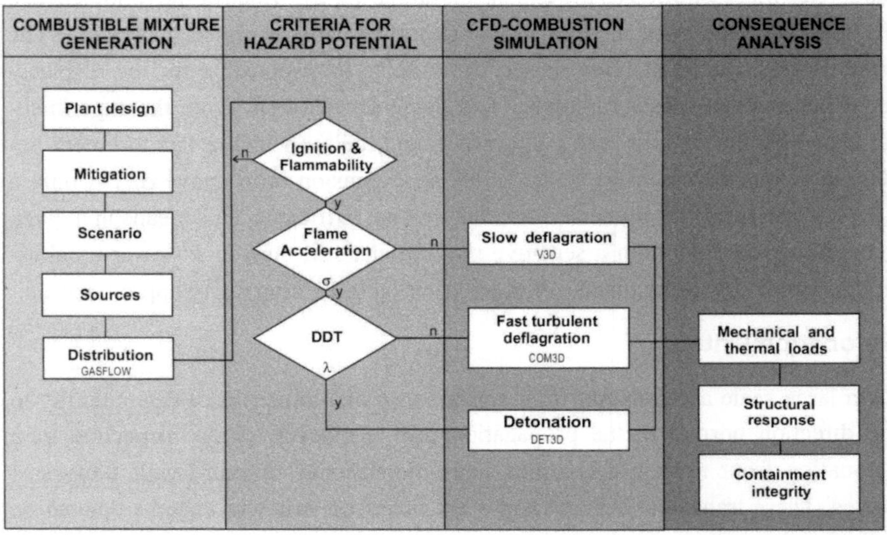

Fig. 1. Schematic representation of the developed analysis procedure for hydrogen related accident scenarios

For the hydrogen analysis we need to know the hydrogen amount, which is released from the reactor primary system. For its evaluation the processes, which take place inside the reactor pressure vessel, must be considered. There are dedicated software packages for integral analysis of in-vessel processes like MELCOR or MAAP4. These codes produce the required input for our containment simulations. Once the location and temporal development of the hydrogen and steam sources are known, the containment analysis starts with a distribution calculation. For this phase the distribution code GASFLOW [20], which was started at Los Alamos National Laboratory, and has been developed further at FZK since 1994, is used. This code solves the main conservation laws together with appropriate equations for turbulence, steam/water phase changes and gas/solid heat transfer. The results of the GASFLOW calculation are time and space resolved distributions of gas concentrations, temperatures, pressure and flow conditions.

The second column in Fig. 1 lists the three criteria for different combustion regimes; each criterion is a prerequisite for the next one, e.g., no flame acceleration without ignition and no detonation without previous flame acceleration. The criteria are repeatedly evaluated during the GASFLOW calculation. Depending on which criteria are fulfilled, the respective combustion code is started.

The third column in Fig. 1 shows the different combustion codes which are used in the containment analysis. For slow deflagrations or standing diffusion flames a code is needed, which allows time integration over prolonged time. Therefore, an implicit integration scheme has been implemented in V3D code [15]. The faster combustion processes need to be calculated for shorter periods and therefore the explicit solvers have been used in the COM3D code. If all three criteria are fulfilled a detonation of the hydrogen-air mixture is possible. In this case the governing equations are reduced into the Euler equations and are solved by the code DET3D [13, 14].

All the combustion codes estimates the thermal and mechanical loads acting on the containment structure. This is shown in the most right column of Fig. 1. Obtained loads are used as input for the structural response calculations, finally resulting in a fully mechanistic statement about the containment integrity. Of course in practice, the containment integrity must be demonstrated for all relevant accident scenarios. Thus the shown analysis procedure must be performed repeatedly.

All the different codes mentioned before must be verified and validated. The formal procedure is quite similar for the different codes, therefore we will focus on one code only, namely the COM3D code, since the turbulent combustion is the most difficult topic of all three considered combustion regimes.

2 The Turbulent Combustion Code COM3D

Turbulent combustion processes are controlled by the interaction between flow field, turbulence and the chemical reactions. The coupling between these processes is strongly non-linear. The governing equations for this situation are the time dependent three-dimensional conservation laws for the mass (1), momentum (2) and energy (3) with the transport equations for the species (4)

$$(\rho)_t + (\rho u_j)_{x_j} = 0 \tag{1}$$

$$(\rho u_j)_t + (\rho u_i u_j)_{x_j} = \rho g_i - p_{x_i} + M_{ij,x_j}, \quad i = 1,2,3 \tag{2}$$

$$(\rho e)_t + ((\rho e + p)u_j)_{x_j} =$$

$$= \rho g_j u_j + u_i M_{ij,x_j} + (\frac{\mu_{tur}}{C_h}(e - \frac{1}{2}u_i u_i \frac{p}{\rho})_{x_j})_{x_j} + B + \rho\varepsilon \tag{3}$$

$$(\rho f_\alpha)_t + (\rho f_\alpha u_j)_{x_j} = \bar{\dot{w}}_\alpha + (\frac{\mu_{tur}}{C_{f_\alpha}} f_{\alpha,x_j})_{x_j} \qquad (4)$$

here

$$f_\alpha = \frac{\rho_\alpha}{\rho}, \quad e = \sum_{\alpha=1}^{N} \frac{f_\alpha}{m_\alpha}(h_\alpha + \Delta h_\alpha^0 - RT) + \frac{1}{2} u_j u_j,$$

$$M_{ij} = \frac{2}{3}\delta_{ij}(\rho k + \mu_{tur} u_{r,x_r}) + \mu_{tur}(u_{i,x_j} + u_{j,x_i}).$$

In addition to the gasdynamic model a turbulence model is required. In COM3D two different turbulence models are available: a standard k-ε model and a renormalization group theory (RNG) k-ε model. The equations for both models include the transport equations of turbulent kinetic energy k (5) and for turbulent kinetic energy dissipation rate ε (6, 6a)

$$(\rho k)_t + (\rho u_j k)_{x_j} = S - \rho\varepsilon + (\frac{\mu_{tur}}{C_k} k_{x_j})_{x_j}, \qquad (5)$$

$$(\rho\varepsilon)_t + (\rho u_j \varepsilon)_{x_j} = \frac{\varepsilon}{k}(C_1 S - C_2 \rho\varepsilon) + (\frac{\mu_{tur}}{C_\varepsilon} \varepsilon_{x_j})_{x_j}, \qquad (6)$$

$$(\rho\varepsilon)_t + (\rho u_j \varepsilon)_{x_j} = \frac{\varepsilon}{k}((C_1 - C_\eta)S - C_2\rho\varepsilon) +$$

$$+ (\frac{\mu_{tur}}{C_\varepsilon} \varepsilon_{x_j})_{x_j} + [C_3 - \frac{2}{3} C_\eta (C_\mu \frac{k}{\varepsilon} u_{j,x_j} + 1)] \rho u_{j,x_j} \varepsilon, \qquad (6a)$$

here

$$S = u_{i,x_i}(-\frac{2}{3}\delta_{ij}(\rho k + \mu_{tur} u_{r,x_r})) + \mu_{tur}(u_{i,x_j} + u_{j,x_i}), \quad \mu_{tur} = C_\mu \frac{k^2}{\varepsilon}$$

and

$$C_\eta = \frac{\eta(1 - \eta/\eta_0)}{1 + \beta\eta^3}, \quad \eta = \frac{k}{\varepsilon}(\frac{1}{2}(u_{i,x_j} + u_{j,x_i})(u_{i,x_j} + u_{j,x_i}))^{1/2},$$

$$C_3 = \frac{-1 + 2C_1 - 1.5(\gamma - 1) + (-1)^\delta \sqrt{6} C_\mu C_\eta \eta}{3}.$$

The values of the model parameters are listed in Table 1. Notice, that one of the advantages of the RNG k-ε model is that the values of the model constants can be evaluated explicitly from the theory in contrast to the standard model.

Table 1. Constants of COM3D k-ε turbulence models

	C_μ	C_1	C_2	C_k	C_ε	β	η_0
Standard k-ε	0.09	1.44	1.92	1.0	1.3	-	-
RNG k-ε	0.0845	1.42	1.68	0.719	0.719	0.012	4.38

In order to close the system of equations a combustion model is required (eq. (4)). In COM3D two combustion models have been implemented. In the first model the mean reaction rate $\bar{\dot{w}}$ is evaluated from

$$\bar{\dot{w}} = \begin{cases} -C'_f \dfrac{\varepsilon}{k} f_\alpha (1 - f_\alpha), & Da >> 1, \\ -k_{chem} f_\alpha \exp(-\dfrac{E_a}{T}), & Da << 1, \end{cases} \tag{7}$$

where $\bar{\dot{w}}$ depends on the Damköhler number $Da = \tau_{turb} / \tau_{chem}$ (correspondingly, characteristic turbulence and chemical time scales). For *Da >> 1*, the chemical reaction rate is infinitely fast and the combustion process is controlled by turbulent mixing. For *Da << 1*, the process is controlled by chemistry and an Arrhenius law is employed. The upper branch in eq. (7) uses an extended 'eddy-break-up' (EBU) model, based on a standard EBU model [18], improved by Said and Borghi [17]. This extension (S_L here denotes the laminar flame speed)

$$C'_f = C_f (1 + 4.4/(1 + 3.2k^{1/2} / S_L))$$

accounts for low turbulence levels and allows to broaden the model applicability range towards slower flames, in comparison with the standard EBU approach.

Since the combustion model (7) uses two different descriptions for the reaction rate it is not well adapted for values of the Damköhler number near 1. The second combustion model, which uses the so-called 'presumed' PDF approach, is intended to cover all values of the Damköhler number, including $Da \approx 1$.

While the 'eddy-break-up' model is fairly old, but well tested, the second COM3D combustion model is relatively new and currently intensively studied. It is based on a statistical description of the reacting media. It introduces probability density functions (PDF) for the model variables, such as species concentrations, temperatures, etc. In the 'presumed' PDF model the shape of the species PDF $P(f_1, f_2, ...f_N)$ is prescribed as a β-function. The mean reaction rate is determined from integration of the reaction rate $\dot{w}_\alpha$ over the PDF:

$$\bar{\dot{w}} = \iint ... \int \dot{w}_\alpha P(f_1, f_2, ...f_N) df_1 df_2 ... df_N \, , \tag{7a}$$

where $P(f_1, f_2, ..., f_N)$ in the case of a one-dimensional PDF is

$$P(f) = \frac{\Gamma(\beta_1 + \beta_2)}{\Gamma(\beta_1)\Gamma(\beta_2)} f^{\beta_1 - 1} (1 - f)^{\beta_2 - 1},$$

and $\dot{w}_\alpha$ could be a one-step Arrhenius reaction law.

The numerical integration of the β-PDF usually causes severe numerical problems due to the singularities of this function and therefore it is often replaced by table look-up of pre-calculated quantities. A recently developed semi-analytical integration method [11] makes the use of this model as fast and computationally efficient as, for example the simple 'eddy-break-up' model. Currently extensive testing of this combustion model is under way.

In addition to the description of the physical models implemented in COM3D some additional information about the numerical methods should be presented. The system of equations (1)-(7a) is solved on a structured equidistant grid allowing complex three-dimensional internal structures. Pre-processing facilities for the generation of complex grids are included in the code suite. A real time graphics application allows to follow the progress of a simulation interactively. Auxiliary tools, which generate local time-dependent data like pressure or 'light sensors' histories, are available.

3 Testing of the Turbulent Combustion Code COM3D

Modern computer codes combine achievements from different branches of science and usually are based upon new or recent developments, therefore 'quality assurance' procedures are strictly required before such a code can be used with trust. The proofing procedure of a newly built code should include thorough testing against well-known results (experimental and theoretical) and an evidence of its ability to solve the problems for which it was designed.

Let us consider how the structure of the code and the choice of models are connected with this testing procedure. Already on the first stage of code planning the initial draft outline of the testing background becomes apparent. The analysis of the code objectives naturally bounds the spectrum of physical phenomena, which should be taken into account, and, therefore, of the physical models, which should be presented in the code. For example, the influence of the conductive heat transfer from burned gas to the bounding walls on fast turbulent flames is negligible, and therefore test cases where this effect is pronounced should be excluded from the testing database.

Further analysis shows, that this first consideration gives only a rough set of physical models. For instance, for the description of fast turbulent flames a turbulence model and a coupled combustion model are required. However, a number of models for the turbulence are currently available, e.g. taken in the order of decreasing models potential: DNS, LES, RNS and different one-equation and algebraic models. The choice of the model must be a compromise between desirable physical authenticity and computational possibilities, which are also dictated by the objectives of the code. For example, in the COM3D simulation for nuclear power plants, the typical computational cell size can hardly be less than 10 cm if computational times should be less than 1 week of processor time. This 'second iteration' brings us to the final choice of physical models. In our case we can afford with the simplest version from the RNS family for the turbulence (k-ε model) and a 'presumed' PDF model (with EBU model as a limiting case) for the combustion.

Now consider what this means for the testing procedure. Once the models are chosen, the set of phenomena, which can be reproduced in the frames of these models, is thus defined, e.g., phenomena like DDT or jet initiation are inherently excluded from the consideration since k-ε theory is certainly inadequate to deal with it. This refined set of phenomena defines possible candidates, which can constitute the experimental and theoretical database to be used in the simulations.

In the rest of this section we make an attempt to formulate the main points of interest during the testing procedure in general and try to illustrate them on the base of our experience. A routine testing procedure usually includes two main parts, namely: code verification and code validation, which are presented in the corresponding sections below.

3.1 Code verification

In the verification phase of CFD code testing it is possible to address the following generic steps:

- numerical algorithm and grid testing,
- test of gasdynamic part,
- test of thermodynamic part,
- code specific tests.

In the first step, special attention should be paid to provide a suitable choice of the numerical scheme. Consider for instance the simulation of turbulent flow. In the frames of the k-ε theory, turbulence generation is determined by the velocity gradients u_{j,x_j}. Therefore for the correct turbulence source predictions, the algorithm has to deliver notably accurate solution of the velocity distribution. From one side, it has to be smooth enough to exclude over-estimation of the turbulence source term; from the other side, it should not flatten the velocity gradients to prevent under-prediction of the turbulence source. TVD (or ENO) [6] schemes currently are the best candidates, which provide second-order accurate essentially non-oscillatory solutions and simultaneously prevent the smearing of discontinuities. The choice of a certain numerical algorithm requires an appropriate selection for the numerical grid type. Only mention, that grid discretisation can affect, among others, the degree of isotropy of the solution, the numerical viscosity of the scheme, and, of course, the overall run-time performance of the code.

In this first, essentially purely numerical, step we include also thorough tests on the completeness and correctness of the code core built. There are, for example, tests on conservation of the total mass, momentum, and energy in the system; stability of the algorithm; achieved level of accuracy, etc.

The second step of code verification includes the proof of the code ability to reproduce simple gas-dynamic problems. The gasdynamic solver is a vital part of a CFD code and it should be checked against a wide range of problems. As a result, the distinct boundaries should be established within which the solver is effective and reliable. Unfortunately, gasdynamics is strongly coupled with the thermodynamics so that these two parts of the code cannot be tested completely separately. At this stage it is often useful to assume the thermodynamic parameters to be constant, and thus to study gasdynamics influenced by the thermodynamics only to a minor degree. As an example from the COM3D verification procedure, a problem with a forward facing step in super-sonic flow can be mentioned [22]. This problem was studied by a number of authors, it is simple in formulation, but rich in features, and can therefore be considered as a good test case.

Table 2. Comparison between COM3D and STANJAN predictions for adiabatic complete combustion of stoichiometric hydrogen-oxygen mixture in a constant volume ($P_0 = 1.013 \cdot 10^5$ Pa, $T_0 = 298.15$ K)

	COM3D	STANJAN	Relative error
Pressure	$13.103 \cdot 10^5$ Pa	$13.991 \cdot 10^5$ Pa	0.85 %
Temperature	5784.6 K	5734.1 K	0.87 %
Density	0.4908 kg/m^3	0.4909 kg/m^3	0.02 %

We include the test of the thermodynamic properties in the verification phase, since the implementation and use of this module is completely clear and requires only the basic testing, which should be addressed mainly in the verification phase. The COM3D code uses for the specification of the material properties the data from thermodynamic tables [19]. For the gas equation of state, a perfect gas approximation is used, which is commonly utilized and works well within a wide spectrum of problems. A simple test for this module is calculation of the complete adiabatic combustion in a constant volume. Table 2 shows a comparison between predictions of the COM3D code and the reference code STANJAN [16].

The last part of the code verification is an overall code testing including all modules of the code. The choice of the problems, which should be modelled in this part, is strongly dependent on the code itself and is defined by the goals of the code. For example, for codes, which are intended to simulate problems with strong shocks, 1D and 3D detonations are a good benchmark. The parameters of 1D detonation waves are known and must be reproduced with good accuracy. A triple-point pattern, which is observed in fully developed detonation, should be reproducible in 3D simulations. Our code is intended to simulate fast turbulent combustion, which is accompanied by relatively strong shocks, therefore such a 'detonation' test seems to be a reasonable choice. For the direct testing of turbulent reactive flow simulations, a handy simple test can scarcely be found, rather a comparison with well-defined simulations of another code could be recommended.

3.2 Experimental Investigations

The experiments available in the literature do only on rare occasions satisfy all demands appearing in a code testing. The best approach would be to set up and perform the experiments, which exhibit all important phenomena at relevant geometrical scales. However, since we cannot perform experiments on the full scale, direct validation is excluded. Instead, our approach is to perform experiments on different possible scales to investigate the models and their scaling potentials and thus eventually demonstrate adequate extrapolation capabilities of the code. In combustion experiments the costs increase rapidly with the size of the experimental facility. Thus, a larger experimental facility means that fewer experiments can be performed. As a result a hierarchy of experiments is called for: basic understanding is provided in small scale experiments, mid scale experiments are used to identify scaling effects for interesting mixtures; and finally a small number of large scale tests is performed in the relevant parameter space, to check if the scaling works as expected.

The database used for the validation of COM3D was collected mainly in four experimental facilities:

- 7.2 m tube at Forschungszentrum Karlsruhe;
- 12 m tube at Forschungszentrum Karlsruhe (see Fig. 2);
- Torpedo tube and Driver tube at Kurchatov Institute;
- RUT facility at Kurchatov Institute.

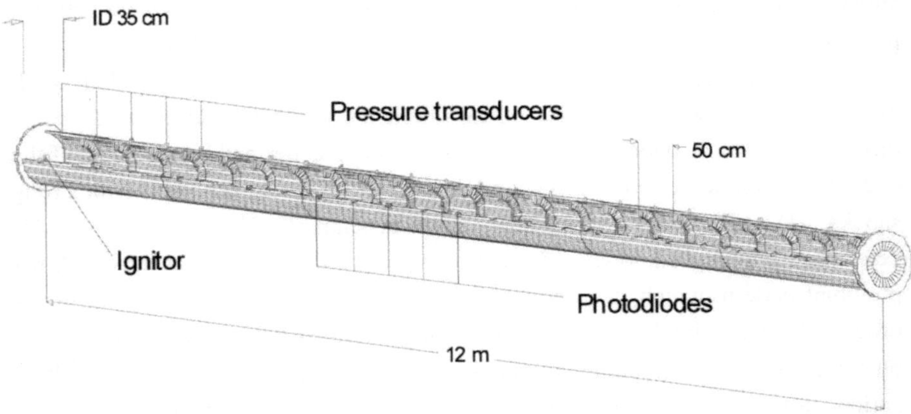

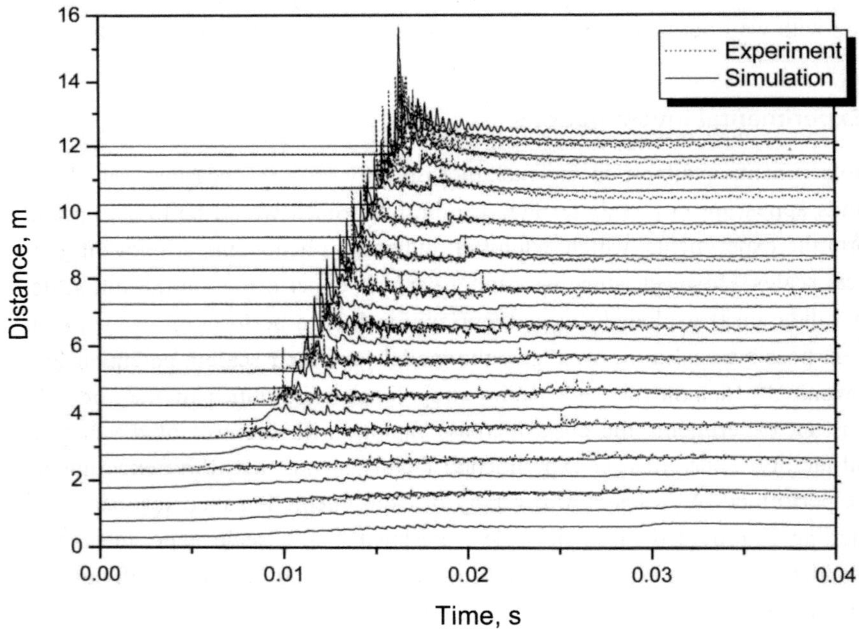

Fig. 2. Comparison of measured and calculated pressure recordings for a combustion experiment in the FZK 12 m tube with blockage ratio 30% and 15% hydrogen-air mixture ($P_0 = 1$ bar, $T_0 = 290$ K).

For the validation process the experimental database derived from tests in these facilities is supplemented by data available in the literature. We consider it as a distinct advantage to be able to perform own investigations, since it allows to carry out the experiments, perfectly fitting to the validation procedure, and simultaneously have a clear understanding of the experimental details and uncertainties.

3.3 Code Validation

In the validation procedure of CFD codes the main efforts are directed to proof the physical models, which accompany the gasdynamic core of the code and interact with it. As mentioned above, the validation process should be performed keeping strictly in mind the objectives of the code. The COM3D code was developed with the aim to predict gasdynamic loads during combustion processes in nuclear power plant containments, therefore the validation should concentrate on pressure wave predictions. An obvious strategy would be first to validate all parts of the code individually and then to evaluate predictions of the code as a whole. However, on this way certain problems immediately appear. For example, even though the turbulence model can be validated separately, the turbulent chemical reaction module apparently cannot be used without turbulence. Therefore the only way is to validate the entire code. Let us consider in some more detail which combustion regimes are expected in the simulations starting from the ignition. Note, that since the k-ε model assumes a developed turbulence, only the processes, which evolve in a fully turbulent flow field, can be included into consideration. For example, ‘self-turbulizing’ flames, which generate turbulence themselves in a significant way cannot be covered by the model and should be excluded as potential test case. Furthermore all experiments, which include events of ignition, re-ignition and quenching, should be taken out of the scope of the validation process. Flame propagation regimes ranging from slow to fast deflagrations need to be adequately reproduced in the simulations and must be presented in the validation database. At later stages of flame acceleration a deflagration-to-detonation transition (DDT) can occur. This event should undoubtedly be excluded from the testing matrix since the limitations of numerical resolution and lack in physical completeness of the models does not allow proper description of this phenomenon.

Thus, in the course of the validation process we concentrate on the flame acceleration and flame propagation during combustion in a wide range of mixture compositions and geometrical situations. Among different possible combustion experiments, which could be carried out on laboratory scale, the most appropriate for the present purpose are the experiments on flame acceleration in combustion tubes with repeated obstacles. Such experiments were performed in numerous configurations by different authors and are understood quite well. Therefore this type of experiments provides a solid foundation for the COM3D validation process [8].

Another important issue for code validation is comparison experimental data with simulation results. It appears that to compare isolated pressure histories taken from individual pressure transducers is not very useful for the reason that in the whole set of transducers some of them could demonstrate extremely good coincidence while others may fail completely. More helpful is to compare x-t diagrams. This type of diagrams shows pressure histories for each transducer, which are located on the plot in accordance with its distance from the ignition point. The analysis of such diagram allows to see immediately how the process developed (i.e. the flame acceleration process), to compare individual pressure recordings, and also the integral agreement between experiment and simulation.

In the experiments with low flame speeds the generated pressure waves are often so weak and smooth that pressure comparison is not conclusive. In these cases it is more convenient to use recordings of another transducers: so-called 'flame-arrival-time' transducers. These transducers measure the light that is emitted within their narrow cone of sight, which is kept small by special collimation. These light detectors see the flame when it passes through the sensitive cone volume. Note that the comparison of light signals is even preferable because it gives explicit information on the flame acceleration; and the flame front position history, which actually defines the generated pressure waves.

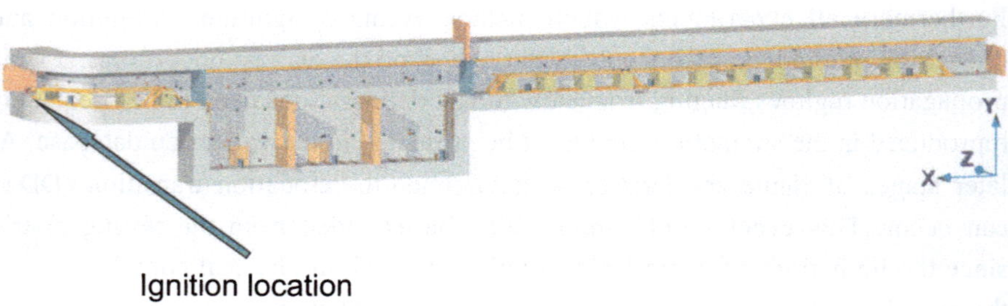

Fig. 3. Model of the RUT facility used in simulation of the combustion experiment HYC02.

When comparing data we must also keep in mind which limitations the measurement devices and the numerical simulations have. For example, pressure transducers have a certain frequency response and thus can only measure signal variations accurately within a certain frequency bandwidth. In the numerical simulation we have pressure value, which is averaged over the volume of the computational cell considered. This size is for large-scale applications much larger than the actual size of a pressure transducer. This influences the extent to which agreement between both data sets can be achieved. We have a similar effect with the photodiode data. These probes measure the light emitted within their cone of sight. In the simulation we approximate these measurements by identifying when the flame enters a set of cells which coincide as good as possible with the photodiode cone. Due to the limited resolution of the computational grid the volumes in the experiment and in the simulation can be quite different. Since we are interested in the time of flame arrival these differences can have a noticeable effect for fast flames.

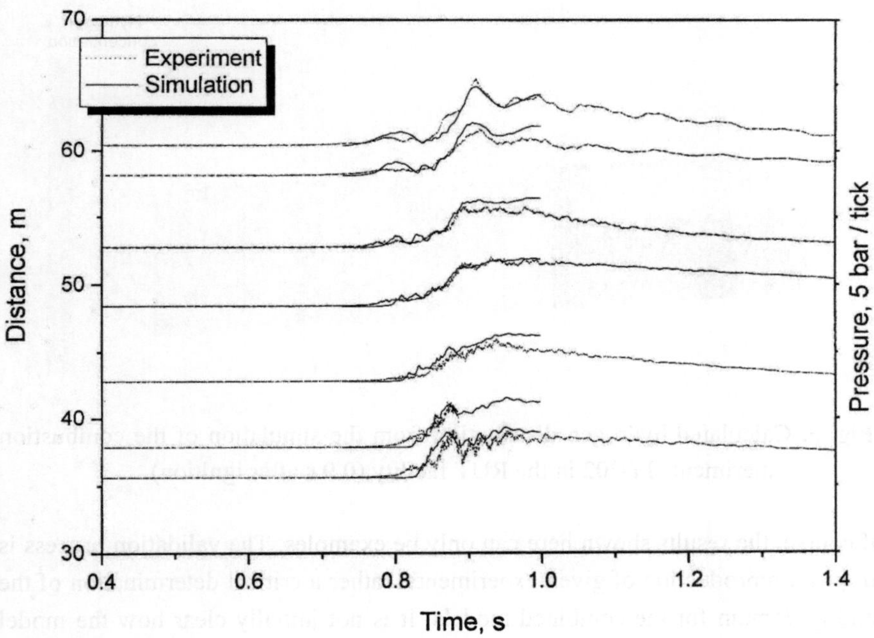

Fig. 4. Comparison of measured and calculated pressure recordings for combustion experiment HYC02 in the RUT facility with 11.5% hydrogen-air mixture. Simulation was performed on CRAY T3E (Research Center Jülich, Germany), total grid included more than 3,000,000 cells, calculation time on 80 processors is about 1 day.

Fig. 2 shows an example for the comparison between a simulation and the corresponding combustion experiment using an x-t diagram. The simulation geometry of the 12 m FZK combustion tube is given at the top the figure. In the lower part the measured pressure histories are shown as dotted lines. The corresponding curves from the simulation are given as solid lines. The agreement between the two sets of curves is quite good. The trajectory of the flame front and major characteristic of the waves are predicted correctly. Notice, that at the end of the validation process the COM3D code does not contain any tuneable parameters.

Similar results for the simulation of a large-scale test in the RUT facility are shown in Fig. 3 and Fig. 4. This example demonstrates that the code models are equally well applicable to turbulent combustion problems on much larger scales (60 m, 500 m^3). In addition Fig. 5 displays the calculated instant hydrogen distribution for the same experiment. Such data are not available from the experiment and provide additional important insight gained from numerical simulations.

Fig. 5. Calculated hydrogen distribution from the simulation of the combustion experiment HYC02 in the RUT facility (0.9 s after ignition).

Of course, the results shown here can only be examples. The validation process is not just a reproduction of given experiments, rather a critical determination of the validity domain for the combined models. It is not initially clear how the model limitations, which may be well-known individually, combine in an integral simulation. Therefore it is required that similar simulations are performed for the whole parameter range of interest. The different simulations will in general show some variation in the accuracy with which the experiments can be reproduced. Deviations are expected to increase as the limits of the validity domain of the models are approached. A cut-off criterion must be used finally to define the parameter range in which acceptable simulation results can be obtained.

The results of the validation calculations [10] are summarized in Fig. 6 as a 'Borghi-diagram'. The diagram axes indicate characteristic length scale and velocity ratios of the combustible mixture. The abscissa is the ratio of a characteristic integral length scale of turbulence to the laminar flame thickness; the ordinate is the ratio of the turbulent fluctuation velocity to the laminar burning velocity. Thick black lines indicate the region of parameters accessed in the validation procedure. The larger region plotted with gray lines is the region which is potentially of interest in industrial applications. Because of their large size, combustion processes on the scale of reactor containment achieve Reynolds numbers larger than in typical combustion applications, which is indicated by the dashed lines. An extension to higher Reynolds numbers should not pose major difficulties to the combustion model, however an extensions to smaller Damköhler numbers would require a different combustion model and thus need to be approached with more caution.

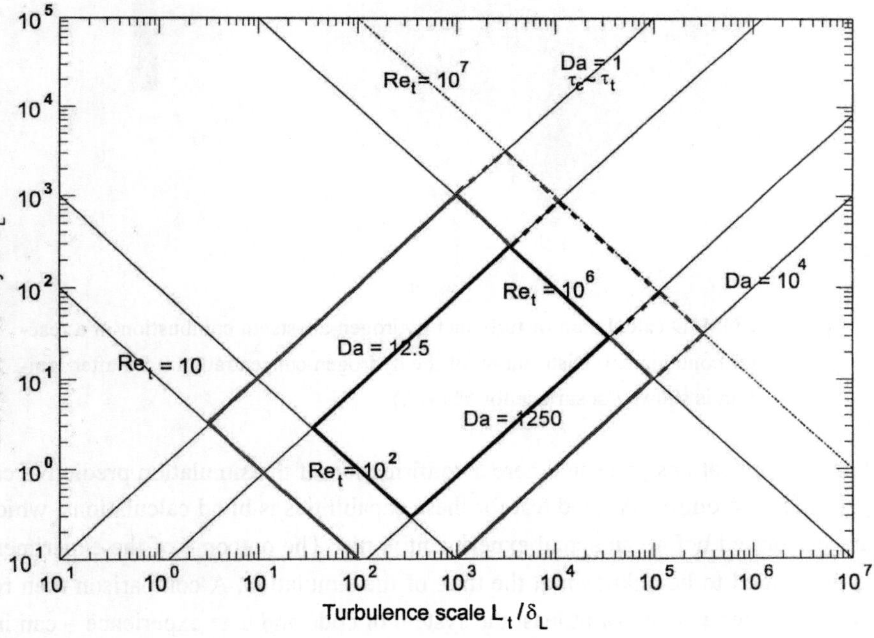

Fig. 6. Range of validity of COM3D turbulent combustion model (eq. (7))

The experience gained during the validation simulations is another important aspect of the whole validation process. Simulations of problems without reference experiments should not be trusted blindly. For such problems it becomes essential that the code user has clear understanding of the ongoing physics. We think that in the current state of combustion simulations the necessary expertise cannot be confined in the code alone, but that the involvement of experienced experts is crucial.

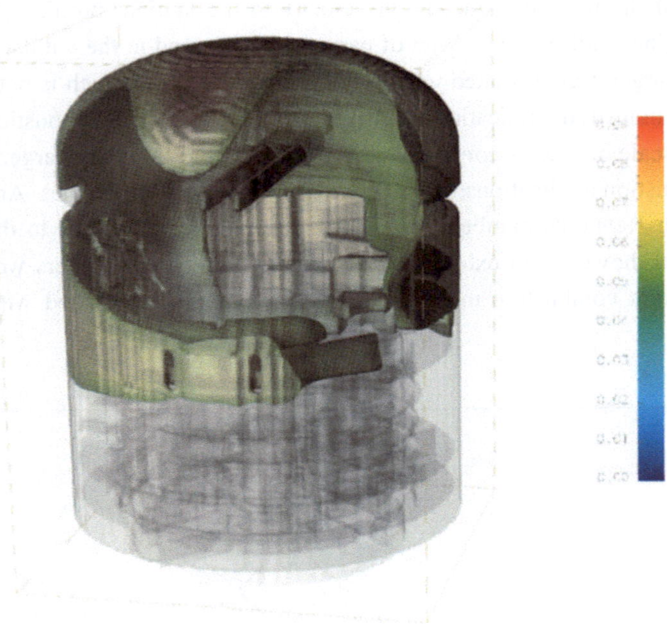

Fig. 7. COM3D calculation of turbulent hydrogen-air-steam combustion in a reactor containment. Distribution of the hydrogen concentration 0.5 s after ignition is shown (isosurface for 5% vol.).

For the applications presented here a confirmation of the simulation predictive capabilities is required. A good test for these capabilities is blind calculations, which are performed before an actual experiment is run. The outcome of the experiment is guarantied to be unknown at the time of the simulation. A comparison then reveals whether a code - or at least the system of code and user experience – can indeed cope with new and yet untested problems. In fact, the numerical results shown in Fig. 3 and Fig. 4 were created during such a blind benchmark exercise.

4 Sample Application of COM3D

We cannot close without giving at least an example of a combustion simulation in a nuclear reactor containment. Fig. 7 gives a snapshot of such simulation in an advanced containment configuration. The building geometry is shown as transparent glass model. The total free volume is about 80,000 m^3. The considered scenario is a so-called 'small break loss of coolant accident'. The hydrogen and steam sources were calculated with MELCOR and then used in the GASFLOW code. The hydrogen and steam concentrations predicted by GASFLOW were used as input to the turbulent combustion code COM3D. The grid used for this calculation had about 2 million cells with a cell size of $(0.4)^3$ m^3. The figure shows the hydrogen cloud with concentrations above 5% at 0.5 s after ignition. A large portion of the hydrogen in the upper 'empty' part has already been burned out. The highest observed pressures are registered on internal walls of the building and not on the outer containment shell. These pressures are well below the specified design values proving containment integrity for this particular accident scenario.

Summary

In the present paper the problems arising in the analysis of hydrogen behaviour during severe accident scenarios in nuclear power plants were described. We have discussed the physical effects, which must be described numerically and have outlined our approach to split the whole accident scenario into different phases, which are dominated by different physical effects. Of the different tools, which are needed to describe the whole accident scenario, one code, namely the turbulent combustion code COM3D, was selected and discussed in more detail. The verification and validation process, which is largely based on comparison to experimental data, was outlined and examples were presented. Finally an example of a full-scale reactor application was briefly discussed.

The results show that explosion process simulations in large-scale environments with CFD tools are possible with todays computer hardware and carefully verified numerical models. Indeed the results indicate that the complex interactions in large-scale three-dimensional geometries cannot be accounted for by other simpler means. CFD models provide information, which is not available from simplified integral descriptions of the physical processes. As soon as dynamical processes become important it is mandatory to resolve these phenomena.

It should be pointed out that the analysis procedure described here is not limited to severe accident studies in nuclear power plants. The same approach has already been used to study distribution and combustion phenomena in applications for a future hydrogen based economy. However, when simulating problems, which are beyond the range of parameters for which a validation has been possible, special care must be taken and the results should not be accepted blindly. At the present state it is not advisable that inexperienced users perform such analyses.

References

1. Bielert U, Breitung W (1999) In: *Annual Meeting on Nuclear Technology ,99*, Deutsches Atomforum e.V., Karlsruhe, pp 159-162
2. Breitung W, Royl P (2000) *Nuclear Engineering and Design* 202: 249-268
3. Dorofeev SB, Kuznetsov M, Alekseev V, Efimenko A, Bezmelnitsyn A, Yankin J, Breitung W (1999) Reports IAE-6127/3, Kurchatov Institute Moscow, Russia, FZKA 6268, Forschungszentrum Karlsruhe, Germany
4. Dorofeev SB (2000) Chapter 3 in Breitung W, Chan C, et al. Report NEA/CSNI/2000/7, OECD Paris. See also ftp://hycom.fzk.de/pub/SOAR.
5. Dorofeev SB, Kuznetsov MS, Alekseev VI, Efimenko AA, Breitung W (2001) *Journal of Loss Prevention in the Process Industries*, 14: 583-89
6. Harten A (1989) *Journal of computational physics* 83: 148-184
7. Kessler G, Breitung W, Eibl J (1992) In: *Proc. 14th Congress on Civilisation through Civil Engineering.* New Delhi, India, pp 365-370
8. Kotchourko A, Breitung W, Veser A, Dorofeev SB (1997) In: *21st Symp. (Int.) on Shock Waves.* Great Keppel Island, Australia, pp 3220.1-3220.6
9. Kotchourko A, Breitung W, Veser A (1999) In: *Annual meeting on nuclear technology'99.* Deutsches Atomforum e.V., Karlsruhe, pp 173-176
10. Kotchourko AS, Dorofeev SB, Breitung W (1999) In: *17th Int. Colloq. on the Dynamics of Explosions and Reactive Systems.* Heidelberg, Germany
11. Kotchourko, Burgeth AB, Dorofeev S and Breitung W (2001) In: *Proc. 18th Int. Colloq. on Dyn. Explos. Reactive Systems.* Seattle, WA, USA, pp 119-123
12. Kumar RK (1985) *Journal of fire sciences, vol* 3, pp 245-262
13. Redlinger R (1999) In: *Jahrestagung Kerntechnik '99*, Inforum GmbH, Karlsruhe, Germany, pp 191-194
14. Redlinger R (2000) *Nachrichten.* Forschungszentrum Karlsruhe, 32: 243-249
15. Redlinger R (2001) http://www.iket.fzk.de/
16. Reynolds WC (1986) STANJAN. Version 3. Dept. of Mechanical Engineering, Stanford University, Palo Alto, CA, USA
17. Said R and Borghi R (1988) In: *22nd Symp. (Int.) on Combustion.* The Combustion Institute, Seattle, WA, USA, pp 569-577
18. Spalding DB (1971) In: *Thirteenth Symposium on Combustion.* The Combustion Institute, Pittsburgh, USA, pp 647-649
19. Stull DR, Prophet H (1971) JANAF thermochemical tables. National Bureau of Standards, NSRDS-NBS37, Ed. 2
20. Travis JR, Spore JW, Royl P, et al. (1998) Reports: LA-13357-M, Los Alamos Nat. Labs, USA, FZKA-5994, Forschungszentrum Karlsruhe, Germany
21. Veser A, Kotchourko A, Breitung W (1999) In: *Annual meeting on nuclear technology'99.* Deutsches Atomforum e.V., Karlsruhe, pp 167-172
22. Woodwoard P, Colella P (1981) In: *Proc. of the 7th Int. Conf. Numer. Methods in Fluid Dyn.,* Stanford/NASA Ames, Springer-Verlag

List of Contributors

Adrian, Gerhard, Dr.
Research and Development, Deutscher Wetterdienst
Kaiserleistraße 42, D-63067 Offenbach
Germany

Bielert, Ulrich, Dr.
Institut für Kern- und Energietechnik, Forschungszentrum Karlsruhe GmbH
Postfach 3640, D-76021 Karlsruhe
Germany

Breitung, Wolfgang, Dr.
Institut für Kern- und Energietechnik, Forschungszentrum Karlsruhe GmbH
Postfach 3640, D-76021 Karlsruhe
Germany

Bromage, Timothy G., Ph. D., Professor
Department of Anthropology, Hunter College, CUNY
695 Park Avenue, New York, N.Y. 10021
U.S.A.

Hertler, Christine, Dr.
Dept. of Vertebrate Paleobiology, Zoological Institute
Johann Wolfgang Goethe - Universität
Siesmayerstraße 70, D-60054 Frankfurt a.M.
Germany

Heusch, Steffen, M.Sc.
SYDRO Consult GbR
Mathildenplatz 8, D- 64287 Darmstadt
Germany

Kazmierczak, Józef, Ph. D., D. Sc., Professor
Institute of Paleobiology, Polish Academy of Sciences
Twarda 51/55, PL-00818 Warszawa
Poland

Kelm, Manfred, Dr.
Institut für Nukleare Entsorgung, Forschungszentrum Karlsruhe GmbH
Postfach 3640, D-76021 Karlsruhe
Germany

Kempe, Stephan, Prof. Dr.
Institut für Angewandte Geowissenschaften, Technische Universität Darmstadt
Schnittspahnstraße 9, D-64287 Darmstadt
Germany

Kienzler, Bernhard, Dr.
Institut für Nukleare Entsorgung, Forschungszentrum Karlsruhe GmbH
Postfach 3640, D-76021 Karlsruhe
Germany

Kotchourko, Alexei, Dr.
Institut für Kern- und Energietechnik, Forschungszentrum Karlsruhe GmbH
Postfach 3640, D-76021 Karlsruhe
Germany

Lempert Marcus, Dr.
Kisters AG
Charlottenburger Allee 5, D-52068 Aachen
Germany

Lützenkirchen, Johannes, Dr.
Institut für Nukleare Entsorgung, Forschungszentrum Karlsruhe GmbH
Postfach 3640, D-76021 Karlsruhe
Germany

Majewski, Detlev
Research and Development, Deutscher Wetterdienst
Kaiserleistraße 42, D-63067 Offenbach
Germany

Ostrowski Manfred W., Prof. Dr.
Institute of Hydraulic and Water Resources Management
Darmstadt University of Technology
Petersenstraße 13, D- 64287 Darmstadt
Germany

Schrenk, Friedemann, Prof. Dr.
Dept. of Vertebrate Paleobiology, Zoological Institute
Johann Wolfgang Goethe - Universität
Siesmayerstraße 70, D-60054 Frankfurt a.M.
Germany

Theis, Christian, Priv.-Doz. Dr.
Institut für Theoretische Physik und Astrophysik der Universität zu Kiel
Olshausenstraße 40, D-24098 Kiel
Germany

Errata

pages 314 and 317: Figure 3 is Figure 4 and Figure 4 is Figure 3

page 320: Figure 5 is missing:

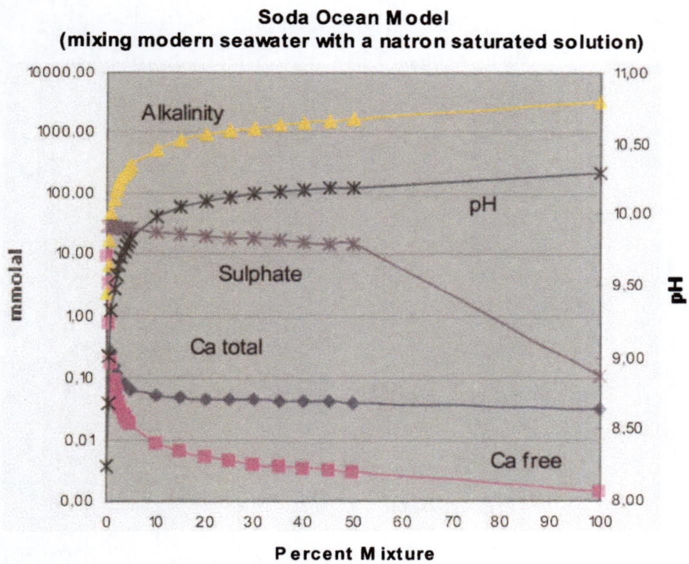

Fig. 5. PHREEQUE mixing experiment of modern seawater (left) with the natron saturated Nemrut Crater Lake Water (line 3, Table 3) (right), illustrating the possible transition of an alkaline ocean to one dominated by halite and gypsum. After linear mixing of constituents, the solutions are adjusted to have a calcite saturation index of +1 and a $pPCO_2$ of 3.5. Note the sharp drop in Ca concentration is accompanied by only a slight increase in alkalinity. Note also that the sulphate concentration would describe an exponential curve between 50 and 100 % if subsamples were calculated for that section of the mixing experiment.

page 318: Line 3 change: “Na+K<<Cl+SO_4” to “Na+K>>Cl+SO_4”;
Table 3 change: “Alkal. mequ” to “Alkal. meq”;
Table 3 change: “Sat” to “SI” (3 times)

MIX
Papier aus verantwortungsvollen Quellen
Paper from responsible sources
FSC® C105338

If you have any concerns about our products,
you can contact us on
ProductSafety@springernature.com

In case Publisher is established outside the EU,
the EU authorized representative is:
Springer Nature Customer Service Center GmbH
Europaplatz 3, 69115 Heidelberg, Germany

Printed by Libri Plureos GmbH
in Hamburg, Germany